Gabelentz and the Science of Language

Gabelentz and the Science of Language

Edited by
James McElvenny

Routledge
Taylor & Francis Group
LONDON AND NEW YORK

First published in 2019 by Amsterdam University Press Ltd.

Published 2025 by Routledge
4 Park Square, Milton Park, Abingdon, Oxon OX14 4RN
605 Third Avenue, New York, NY 10158

Routledge is an imprint of the Taylor & Francis Group, an informa business

ISBN: 9789462986244 (hbk)
ISBN: 9781041179955 (pbk)
ISBN: 9781003696285 (ebk)
NUR 610

Cover illustration: © Kota Ezawa

Cover design: Coördesign, Leiden

DOI 10.5117/9789462986244

Contents

1 Introduction

James McElvenny

Abstract

This chapter begins by introducing Georg von der Gabelentz and his place in the history of linguistics. It then outlines the content of this volume and its relation to existing Gabelentz scholarship.

The last decades of the nineteenth century represent a crucial period in the history of linguistics. This is the era in which the historical-comparative method, the flag-bearer of the new scientific language study of the nineteenth century, reached the peak of its institutional dominance and its most extreme form in the Neogrammarian school. But it is also the eve of the structuralist revolution, whose mounting challenges overwhelmed the historical paradigm and swept it away shortly after the turn of the century.

An outlier in this intellectual environment is the German sinologist and general linguist Georg von der Gabelentz (1840–1893). As Professor of East Asian Languages first at the Neogrammarian stronghold of the University of Leipzig from 1878 to 1889 and then Professor of Sinology and General Linguistics in Berlin from 1889 until his death in 1893, Gabelentz was present at the chief centres of linguistic scholarship of the time. His work was marked, however, by an adherence to the seemingly antiquated doctrines of Wilhelm von Humboldt (1767–1835). Reviewing the posthumous second edition of Gabelentz's (2016 [1891]) magnum opus, *Die Sprachwissenschaft* (The Science of Language), the Indo-Europeanist Ludwig Sütterlin (1863–1934) summed up the contemporary reception of Gabelentz when he remarked: 'Even though its first edition came out as recently as 1891 and its second edition in 1901, Gabelentz's book seems to us like a remnant of a former time: with him dies a point of view that in the end was established by Wilhelm von Humboldt' (Sütterlin, 1904, p. 319).[1]

1 Original: 'Das Buch von der Gabelentzens mutet uns, obwohl seine erste Auflage erst 1891 erschienen ist und die zweite 1901, schon wie ein Ueberbleibsel aus vergangener Zeit an;

McElvenny, James (ed.). *Georg von der Gabelentz and the Science of Language*.
Taylor & Francis Group, 2019
DOI: 10.5117/9789462986244/CH01

But Gabelentz was not simply old-fashioned: he sought to update the Humboldtian programme as a means to overcoming the perceived limitations that linguistics had acquired through the course of the nineteenth century. He hoped to revive the study of 'general linguistics', which he conceived as the broad investigation on the Humboldtian model of the human capacity for language (see Elffers, 2012). This programme was opposed to the narrow focus of the Neogrammarians, whose work concerned itself almost exclusively with historical sound change in Indo-European languages. Although some Neogrammarian theoreticians – such as Hermann Paul (1846–1921), whom Gabelentz (2016 [1891], p. 143) cited favourably – allowed themselves to wander in the realm of general linguistics, Gabelentz criticized the historical-comparative school for its tendency to treat its restricted technical concerns as the totality of scientific linguistics:

> Most of us have limited our research to one or another language family, and the genealogical-historical school has demonstrated such brilliant progress that we should not begrudge them a certain amount of self-satisfaction. Nothing seemed more reasonable than to say: in linguistics progress occurs entirely and exclusively within this school; those who remain outside it may call themselves philologists, philosophers of language, even language experts or polyglots, or whatever they like, but they must not pretend that they are linguists and that their subject is linguistics. Whoever speaks like that confuses the small field that he is ploughing with the meadows of a large community, and, to use a Chinese analogy, thinks like someone who sits in a well and maintains that the sky is small.
> (Gabelentz, 2016 [1891], p. 12)[2]

Historical-comparative linguistics still features prominently in Gabelentz's *Sprachwissenschaft*, but is treated only in the third 'book', or major section,

es stirbt mit ihm eine Betrachtungsweise aus, die sich in letzter Linie von W. von Humboldt herschreibt.'

2 Original: 'Die Meisten von uns haben ihre Arbeit auf die Erforschung der einen oder anderen Sprachfamilie beschränkt, und die genealogisch-historische Schule hat so glänzende Fortschritte zu verzeichnen, dass ihr eine gewisse Selbstgenügsamkeit nicht zu verdenken ist. Nichts lag näher, als zu sagen: Der Fortschritt der Sprachwissenschaft ist ganz und ausschliesslich bei dieser Schule; die draussen mögen sich Philologen, Sprachphilosophen, wohl auch Sprachkundige, Polyglotten nennen, oder wie es ihnen sonst beliebt, sie sollen nur nicht sich für Linguisten, nicht ihre Sache für Sprachwissenschaft ausgeben. Wer so spricht, der verwechselt den kleinen Acker, den er pflügt, mit der Flur einer grossen Gemeinde und urtheilt, um mich eines chinesischen Vergleiches zu bedienen, wie einer, der im Brunnen sitzt und behauptet, der Himmel sei klein.'

of this work. The other three 'books' deal with the place of linguistics as a science among other sciences (Book 1); the study and description of individual languages (Book 2); and general linguistics, covering mostly the nature of human language and cross-linguistic typological comparison (Book 4). The broad themes Gabelentz addresses and their arrangement lead Morpurgo Davies (1998, pp. 299-300) to attribute an 'inescapable air of modernity' to Gabelentz's book in her survey of late nineteenth-century theoretical works in linguistics, and it is these qualities that make Gabelentz a key transitional figure from the nineteenth to the twentieth century. His critique of the Neogrammarian mainstream and efforts to modernize traditional linguistic thought provide an invaluable window onto the field in this period. In addition, his own theoretical and technical innovations are frequently cited as anticipating many later ideas, including modern approaches to typology, grammaticalization and structuralist theory (see McElvenny, 2017).

This volume brings together a selection of texts that illuminate Gabelentz's life and work, and his place in the scholarly environment of his time. There is already a considerable body of secondary literature on Gabelentz, most of it in German (e.g. the classic Richter & Reichardt, 1979, or the more recent Gimm, 2013). In the past few years, the Berlin Ost-West-Gesellschaft für Sprach- und Kulturforschung (East-West Society for linguistic and cultural research) has spurred on research in this area with a number of anthologies containing previously unpublished archival materials alongside old and new essays on Gabelentz (e.g. Ezawa & Vogel, 2013; Ezawa et al., 2014). This volume is a further contribution to this genre which, through the choice of English as the language of publication, is intended to make Gabelentz scholarship more accessible to an international audience. The contents of this volume are quite heterogeneous, ranging from biography to translation to academic exposition, and have been deliberately chosen to provide a representative sample of some of the diverse directions research into Gabelentz and his work has taken. Each chapter is self-contained and can be read in isolation, although points of contact between the chapters are highlighted through cross-references.

Vogel and McElvenny open the volume in Chapter 2 with a biographical essay concentrating on the role played by Gabelentz's family background and his upbringing in shaping his linguistic and Sinological interests. Gabelentz's father, Hans Conon von der Gabelentz (1807–1874), was himself a renowned language scholar who published several studies of 'exotic' languages, wrote pioneering works on linguistic typology and, through his library at the family estate in Poschwitz, assembled and made available a major research resource for European linguists at the time. The Gabelentz children were raised in a world of languages and were encouraged to partake in their father's scholarly passion: linguistics and language study were very much a family affair. This chapter explores the interplay between Gabelentz's family and his academic career, drawing principally on personal correspondence, manuscript biographies penned by Gabelentz's sister, Clementine von Münchhausen (1849–1913), and oral tradition within the family.

In Chapter 3, the focus of the volume shifts to Gabelentz's contributions to linguistic theory with Elffers's account of his role as a pioneer of information structure. Alongside such figures as Henri Weil (1818–1909), Hermann Paul and Philipp Wegener (1848–1916), Gabelentz is widely considered to be among the first generation of theorists to address questions of information structure. Beginning with his earliest linguistic publications on 'comparative syntax' and continuing throughout his career, Gabelentz explored the discourse effects of word order through an empirical investigation of a wide variety of languages. This chapter undertakes a detailed examination of the evolution of Gabelentz's ideas in this area, their place in the contemporary intellectual environment and their influence on later scholarship. In particular, it shows how Gabelentz's account was embedded within the linguistic component of *Völkerpsychologie* developed by H. Steinthal (1823–1899), in many respects an offshoot of the Humboldtian tradition.

Chapter 4 turns to a puzzling but revealing episode in Gabelentz's career, with Hurch and Purgay's exploration of his Basque-Berber studies. Discussions of Gabelentz's *œuvre* tend to omit the fact that he attempted in some of his later writings to demonstrate a genealogical connection between the Basque and Berber languages. Based on an examination of Gabelentz's texts, unpublished notes and correspondence, as well as several contemporary reviews, this chapter examines Gabelentz's proposal and what it shows about his theoretical views vis-à-vis historical-comparative linguistics and his place in the linguistic community of the time. It shows how Gabelentz's critique of the prevailing historical-comparative approach led him to abandon all established methods and draw wildly implausible conclusions. Even the most vehement critics of the Neogrammarians, such as Hugo Schuchardt (1842–1927), looked on uncomprehendingly at the turn Gabelentz had taken.

In Chapter 5, Willems investigates the philosophy motivating Gabelentz's perspective on the study of language as reflected in *Die Sprachwissenschaft*. The dominant philosophy underlying most linguistic work in Gabelentz's day looked to a psychological conception of language. In his own work, Gabelentz of course drew on and responded to this conception, but his *Sprachwissenschaft* nevertheless strikes another chord, which sets it apart from contemporary sources. In this chapter, Willems argues that Gabelentz's book is particularly noteworthy for its wide-ranging bottom-up approach to linguistic phenomena and its propensity to conceive of language both as object and instrument of enquiry in a sense akin to the 'reflexive' stance that would later become a hallmark of Edmund Husserl's (1859–1938) phenomenology. In particular, Willems discerns the 'phenomenological mindset' in the way a different philosophy of the language sciences and theory of meaning take shape in *Die Sprachwissenschaft*.

Chapter 6 closes the volume with an English translation of 'Content and Form of Speech', one of the core theoretical chapters of *Die Sprachwissenschaft*, which addresses the notion of 'form' in language and speech, and its opposites, 'content' and 'matter'. 'Form' is a key concept in much nineteenth-century linguistic scholarship, which was employed in various senses to a number of different ends, playing a particularly important role in Humboldtian approaches to typology. In this chapter, Gabelentz provides a wide-ranging survey of existing views on 'form' in language – including extensive quotations from other authors – and develops his own position. This text is therefore useful not only as a statement of Gabelentz's own thinking, but also as an overview of opinions in the field at the time. The translation is accompanied by a short introduction and notes that contextualize the text and help the present-day reader to follow the historical debates.

Works cited

Elffers, Els. 2012. 'The rise of general linguistics as an academic discipline: Georg von der Gabelentz as a co-founder'. In: *The making of the humanities. Vol. II: From early modern to modern disciplines*, Rens Bod, Jaap Maat & Thijs Weststeijn (eds.), pp. 55-70. Amsterdam: Amsterdam University Press.

Ezawa, Kennosuke & Annemete von Vogel (eds.). 2013. *Georg von der Gabelentz: ein biographisches Lesebuch*. Tübingen: Narr.

Ezawa, Kennosuke, Franz Hundsnurscher & Annemete von Vogel (eds.). 2014. *Beiträge zur Gabelentz-Forschung*. Tübingen: Narr.

Gabelentz, Georg von der. 2016 [1891]. *Die Sprachwissenschaft, ihre Aufgaben, Methoden und bisherigen Ergebnisse*, Manfred Ringmacher & James McElvenny (eds.). Berlin: Language Science Press.

Gimm, Martin. 2013. *Georg von der Gabelentz zum Gedenken: Materialien zum Leben und Werk*. Wiesbaden: Harrassowitz.

McElvenny, James. 2017. 'Georg von der Gabelentz'. *Oxford Research Encyclopedia of Linguistics*. http://linguistics.oxfordre.com/view/10.1093/

acrefore/9780199384655.001.0001/acrefore-9780199384655-e-379, accessed 2 August 2018.

Morpurgo Davies, Anna. 1998. *History of Linguistics*, vol. IV: *Nineteenth-century linguistics*. London: Longman.

Richter, Eberhardt & Manfred Reichardt (eds.). 1979. *Hans Georg Conon von der Gabelentz: Erbe und Verpflichtung*. Berlin: Akademie der Wissenschaften der DDR.

Sütterlin, Ludwig. 1904. Review of 1901 second edition of Gabelentz (2016 [1891]). *Literaturblatt für germanische und romanische Philologie* 25, pp. 319-320.

About the author

James McElvenny is an intellectual historian specialising in the history of linguistics. He is currently Newton International Fellow in the Department of Linguistics and English Language at the University of Edinburgh.
james.mcelvenny@mailbox.org

2 The Gabelentz family in their own words

Annemete von Vogel and James McElvenny

Abstract
This chapter presents a biographical essay concentrating on the role played by Gabelentz's family background and his upbringing in shaping his linguistic and Sinological interests. Gabelentz's father, Hans Conon von der Gabelentz (1807–1874), was himself a renowned language scholar who published several studies of 'exotic' languages, wrote pioneering works on linguistic typology and, through his library at the family estate in Poschwitz, assembled and made available a major research resource for European linguists at the time. The Gabelentz children were raised in a world of languages and were encouraged to partake in their father's scholarly passion: linguistics and language study were therefore very much a family affair. This chapter explores the interplay between Gabelentz's family and his academic career, drawing principally on personal correspondence, manuscript biographies penned by Gabelentz's sister, Clementine von Münchhausen (1849–1913), and oral tradition within the family.

Keywords: Hans Conon von der Gabelentz, Humboldtian tradition, typology, Poschwitz library

1 Introduction

The future career of Georg von der Gabelentz (1840–1893) as sinologist and general linguist was in many ways determined by birth. Georg's father, Hans Conon von der Gabelentz (1807–1874), was himself a renowned linguist and created for his children an environment in which they were surrounded by languages – classical and modern, familiar and exotic. Hans Conon eagerly encouraged Georg's early enthusiasm for language study, ultimately inspiring

McElvenny, James (ed.). *Georg von der Gabelentz and the Science of Language*. Taylor & Francis Group, 2019
DOI: 10.5117/9789462986244/CH02

and shaping his son's academic ambitions. Throughout his life, Georg stayed in close contact with his family – his father, mother and siblings – with whom he shared his intellectual passions. In turn, Georg continued the family succession by fostering the linguistic interests of his nephew, Albrecht Graf von der Schulenburg (1856–1902).

In this chapter, we look into the interplay between Georg's linguistics and his family background as revealed in the correspondence and personal reminiscences of Georg and his family members. Our main sources are unpublished correspondence held in the Gabelentz family collection of the Thüringisches Staatsarchiv (Thuringian State Archives) in Altenburg;[1] a manuscript biography of Georg (Münchhausen, 2013 [1913]) by his sister Clementine von Münchhausen (née von der Gabelentz; 1849–1913); a speech Georg gave to the Königlich-Sächsische Akademie der Wissenschaften (Royal Saxon Society of Sciences) about Hans Conon's scholarly achievements and the upbringing he afforded his children (Gabelentz, 1886);[2] and family lore passed down orally to the principal author, Annemete von Vogel, great-granddaughter of Clementine.

We begin in Section 2 with an examination of Hans Conon's contributions to linguistic scholarship and his influence on Georg. We then turn in Section 3 to Georg's childhood, education and his initial career in law and government administration. Finally, in Section 4, we look at Georg's move into academia as professor in Leipzig and Berlin, and the composition of his two major works, the *Chinesische Grammatik* (Chinese Grammar) (1881) and *Die Sprachwissenschaft* (The Science of Language) (2016 [1891]).

2 The influence of Hans Conon

'In terms of a great part of my views on the philosophy of language, I do not know how much they are essentially my own and how much they come from my father. Perhaps more often than I am aware the core idea is his and its elaboration mine' (Gabelentz, 1886, p. 233).[3] So was Georg's assessment

1 The correspondence quoted in this chapter is kept in the Thüringisches Staatsarchiv, Altenburg, under the signature 'Familienarchiv von der Gabelentz, Nr. 719'. For detailed information on the Gabelentz family collection, see Emig (2013).

2 Draft versions of Georg's speech are preserved in his notebook, edited and published by Manfred Ringmacher as Gabelentz (2011 [1879], pp. 370-382).

3 Original: 'Von einem grossen Theile meiner sprachphilosophischen Anschauungen weiss ich nicht, wieviel im Grunde mir selbst eigen, wieviel mir von meinem Vater überkommen ist. Öfter als ich mir bewusst bin, mag der Hauptgedanke ihm, die Ausgestaltung mir angehören.'

of the influence of his father Hans Conon on his own linguistic views. Hans Conon imbued his children with a love of languages and linguistic study. Georg (ibid., pp. 239-240) recounts how as an eight-year-old learning English he noticed the correspondence of English *th* and German *d* and, on asking his father about it, received an introduction to sound laws. Over the following years he was fed a steady stream of grammars and theoretical works in linguistics. But Georg was not the only child to be inducted into his father's science: Clementine produced a manuscript grammar of the eastern Himalayan language Lepcha, commenting that she 'knew how to do such work even at the age of eighteen', owing to her father's former instruction (Münchhausen, 1907, p. 6).[4]

Hans Conon's own linguistic interests emerged at a young age and were cultivated throughout his life. It is reported that, when asked by his grandmother what he wanted to do when he grew up, he replied: 'I want to learn all languages' (Gabelentz, 1886, pp. 219-220). Although too shy and thoughtful to be a consummate polyglot, as a child Hans Conon spent much of his free time immersed in his Greek grammar, took elective classes in Hebrew at his gymnasium (secondary school), and studied Arabic privately (ibid., p. 234). His fascination for Chinese was also awoken at a young age when his teacher showed him the Chinese character 鹿 *lù* 'deer' in a book. At sixteen, he began studying Chinese with the aid of Jean-Pierre Abel-Rémusat's 1822 *Élémens de la grammaire chinoise* (Elements of Chinese grammar), which was lent to him by his friend Hermann Brockhaus (1806–1877), a son of the prominent publishing family and himself later a professor of Sanskrit. Hans Conon continued his studies of Chinese privately during his time at university (ibid., p. 225).

As the scion of an old aristocratic family in the small central German duchy of Saxe-Altenburg, Hans Conon was expected to pursue a career in public service. He dutifully studied law and public administration (*Kameralistik*) and worked in a number of high-profile roles in the government of the duchy, including premier and president of the parliament. Alongside his professional obligations, Hans Conon found time for his own diverse scholarly activities, which extended from research into the local history of his region (see Gabelentz, 1886, pp. 222-223) to his much more widely received philological and linguistic work. On the philological front, he compiled a scholarly edition of the 'Ulphilas' Gothic Bible (Gabelentz & Loebe, 1836–1846) and translated the classic Chinese novel 金瓶梅 *Jīnpíngméi*

4 The quotation attributed to Clementine is: 'Wie so etwas zu bearbeiten sei, wusste ich mit 18 Jahren allein.'

into German (first published as Gabelentz, 2005–2013). In a more linguistic vein, he wrote grammars of such languages as Manchu (1832) and Dakota (1852), and undertook one of the earliest historical-comparative studies of the languages of Melanesia (Gabelentz, 1861a and 1873). A high point of his theoretical work is the essay 'Über das Passivum' (On the passive) (1861b), in which he compared the voice systems from a diverse selection of languages (Gabelentz, 1886, p. 229). This essay was conceived in the tradition of language study pioneered by Wilhelm von Humboldt (1767–1835) and was modelled on Humboldt's (1907 [1827]) 'Über den Dualis' (On the dual).[5]

It is this Humboldtian strand of Hans Conon's thought that was most influential on Georg. The ultimate aim of Georg's later conception of general linguistics was to take up the Humboldtian challenge of capturing the mutual relations between languages and supposed national characters through an empirical investigation of structural diversity in the world's languages (Gabelentz, 2016 [1891], p. 502; McElvenny, 2017, pp. 6-7). The culmination of these efforts was Georg's proposal for 'typology' (*Typologie*) in linguistics, a programme that progressively took shape throughout Georg's career, but which he only named and gave definite outlines in his very last, posthumous publication, 'Typologie: eine neue Aufgabe der Linguistik' (Typology: a new task of linguistics) (Gabelentz, 1894a; see Plank, 1991).

In addition to his own research, Hans Conon created a significant resource for philology in Europe in the private library he established at Schloss Poschwitz, the family residence in Altenburg. In the library, which he continually expanded, Hans Conon assembled one of the largest collections of grammars, dictionaries and linguistic treatises in the world. Colleagues would come to Altenburg to consult the library's collections or borrow volumes through the post. Hans Conon's pride in his library is evident in an anecdote Clementine (Münchhausen, 1910, p. 71) tells. Hans Conon was overjoyed when he was able to supply a colleague with a volume that could not be found in any of the great libraries of Vienna, Paris, Berlin or London:

> One morning I came [...] into his [Papa's] study and saw from his beaming face that he was extremely happy. He told me: 'I got a letter from Vienna today, [...] from a man who requested a book. [...] They did not have the book in Vienna and had told him to look in Paris; in Paris they did not have it either and told him to look in Berlin; they didn't have it either,

5 In addition to his publications, Hans Conon maintained an extensive correspondence with many leading scholars of his day, which is preserved in the Gabelentz family archive in Altenburg and has in part been published (see Walravens, 2008; 2013; 2015).

> maybe it was in London, and in London they wrote to him, if someone has it, it will be me. […] It [the book] is already in the post.'[6]

References to the invaluable Poschwitz library can be found throughout the linguistic literature of the nineteenth century. For example, August Schleicher (1821–1861), a leading linguist of mid-century, thanked Hans Conon, along with a number of other contemporary scholars, for his assistance in providing sources for his typological studies (Schleicher, 1865, pp. 499-500).

After Hans Conon's death in 1874 the library passed into Georg's stewardship, and was further expanded by him, as discussed below. The library continued to exist in the care of the family until 1945, in which year the majority of its volumes were requisitioned by Soviet troops and taken to Russia as war reparations.[7]

3 Georg's childhood and youth, legal career, and turn to academia

Georg very much followed in his father's footsteps. Like his father, Georg keenly applied himself to learning a number of languages both at home and at school. At home, the children had from 1857 to 1859 an English governess, Elisabeth Foster, who gave them lessons in her native language. At the Altenburg gymnasium, Georg studied French and the standard classical languages, Greek and Latin (Münchhausen, 2013 [1913], pp. 94-95). Chinese also caught his attention early on: Clementine (ibid., p. 94) comments that '[a]longside his school work, during his gymnasial years Georg learnt Chinese: I remember with what tender love he would say the name "Abel-Rémusat",

6 Original: 'Da kam ich eines Morgens […] in seine [Papas] Stube und merkte an seinem strahlenden Gesicht, daß er eine Freude gehabt hatte. Er erzählte: Da habe ich heute einen Brief bekommen aus Wien, von einem Herrn, […] der fragte nach einem Buch. […] In Wien hätten sies nicht, hätten ihn nach Paris verwiesen, die hattens auch nicht, verwiesen ihn nach Berlin, die hattens wieder nicht, vielleicht wärs in London, und in London haben sie ihm geschrieben, wenns einer hätte, wäre ich's. […] 's ist schon auf der Post.'

7 On a visit to Russia in 2011, one of the authors (McElvenny) located volumes containing the Poschwitz library bookplate in the Rudomino Foreign Literature Library in Moscow. Although the books were kept in the rare book collection, there did not seem to be any specific indication of their origin in the catalogue. Further research is required to establish how many of the volumes from Poschwitz are in the Foreign Literature Library and whether there may be volumes in other libraries.

whose work he chiefly used.'[8] With his sister Margarete, Georg read the Finnish national epic *Kalevala* and practised reading it aloud with friends. Georg had a clear talent for foreign languages. Clementine (loc. cit.) recounts how he once corrected his French teacher in class, that he wrote Greek 'in a beautiful flowing hand' (*in schöner fließender Hand*), and that he was asked to deliver the graduation speech for his gymnasium class because of the 'faultless English' (*tadelloses Englisch*) he had learnt from his governess.

Georg's professional career initially also mimicked that of his father in its contours. At the universities of Jena and Leipzig he studied law and public administration, while continuing to occupy every spare minute with the study of languages. His free time, it must be observed, was somewhat limited by his active participation in the delights of student life, most notably fencing, drinking and all manner of pranks on the hapless townspeople (see Münchhausen, 2013 [1913], pp. 96-98). From 1863 to 1869 he progressively completed his state-mandated law examinations and worked in a variety of roles at courts in Dresden, Leisnig and Chemnitz in the Kingdom of Saxony. When Alsace was annexed by the newly founded German Empire in 1871, Georg volunteered to serve in the imperial administration there. His excellent French aided him in his interactions with the local population, which was in many quarters quite hostile to the German occupiers (ibid., pp. 99-107).

Throughout this period, Georg maintained a range of scholarly interests. In Dresden he participated in the activities of the local geographical society and in Leisnig he co-founded an archaeological society (*Altertumsverein*) (see Münchhausen, 2013 [1913], p. 99). During his time in Chemnitz, he regularly travelled home to Poschwitz on the weekend, and would study the Sanskrit grammar of Pāṇini while on the train. It was in these years that he produced his first major theoretical publication in linguistics, the 1869 'Ideen zu einer vergleichenden Syntax' (Ideas on a comparative syntax). In this essay, Georg sought to open up the realm of syntax to cross-linguistic comparison, on the model of existing efforts that compared morphological structures across languages (Georg cites in particular Schleicher, 1859). It was here that Georg first introduced his terms 'psychological subject' and 'psychological predicate', which were addressed at contemporary debates on the nature of subject and predicate and paved the way for future work on information structure (see Elffers, this volume).

8 Original: 'Neben seinen Schularbeiten hat Georg schon auf dem Gymnasium Chinesisch getrieben, ich entsinne mich, mit welch zärtlicher Liebe er allein den Namen Abel Rémusat aussprach, dessen Werk er wohl hauptsächlich benutzte.'

The 1870s represented a caesura in Georg's life. In 1872 he married – against the advice of his family – his distant cousin Alexandra von Rothkirch und Trach (1854–1925). Alexandra, quite young and reputed for her beauty, had a very liberal upbringing at the Russian court in St Petersburg, where she had lived with her parents. These qualities were not those of a future faithful wife, thought Georg's family – and their fears would ultimately be confirmed. Georg and Alexandra had two children, the sons Albrecht (1873–1933) and Wolf Erich (1884–1914). Georg was very attached to Wolf Erich, who was born mentally handicapped, and made special provision for the boy in his will.

By 1874, Georg had resolved to abandon his legal career, which he described as a 'marriage of convenience' (*Vernunftheirat*) in a letter to Clementine on 23 April 1874, and to pursue his 'first love' (*premier amour*) of language study professionally. A month before, on 18 March 1874, he wrote to Clementine at length (Münchhausen, 2013 [1913], p. 117):

> I'm getting increasingly tired of this office work. Low pay, poor chances of promotion – that's what you get toiling *pro bono publico*, for the good public.[9] If I manage to get better known in the scholarly world then you will be able to see me changing course and standing behind a lectern at the university in Leipzig. Germany's biggest university still has no professorship for Chinese language; of course it also depends on the salary, since putting up with 900 Reichstaler forever is against my principles.[10] But I would have so much more time, and I could achieve so much more! The plan is still in the early stages, but it's already circulating in my mind.[11]

9 Gabelentz's rendering of *pro bono publico* as 'fürs gute Publicum' (for the good public) is a pun: *pro bono publico* of course means 'for the public good' but from a purely grammatical point of view could equally mean 'for the good public', describing his public service.

10 900 Reichstaler in 1874 had approximately the equivalent purchasing power of €17,100 in 2017.

11 Original: 'Den Büreaudinst bekomme ich immer mehr satt. Geringer Gehalt, schlechtes Avancement, das hat man dafür, daß man 8 Stunden täglich sich *pro bono publico* – fürs gute Publicum – abmüht. Gelingt es mir mich in der Gelehrtenwelt bekannter zu machen als ich heute bin, so kannst Du mich noch umsatteln und auf einem Katheder der Leipziger Universität stehen sehen. Deutschlands größte Hochschule hat ja noch keinen Lehrstuhl der chinesischen Sprache, freilich kommt es dabei mit auf den Gehalt an, denn ewig mit 900 rt. [Reichstaler] fürlieb nehmen ist wider meine Grundsätze. Aber was gewänne ich an Zeit, und was könnte ich dann schaffen! Der Plan ist eben noch ein unreifer, aber er geht mir doch schon recht im Kopfe herum.'

Over the next two years Georg worked tirelessly on further linguistic texts, most notably a second, much longer instalment of his 'Vergleichende Syntax' (Gabelentz, 1875), and his doctoral dissertation, a scholarly edition of the Chinese classic 太極圖 Tàijítú by 周敦頤 Zhōu Dūnyí (1017–1073), which was accepted at the university of Leipzig in 1876 (Gabelentz, 1876).

1874 is also the year in which Hans Conon died, leaving Georg the family estate at Poschwitz and its attached library. Georg keenly accepted his custodianship of the library and continually expanded its collection until the end of his life. In his correspondence with his mother Henriette (née von Linsingen; 1813–1892) after this date, he frequently reports on his new acquisitions. On 30 November 1877 he writes:

> From Brockhaus's auction I have purchased 30 volumes for a total of 21 Mark;[12] among them is the same copy of *Rémusat*'s Chinese grammar which Brockhaus lent Papa when he was in the gymnasium and from which he got his first knowledge of Chinese. I've given the book to [brother] Albert for his birthday. From Russia and South America, too, I have received interesting additions, all gifts; the same from Holland and Vietnam. A friendly merchant from Hamburg, G[eorg] Repsold [1804–1885], who I otherwise don't know at all, appears to have taken on the task of sending me books about the languages of Brazil. Not long ago he gave me a new edition of *Mamiani*'s Kìriri-Grammatik, which you copied out for Papa and he translated and published, and already he holds out the prospect for me of a new *Tupí* grammar.[13]

On 31 January 1880 Georg reports on a further order from Japan of around 500 volumes for 360 Mark,[14] many of which he sold on to the university library in Leipzig 'with a heavy heart' (*mit schwerem Herzen*), in order to recoup some

12 21 Mark in 1877 had roughly the equivalent purchasing power of €150 in 2017.

13 Original: 'Aus Brockhaus' Auktion habe ich für zusammen 21 Mkoo [Mark] 30 Bände Bücher erstanden; darunter dasselbe Exemplar von *Rémusat*'s chinesischer Grammatik, welches Brockhaus als Gymnasiast dem Papa geliehen, und woraus dieser seine ersten chinesischen Kenntnisse geschöpft hat. Ich habe das Buch Albert zu seinem Geburtstage geschenkt. Auch aus Rußland und Südamerika sind mir wieder interessante Neuigkeiten zugegangen, Alles Geschenke; desgleichen aus Holland und Wien. Ein freundlicher Hambuger Kaufmann, G. Repsold, den ich sonst gar nicht kenne, scheint es sich zur Aufgabe gemacht zu haben, mir Bücher über die Sprachen Brasiliens zugehen zu lassen. Unlängst schenkte er mir eine neue Ausgabe von *Mamiani*'s Kìriri-Grammatik, die Du dem Papa abgeschrieben und der Papa übersetzt herausgegeben hat, und schon stellt er mir wieder eine neue *Tupí*-Grammatik in Aussicht.'

14 360 Mark in 1880 had roughly the purchasing power of €2,500 in 2017. This would have represented around two months' salary for a young lawyer like Georg. Since, as Georg reports in

of the cost. At the end of the year, on 3 December 1880, he wrote, echoing his father's pride in the superiority of their family library over several major public collections: 'Just think, Börries [the husband of Clementine] gave me the Uslarian books[15] as a present just as the Berlin library ordered them for itself – once again a triumph for the Poschwitz library!'[16] On 1 May 1882 he reports another book package from Japan, containing 40 volumes.

4 Professor in Leipzig and Berlin

In 1878, two years after receiving his doctorate, Georg stepped into his first academic role, Extraordinary Professor of East Asian Languages at the University of Leipzig. The next decade and a half until his death proved to be the most productive period in his life, in which he published his two major works, in Sinology the 1881 *Chinesische Grammatik* and in general linguistics the 1891 *Die Sprachwissenschaft*. In 1889 he reached the height of his career with a call to a coveted ordinary professorship in Berlin.

Immediately on being called to the professorship in Leipzig, Georg was inundated with academic tasks. On 1 April 1879 he wrote to his mother:

> At the moment I am fully occupied. A work on the Melanesian languages [Gabelentz, 1879] has to be finished, then my inaugural lecture has to be written. And now [Leipzig publisher] T.O. Weigel wants a comprehensive Chinese grammar from me. Such a request was indeed something that I did not expect and is a great piece of luck for me.[17]

This commission from Weigel became the *Chinesische Grammatik*. As the excerpt above attests, Georg was delighted with this opportunity to make his mark on the field of Sinology. The progress of this book is

his letter, he received 200 Mark from the university for the books he sold on to them, he ended up paying only 160 Mark, around 60% of a young lawyer's monthly income at the time.

15 The 'Uslarian books' refer to a series of works by the Russian language scholar and army officer Peter Karlowitsch von Uslar (1816–1875). His works were translated into German and published between 1863 and 1873 in the *Mémoires* of the Imperial Academy of Sciences of St. Petersburg by Franz Anton Schiefner (1817–1879).

16 Original: 'Denke Dir, eben hat mir Börries die Uslar'schen Bücher geschenkt, als die Berliner Bibliothek sie für sich bestellte, – wieder ein Triumph der Poschwitzer Bibliothek!'

17 Original: 'Vor der Hand bin ich aber hier voll in Anspruch genommen. Eine Arbeit über die melanesischen Sprachen will fertig gemacht, dann die Antrittsvorlesung geschrieben werden. Und nun wünscht T. O. Weigel von mir eine ausführliche chinesische Grammatik. Ein solcher Antrag war in der That nicht zu erwarten und ist ein wahres Glück für mich.'

charted through his correspondence with his mother over the following months and years. On 3 December 1880 he wrote to her: 'The first proofs of my grammar now lie before me; it will be an impressive book in every way.'[18] On 10 October 1881, he wrote from his country house in Lemnitz: 'In the peace and quiet here I've finally got my book ready for printing, and now I can breathe more freely again.'[19] Printing was finished at the end of November that year and at the beginning of December Georg expected his first royalties from Weigel. On 17 March 1882 he wrote to his mother: 'Up to today I have read three reviews of my book filled with praise, and my publisher is also happy with the number of orders – already more than 260 a few weeks ago.'[20]

The *Grammatik* did not quite bring Georg all that he had hoped for, however. He had expected that on the appearance of the book he would be promoted to Ordinary Professor in Leipzig, a position accompanied by a higher and more secure income. A promotion was forthcoming in 1882, but only to Ordinary Honorary Professor. A motif marking the correspondence between Georg and his mother over the next decade are Georg's frequent requests for money. In these years, Georg was able to support himself and his family with the rental income generated by the estate at Poschwitz, which his mother managed on his behalf.

The ordinary professorship came only in 1889, when Georg was made a member of the Königlich-Preußische Akademie der Wissenschaften (Royal Prussian Academy of Sciences) and called to the Friedrich-Wilhelms-Universität in Berlin as Ordinary Professor of Sinology and General Linguistics. In a letter to Clementine on 31 July 1889 (Münchhausen, 2013 [1913], p. 124), Georg described his induction into the academy and call to Berlin as the 'highest distinction that a scholar can expect from his peers and for the rest of my life a fundamental improvement of my financial situation'.[21] But professional success contrasted with looming personal crisis. Georg's brother Albert had been collecting evidence of his wife Alexandra's infidelity, with which he confronted Georg, urging him to file for divorce. The divorce

18 Original: 'Von meiner Grammatik liegen mir die ersten Druckbogen vor; es wird ein in jeder Hinsicht stattliches Buch werden.'

19 Original: 'In der hiesigen ungestörten Ruhe ist mein Buch endlich druckfertig geworden, und nun kann ich wieder freier aufathmen.'

20 Original: 'Von meinem Buche habe ich bis heute drei durchweg sehr lobende Kritiken zu lesen bekommen, und auch mein Verleger ist mit der Zahl der Bestellungen – vor einigen Wochen schon über 260 – sehr zufrieden.'

21 Original: 'Die höchste Auszeichnung, die ein Gelehrter von Seinesgleichen erwarten kann und für den Rest meines Lebens eine wesentliche Besserung meiner Vermögenslage.'

proceedings, under which Georg suffered immensely, were completed in 1891 (ibid., p. 125). Shortly afterwards, Georg's fortune improved, when he met his second wife, the young widow Gertrud von Adelebsen (née von Oldershausen; 1858–1904). She gave birth to another son, Hanns-Conon, in 1892 (d. 1977). This good fortune did not last long, however: Georg died the following year (ibid., pp. 126, 130, 140).

Invigorated by his ordinary professorship in Berlin, Georg was able to complete *Die Sprachwissenschaft*, his chief theoretical statement on general linguistics, on which he had been working for many years. He held great hopes for the book's wide reception and commercial success, and his perfectionism drove him to immediately start preparing revisions for the anticipated second edition. On 30 April 1891, he wrote to Clementine (Münchhausen, 2013 [1913], p. 126):

> My book will now probably appear next week and, if the signs do not deceive, it will quickly sell out. For the second edition I already have a number of additions and corrections. Hopefully it will deliver what it promises and will become a goldmine for me. I could really use that this year, since [the expenses arising from] the change of tenants [on the Poschwitz estate] has cost me many thousands.[22]

After the book came out, he wrote to Clementine on 18 May 1891: 'I will send you a fresh copy of my book from Berlin. Maybe we'll see a second edition within the next year, which should be an expanded and improved one – including what I have already added and changed.'[23] On 12 June 1891 (Münchhausen, 2013 [1913], p. 126), he added: 'In the meantime I have prepared quite a lot for the second edition, all in all a good 30 printed pages. From all sides I receive letters of recognition and delight, but of course there will be no lack of criticism.'[24] His revisions are a recurring topic in their

22 Original: 'Mein Buch wird nun wohl in nächster Woche herausgegeben werden und, wenn die Anzeichen nicht trügen, sehr bald vergriffen sein. Zur zweiten Auflage habe ich schon eine Menge Nachträge und Verbesserungen. Hoffentlich hält es, was es verspricht, und wird mir zur Goldgrube. Gerade in diesem Jahr kann ich eine solche sehr gut gebrauchen, denn der Pachtwechsel kostet mich viele Tausende.'

23 Original: 'Ein properes Exemplar meines Buches werde ich Dir von Berlin aus zuschicken, Vielleicht erleben wir in Jahresfrist die zweite Auflage, die eine vermehrte und verbesserte werden soll, – schon mit dem was ich bis jetzt nachgetragen und verändert habe.'

24 Original: 'Dazwischen habe ich doch sehr viel für die zweite Auflage vorgearbeitet, Alles in Allem wohl gut 30 Seiten Druck. Anerkennende, zuweilen entzückte Briefe erhalte ich von allen Seiten; an Widerspruch wird es aber auch nicht fehlen.'

correspondence over the next year.[25] On 6 May 1891 he had already noted 'and [I] always find new material for the second edition' (*...und finde immer Neues für die zweite Auflage*) (ibid., p. 134), and on 24 March 1892 he writes: 'Along the way my *Sprachwissenschaft* gets a few new scraps' (*Nebenbei fällt mancher neue Brocken für meine Sprachwissenschaft ab*) (ibid., p. 130).

Georg never saw the second edition of *Die Sprachwissenschaft*. It appeared in 1901, a decade after the first edition and eight years after his death. Albrecht von der Schulenburg, the son of Georg's sister Margarete, completed Georg's revisions and saw the book through the press. Albrecht was also responsible for posthumously publishing a study by Georg positing a genealogical relationship between the Basque and Berber languages (Gabelentz, 1894b), which was very poorly received in the scholarly community (see Hurch & Purgay, this volume). Frequent mentions of Albrecht's presence at Poschwitz, Lemnitz and Berlin in Georg's correspondence with Clementine in the last years of his life (Münchhausen, 2013 [1913], pp. 125-128) attest to the close relationship that developed between uncle and nephew. Albrecht shared Georg's Sinological and linguistic interests: he had been Georg's student in Berlin and was himself proficient in Chinese. He began an academic career in linguistics, as a lecturer (*Privatdozent*) in Munich, but his untimely death in 1902 prevented him from taking up a call to an ordinary professorship in Göttingen. It remains an enduring question in Gabelentz scholarship as to how faithfully Albrecht's revisions of and additions to *Die Sprachwissenschaft* reflect his uncle's intentions and to what extent Albrecht included his own innovations.

Works cited

Abel-Rémusat, Jean-Pierre. 1822. *Élémens de la grammaire chinoise*. Paris: Imprimerie Royale.

Emig, Joachim. 2013. 'Das Familienarchiv v. d. Gabelentz im Thüringischen Staatsarchiv Altenburg'. In: Ezawa & Vogel (2013), pp. 311-322.

Ezawa, Kennosuke & Annemete von Vogel (eds.). 2013. *Georg von der Gabelentz: ein biographisches Lesebuch*. Tübingen: Narr.

Gabelentz, Georg von der. 1869. 'Ideen zu einer vergleichenden Syntax. Wort- und Satzstellung'. *Zeitschrift für Völkerpsychologie und Sprachwissenschaft* 6, pp. 376-384.

25 Many of the notes Georg made in preparing the first and second editions of *Die Sprachwissenschaft* can be found in his notebook, published as Gabelentz (2011 [1879]).

Gabelentz, Georg von der. 1876. *Thai-ki-thu, des Tscheu-Tsï Tafel des Urprinzipes, mit Tschu-Hi's Commentare nach dem Hoh-pih-sing-li*. Dresden: Commissions-Verlag.

Gabelentz, Georg von der. 1879. 'Malaiisch-polynesische und melanesische Sprachen und Literaturen'. *Zeitschrift der Deutschen Morgenländischen Gesellschaft*, supplement to vol. 33, pp. 39-44.

Gabelentz, Georg von der. 1881. *Chinesische Grammatik, mit Ausschluss des niederen Stiles und der heutigen Umgangssprache*. Leipzig: T. O. Weigel.

Gabelentz, Georg von der. 1886. 'Hans Conon von der Gabelentz als Sprachforscher'. *Berichte über die Verhandlungen der königlich-sächsischen Gesellschaft der Wissenschaften zu Leipzig, philologisch-historische Classe* 38, pp. 217-241.

Gabelentz, Georg von der. 1894a. 'Hypologie [Typologie], eine neue Aufgabe der Linguistik'. *Indogermanische Forschungen* 4, pp. 1–7.

Gabelentz, Georg von der. 1894b. *Die Verwandtschaft des Baskischen mit den Berbersprachen Nord-Africas*, Albrecht Graf von der Schulenburg (ed.). Braunschweig: Sattler.

Gabelentz, Georg von der. 2011 [1879]. 'Zur allgemeinen Sprachwissenschaft'. In: Wilfried Kürschner (ed.), *Miscellanea Linguistica*, pp. 225–394. Frankfurt am Main: Peter Lang.

Gabelentz, Georg von der. 2016 [1891]. *Die Sprachwissenschaft, ihre Aufgaben, Methoden und bisherigen Ergebnisse*. Leipzig: T.O. Weigel Nachfolger.

Gabelentz, Hans Conon von der. 1832. *Élémens de la Grammaire Mandchoue*. Altenburg: Comptoire de la Littérature.

Gabelentz, Hans Conon von der & Julius Loebe. 1836–1846. *Ulfias*, vols. 1 & 2. *Veteris et Novi Testamenti Versionis Gothicae*. Altenburg & Leipzig: Libraria Schnuphasiana.

Gabelentz, Hans Conon von der. 1852. *Grammatik der Dakota-Sprache*. Leipzig: F.A. Brockhaus.

Gabelentz, Hans Conon von der. 1861a & 1873. 'Die melanesischen Sprachen'. *Abhandlungen der Königlich-Sächsischen Akademie der Wissenschaften zu Leipzig, philologisch-historische Classe* 3, pp. 1-266.

Gabelentz, Hans Conon von der. 1861b. 'Über das Passivum'. *Abhandlungen der Königlich-Sächsischen Akademie der Wissenschaften zu Leipzig, philologisch-historische Classe* 3, pp. 449-546.

Gabelentz, Hans Conon von der. 2005–2013. *Jin Ping Mei: chinesischer Roman, erstmalig vollständig ins Deutsche übersetzt*, 10 vols., Martin Gimm (ed.). Berlin: Staatsbibliothek zu Berlin.

Humboldt, Wilhelm von. 1907 [1827]. 'Über den Dualis'. In: Albert Leitzmann (ed.), *Wilhelm von Humboldt, Gesammelte Schriften*, vol. 6, part 1, pp. 4-30. Berlin: Behr.

McElvenny, James. 2017. 'Grammar, typology and the Humboldtian tradition in the work of Georg von der Gabelentz'. *Language and History* 60:1, pp. 1-20.

Münchhausen, Börries. 1907. *Clementine von Münchhausen, geb. v. d. Gabelentz. Anlagen und Talente, Entwicklung u. Studien.* MS, in family possession.

Münchhausen, Clementine von. 1910. *Hans Conon v. d. Gabelentz geschildert von C. v. Münchhausen geb. v. d. Gabelentz.* MS, in family possession.

Münchhausen, Clementine von. 2013 [1913]. 'H. Georg v. d. Gabelentz. Biographie und Charakteristik', Annemete von Vogel (ed.). In Ezawa & Vogel (2013), pp. 85-171.

Plank, Frans. 1991. 'Hypology, typology: The Gabelentz puzzle'. *Folia Linguistica* 25, pp. 421–458.

Schleicher, August. 1859. 'Zur Morphologie der Sprachen'. *Mémoires de l'Académie Impériale des Sciences de St.-Petersbourg* 1.7, pp. 1-38.

Schleicher, August. 1865. *Die Unterscheidung von Nomen und Verbum in der lautlichen Form.* Leipzig: Hirzel.

Walravens, Hartmut (ed.). 2008. *'Freilich lag in den zu überwindenden Schierigkeiten ein besonderer Reiz ...': Briefwechsel der Sprachwissenschaftler Hans Conon von der Gabelentz, Wilhelm Schott und Anton Schiefner, 1834–1874.* Wiesbaden: Harrassowitz.

Walravens, Hartmut (ed.). 2013. *'... Ihr ewig dankbarer B. Jülg': Briefwechsel der Sprachwissenschaftler Bernhard Jülg (1825–1886) und Hans Conon von der Gabelentz (1807–1874).* Wiesbaden: Harrassowitz.

Walravens, Hartmut (ed.). 2015. *Hans Conon von der Gabelentz (1807–1874) und sein Umkreis: linguistische Briefwechsel mit Hermann Brockhaus, Heinrich Ewald, Karl Friedrich Neumann, August Friedrich Pott, Wilhelm Radloff und Karl Andree.* Wiesbaden: Harrassowitz.

About the authors

Annemete von Vogel is a descendant of the Gabelentz family and has co-edited several volumes on Georg von der Gabelentz which focus on his family background.
annemete.v.vogel@gmx.de

James McElvenny is an intellectual historian specialising in the history of linguistics. He is currently Newton International Fellow in the Department of Linguistics and English Language at the University of Edinburgh.
james.mcelvenny@mailbox.org

3 Georg von der Gabelentz as a pioneer of information structure

Els Elffers

Abstract

In 1869, Gabelentz introduced the terms 'psychological subject' and 'psychological predicate', which are generally regarded as predecessors of information structural terms such as 'theme' and 'rheme', and 'topic' and 'comment'. Together with some other linguists – Henri Weil (1818–1909), Philipp Wegener (1848–1916), Hermann Paul (1846–1921) – Gabelentz belonged to the first generation of theorists of information structure. This chapter focuses on Gabelentz's specific contribution to this area. Unlike his fellow pioneers in this area, Gabelentz developed his ideas exclusively in the context of questions about word order. He actively tried to apply them to a wide variety of languages, and to embed them in the framework of *Völkerpsychologie*. These contextual factors explain many unique characteristics of Gabelentz's variety of information structure, for example his exclusive focus on what is nowadays called the 'aboutness' aspect of information structure.

Keywords: psychological subject, psychological predicate, information structure, word order, 'aboutness', *Völkerpsychologie*

1 Introduction

Historians of linguistics and theorists of information structure are in complete agreement: Georg von der Gabelentz was a pioneer of information structure.[1] He counts among the first scholars who recognized that parts of

1 Cf., e.g., McElvenny's (2017b) handbook article about Gabelentz: 'Gabelentz's solution to the subject-predicate problem [...] fed directly into the development of present-day conceptions of information structure.' In Krifka's (2007, p. 40) introduction to information structure, the

McElvenny, James (ed.). *Georg von der Gabelentz and the Science of Language*. Taylor & Francis Group, 2019
DOI: 10.5117/9789462986244/CH03

sentences can be categorized in terms of their role in the communication between the speaker and the listener, the area which is today labelled 'information structure'.[2]

In Gabelentz's (1869) first publication on the subject, he discusses the sentences *Napoleon wurde bei Leipzig geschlagen* (lit. 'Napoleon was near Leipzig defeated') and *Bei Leipzig wurde Napoleon geschlagen* (lit. 'Near Leipzig was Napoleon defeated'). Both sentences have the same grammatical structure and their factual content is identical. But, as Gabelentz claims, there is a difference: in the first sentence, the speaker talks about Napoleon, in the second sentence about Leipzig's surroundings.[3]

Gabelentz accounts for this difference by saying that the first sentence element (in the first sentence *Napoleon*, in the second sentence *bei Leipzig*) always indicates what the speaker 'wants the listener to think about' (*worüber [der Redende] den Angeredeten denken lassen will*); he introduces the term 'psychological subject' (*das psychologische Subject*) to refer to this element. The subsequent parts of the sentence together constitute what is dubbed the 'psychological predicate' (*das psychologische Prädicat*): this element represents what the listener 'must think about' the psychological subject (*was [der Angeredete] darüber denken soll*) (Gabelentz, 1869, p. 378). The psychological subject and predicate may be identical to the grammatical subject and predicate, but they often differ from each other. In our second sentence they are different, because in both sentences the grammatical subject is *Napoleon*.

Gabelentz's analysis of these sentences in terms of psychological subject and predicate corresponds to analyses by present-day investigators of information structure. Their division of sentences into a 'topic' and a 'comment' or a 'theme' and a 'rheme' is based upon similar observations of communicative functions of sentence elements.

Alongside Gabelentz's ideas about psychological subject and predicate, more or less comparable ideas were presented, in partially different terms, by contemporary linguists such as Henri Weil (1818–1909), Hermann Paul (1846–1921) and Philipp Wegener (1848–1916). The information-structural insights of these nineteenth-century linguists are together generally regarded as anticipating more comprehensive theories of information structure,

present-day concepts "topic" and "comment" are said "to refer to what has been introduced into linguistic thinking as "psychological subject" and "psychological predicate" by von der Gabelentz (1869) [...].'

2 The term 'information structure' was introduced in Halliday (1967).

3 '[...] In dem einen Falle ist es Napoleon, in dem andern die Gegend bei Leipzig, von der ich reden [...] will [...]' (Gabelentz, 1869, p. 380).

which were developed in the course of the twentieth century, first in Prague structuralism, later in paradigms such as generative, functional and cognitive linguistics.[4] The terms 'psychological subject' and 'psychological predicate' were soon abandoned. 'Theme' and 'rheme' became their first successors in Prague structuralism. In later movements other terms were introduced, such as 'topic/comment', 'old or given information/new information', 'presupposition/focus'.

Just like their predecessors, these terms have a variable and partially dissimilar content, depending upon which linguistic phenomena are in focus (word order, accentuation, specific particles [as in Japanese], specific constructions, such as passive or [pseudo-]cleft constructions) and upon the theoretical frameworks in which they are embedded. In modern studies, the term 'information structure' covers the whole area.

The received view of the above-mentioned nineteenth-century linguists as joint precursors of twentieth-century information-structural approaches was already present in early Prague structuralism. Mathesius (1964 [1928], p. 67) identifies his own concepts 'theme' and 'enunciation' (later: 'rheme') with 'what is usually called the psychological subject and the psychological predicate, respectively'. Firbas (1974, pp. 10-11) discusses Mathesius's sources of inspiration and claims that 'in the last quarter of the nineteenth century, they [problems of 'theme' and 'rheme'] were dealt with by German scholars, for instance by G. von der Gabelentz, H. Paul and Ph. Wegener. Most of them used the terms 'psychological subject' and 'psychological predicate [...]'. Firbas also mentions Weil's work as especially important for Mathesius.

In this chapter, I shall focus on Gabelentz's ideas about information structure *avant la lettre*. In contrast to the bird's eye perspective of the received view, my perspective will be 'from within', so that specific details of Gabelentz's view and their specific intellectual context come into the picture.

In Section 2, I discuss information-structural thought in the pre-Gabelentz period. Although Gabelentz was a pioneer, he was not the very first to set foot in the area. In Sections 3-5, I present Gabelentz's ideas on psychological subject and predicate. Section 3 deals with their first tentative introduction in the 1869 article 'Ideen zu einer vergleichenden Syntax. Wort- und Satzstellung' (Ideas on comparative syntax. Word and sentence ordering), in the *Zeitschrift für Völkerpsychologie und Sprachwissenschaft*, and its

4 For example, according to Sgall (2003, p. 279) recent information structural insights are based upon 'the pioneering inquiry by H. Weil and further discussions by G. von der Gabelentz, H. Paul, P. Wegener. [...].'

sequel, published in two parts in 1875 in the same journal, entitled 'Weiteres zur vergleichenden Syntax. Wort- und Satzstellung' (More on comparative syntax. Word and sentence ordering). Section 4 deals with the use of the concepts 'psychological subject' and 'psychological predicate' as descriptive tools in Gabelentz's *Chinesische Grammatik* (Chinese grammar) (1881) and his subsequent discussion with its reviewer Franz Misteli (1841–1903) in 1887 in the *Zeitschrift für allgemeine Sprachwissenschaft*. Section 5 describes how the concepts are reflected in the broader context of Gabelentz's handbook *Die Sprachwissenschaft, ihre Aufgaben, Methoden und bisherigen Ergebnisse* (The Science of Language, Its aims, methods and results to date) (2016 [1891]). In Section 6, all these findings are analysed in more detail. Section 7 goes briefly into the reception of Gabelentz's information-structural ideas.

2 Predecessors

Identifying the earliest information-structural ideas is no straightforward matter. For example, a Dutch treatise on logic by Hendrik Laurensz. Spieghel (1549–1612) seems to foreshadow such ideas. Discussing subject and predicate, Spieghel (1962 [1585], p. 99) casually mentions some sentences in which the grammatical subject is not the 'real' subject being spoken about. For example, in the sentence 'Trouble follows the increasing money', 'money' is the subject of the proposition, 'because this is what is actually talked about' (*De zórghe vólght het anwassende gheld. Hier is gheld t' onderwurp des voorstels: want men eyghentlyck daarvan spreeckt* [...]). There is no telling how unique this observation was at that time.

Nevertheless, we can be sure that the first more systematic ideas about information structure were developed in nineteenth-century Germany and France. The *Frankfurter Gelehrtenverein* (Frankfurt society of scholars) proved fertile soil for discussions among grammarians such as Karl Ferdinand Becker (1775–1851) and Simon Herling (1780–1849) about the function of word order and accentuation. In Becker's work, some preliminary information-structural ideas can be found. Works of Becker and Herling inspired the German-born French linguist Weil in his extensive study of word order, which anticipates Gabelentz's ideas. H. Steinthal (1823–1899) knew Becker's works very well and criticized them severely, but these very works may also have inspired him to develop his own tentative information-structural ideas.

Before I discuss these three predecessors of Gabelentz, two conceptual distinctions must be introduced.

a) 'Speech-act scheme' and 'participant scheme'

Traditionally, the grammatical concepts 'subject' and 'predicate' were related to (i) a semantic structure in which the predicate (= verb + objects and adjuncts) conveys information about the subject, (ii) a semantic structure in which the subject is a participant (generally the initiator) in the process referred to by the predicate (= verb), alongside other participants (e.g. the direct object and the indirect object). Elffers (1991) introduced the terms 'speech-act scheme' and 'participant scheme' to refer to these respective structures, which are often intermingled in grammatical works. Only the speech-act scheme is relevant to applications of the concepts (usually after a 'redoubling' of the concepts into 'grammatical' and 'logical' or 'psychological' sub-concepts) to information-structural phenomena.[5]

b) 'Aboutness' and 'discourse anchoring'

In many recent studies, information-structural phenomena are classified in various groups. The main bipartition is made between 'aboutness' and 'discourse anchoring' (cf. Heusinger, 1999, p. 102). 'Aboutness' is primarily related to constituent order and is often described in terms of 'topic' and 'comment'. 'Discourse anchoring' is mainly related to accentuation and is often described in terms of 'given information' and 'new information'. For Gabelentz, 'aboutness' is clearly the underlying idea of his psychological subject and predicate. He analyses, for example, the sentence *Napoleon wurde bei Leipzig geschlagen* in terms of 'aboutness', which he relates to constituent order. When, in the same sentence, *geschlagen* or *bei Leipzig* is highlighted by a strong accent, there is also a difference in 'discourse anchoring': the different accentuations indicate that different material is presented as 'given' and 'new', in relation to the speech situation. There is certainly some interaction between 'aboutness' and 'discourse anchoring', but the distinction between them, after a long period of (near-)identification, is a valuable milestone in the development of theories of information structure.[6]

When we look at Gabelentz and his predecessors, taking into account both distinctions, we have to conclude that Gabelentz had no predecessors in the strict sense of the word. Nobody before Gabelentz redoubled the concepts 'subject' and 'predicate' according to the speech-act scheme in

5 In the rest of this article, I will use the verb 'redouble' and its derivations to refer to the division of the categories 'subject' and 'predicate' in, on the one hand, 'grammatical' subject and predicate, and, on the other hand, 'logical' or 'psychological' subject and predicate.

6 Halliday (1967) and Chafe (1976) are seminal articles in which the distinction is taken into account.

order to account for constituent order phenomena in terms of 'aboutness'. But there were (i) earlier redoublings of subject and predicate according to the participant scheme (e.g. Becker, ^{2}1842 [1836], Part 1); (ii) earlier redoublings of subject and predicate according to the speech-act scheme in terms of discourse anchoring (Becker, ^{2}1842 [1836], Part 2; Steinthal, 1855); and (iii) earlier independent 'aboutness' concepts, unrelated to subject and predicate, which account for word order phenomena (Weil, 1879 [1844]).

Ad (i): The first redoublings of subject and predicate can be found in eighteenth and nineteenth-century grammatical treatises. They were in line with the participant scheme. 'Logical' subjects and predicates were distinguished from their 'grammatical' counterparts in order to account for a restricted set of phenomena: active-passive sentence pairs (same logical subject, different grammatical subjects) and impersonal sentences (no/empty grammatical subject, logical subject in various grammatical roles). For example, Becker (^{2}1842 [1836], Part 1) adopts this type of analysis. He discusses the impersonal sentence *Es fehlt an Wein* ('Wine is lacking', lit. 'It lacks of wine'), which has *an Wein* as its alleged logical subject.

Ad (ii): In Becker (^{2}1842 [1836], Part 2) a remarkable change has implicitly come about: the difference between grammatical and logical subject and predicate is now reinterpreted in line with the speech-act scheme.[7] The same sentence *Es fehlt an Wein* is now denied a logical subject; *an Wein* is a part of the logical predicate. The logical subjects of active-passive counterparts, still identical in Part 1, are now different and identical to the grammatical subjects.

In Steinthal (1855, pp. 199-200) a greater variety of 'speech act' redoublings is presented. Differences between 'grammatical' and 'logical' subject and predicate are presented as arguments for separating grammar from logic.[8] Steinthal's examples are of the 'discourse anchoring' type, and accordingly provided with preceding contexts. For example, after the question *How did the patient sleep?*, the answer *He slept well* has *he slept* as its logical subject and *well* as its logical predicate.

Ad (iii): Weil (1879 [1844]) presents a book-length theory of word order. He argues that word order is entirely unrelated to subject and predicate.

7 Becker's re-analyses fit in with a general trend of re-analysing impersonal sentences along these lines. This trend, in turn, fits in with a trend of growing 'realism' in grammar, which involved closer examination of what sentences actually convey. Cf. Elffers (1991, sections 8.2 and 9.5.1).

8 The independence of grammar from logic is the central theme of Steinthal (1855). Becker is severely criticized in the book for his alleged 'logicism' in grammar.

Weil distinguishes the speaker's *marche syntaxique* (syntactic course) from his *marche des idées* (course of ideas). The latter *marche* predicts word order. Sentences begin by mentioning the *point de depart* (point of departure) or *notion initiale* (initial notion), about which the speaker wants to convey information. This information follows in the second element, the *but du discours* or *but de l'énoncé* (target of the discourse/ target of the expression).

Weil's and Gabelentz's theories are fundamentally different in that only the latter appeals to (redoubled) 'subject' and 'predicate' to account for observations of the communicative function of constituent order. But the observations themselves are similar. They resemble each other to a degree that caused Weil to remark that Gabelentz, in his two articles about comparative syntax (1869; 1875), 'did not say anything that I had not explained twenty-five years before him' (Weil, 1879 [1844], p. viii). This statement is exaggerated, if only because of Gabelentz's much wider and much more detailed coverage of language types in his examples and theoretical expositions. But the overall similarity between the two theories is undeniable.

In summary: there were some information-structural ideas before Gabelentz's first contributions in the area. Part of these ideas appeal to the traditional concepts 'subject' and 'predicate' in their speech-act variety, by redoubling them. The very idea of redoubling subject and predicate could be triggered by earlier redoublings according to the participant scheme.

Did Gabelentz know about these predecessors? His texts do not contain any indication. Weil is not mentioned at all. Becker is only mentioned once in *Die Sprachwissenschaft*, in a passage added in the posthumous second edition, as an example of outdated 'philosophical grammar', without any reference to his ideas about subject and predicate (Gabelentz, 2016 [1891], p. 17). Steinthal's views are amply discussed throughout Gabelentz's works, but his information-structural ideas are not mentioned.[9] Despite this lack of evidence, it is of course possible that Gabelentz read relevant texts of his predecessors. We shall not speculate on this issue here.[10]

9 In the introduction to his 1869 article, Gabelentz is emphatic in presenting its content as the results of his own efforts. At the same time he eschews the pretension of being their very first discoverer (*erster Entdecker*) and mentions the numerous sharp observations that can be found throughout Steinthal's work. However, explicit references to Steinthal are entirely lacking.

10 Knobloch (1988, p. 336) suggests that Gabelentz preferred not to mention the naturalized Frenchman Weil and to praise Steinthal because the article was published in a journal of which the non-Francophile Steinthal was an editor. Knobloch himself calls this idea speculation.

3 Gabelentz (1869 and 1875). Introduction of 'psychological subject' and 'psychological predicate'

3.1 First explorations: 'Ideen zu einer vergleichenden Syntax. Wort- und Satzstellung' (1869)

'Fragmentary' (*fragmentarisch*), 'sketchy' (*skizzenhaft*), 'uncertain' (*precär*), 'imperfect' (*mangelhaft*), 'a first' (*ein Erstling*): all these qualifications are applied by Gabelentz, in the first section of his 1869 article, to convince his readers of the modesty of these first explorations. These explorations did not involve information structure in the first place. Gabelentz wanted to explore, in conformity with the title of the article, the order of sentence constituents in various language types. Information-structural ideas are his answer to questions about order phenomena.

The answer is simple, at least for declarative main clauses.[11] Gabelentz (1869, p. 378) asks: 'What are we aiming at when we speak to another person? We want to evoke a thought in him. I believe this is something twofold: first, we direct his attention (his thought) toward something; secondly, we let him think this or that about this something'.[12] These elements are called 'psychological subject' and 'psychological predicate', respectively.

The natural order is: first the subject, second the predicate. For the grammatical sub-concepts this is, according to Gabelentz, a general trend in all languages known to him. For the psychological sub-concepts it must be a law without exception. The psychological subject can be omitted, however, when it is superfluous because it is already present in the listener's thought, as in *Was! Schon wieder da?* (What! Here again?), *Ja, und alles besorgt* (Yes, and everything fixed), where *du bist* (you are) and *ich habe* (I have) are omitted (Gabelentz, 1869, p. 379).

Gabelentz amply demonstrates the difference between grammatical and psychological subject and predicate. For example, adverbial constituents

11 Gabelentz disregards rhetorically motivated orders, inversions in relative sentences, imperatives and questions, and irregular orders due to, e.g., repetitions when the listener did not understand us (Gabelentz, 1869, p. 377). In the other texts we will discuss, similar restrictions are mentioned. Gabelentz prefers proverbs and other autonomous expressions as material for analysis. All this suggests some awareness of interaction between 'aboutness' and 'discourse anchoring' phenomena.

12 Original: 'Was bezweckt man nun, indem man zu einem andern etwas spricht? Man will dadurch einen Gedanken in ihm erwecken. Ich glaube, hierzu gehört doppeltes: erstens, dass man des Andern Aufmerksamkeit (sein Denken) auf etwas hinleite, zweitens, dass man ihn über dieses etwas das und das denken lasse [...].'

may fulfil both roles: psychological subject or (part of the) psychological predicate. Here the above-mentioned sentences *Napoleon wurde bei Leipzig geschlagen* and *Bei Leipzig wurde Napoleon geschlagen* are presented as instructive examples.

The grammatical predicate can be psychological subject, as well as psychological predicate. In German, the former is only possible in some regional varieties (e.g. *Kommt ein Vogel geflogen* [lit. 'Comes a bird flying']). In standard German and many other languages that do not allow an initial verb in declarative sentences, a 'surrogate' construction is available. A pronoun or adverb is added, as in the German sentence *Es drückt mich der Stiefel* (lit. 'It hurts me the boot'). This type of construction allows the grammatical predicate to become psychological subject, despite word order restrictions. Gabelentz observes that the sentence is an adequate answer to the question *What is wrong with you?,* whereas *Der Stiefel drückt mich* (The boot hurts me) can be an answer to *What is the matter with your boot?* (Gabelentz, 1869, p. 381).

'Surrogates' are also appealed to in Chinese which, according to Gabelentz, does not allow for a preposed direct object without an additional particle. This particle makes the constituent a pseudo-adverbial. In this way, the function of psychological subject becomes possible. Other 'preposing' mechanisms, such as passive and cleft constructions, are mentioned in this context.

Gabelentz's main bipartition of sentences in a psychological subject and predicate is not exhaustive, however. Both elements can be complicated by internal 'secondary predicates'; for example, genitives, possessives, adverbs, adjectives or participles. Their function is to reveal to the listener how to imagine the main elements in more detail. When postposed, as in Alifuru *tou sakit essa* (man sick a = a sick man), attributive modifications of nouns are examples of secondary psychological predicates. When preposed, as in Hungarian *jó bór* (good wine), these modifications have 'lost their predicative character' and have become 'parenthetical elements' in a complex constituent.

Gabelentz analyses finite verbs and their personal suffixes as psychological subjects and psychological predicates respectively.[13] Although Gabelentz does not draw this conclusion, this analysis involves another type of complex psychological subjects and predicates.

13 In Gabelentz's day, personal suffixes of finite verbs were generally regarded as (remnants of) grammatical subjects.

Of course, the linguist Gabelentz knew about grammatical restrictions on constituent order, which confine the free employability of constituent order as a means to express information structure. However, he does not at all thematize the issue in this article.[14] Some clues to his ideas about fixed constituent order may be drawn from a remark about the general character of his investigation: it belongs primarily to *Völkerpsychologie* (ethnopsychology). Gabelentz wants to relate order differences between languages to characteristics of their speakers' thought: 'The way a people arranges its concepts, its thoughts, is the way it arranges its sentences' (*Wie ein Volk seine Begriffe, seine Gedanken ordnet, so ordnet es seine Sätze*) (Gabelentz, 1869, p. 377). This suggests that Gabelentz aims at psychological explanations of fixed constituent order. Given his general law, such explanations should appeal to alleged psychological inclinations of language users to attribute the role of psychological subject or predicate to some specific type of constituent.

However, the article's content is not in agreement with this idea. Only one order phenomenon is very briefly explained in terms of psychology; not *Völkerpsychologie*, however, but general human psychology. The explanation is, moreover, mainly historical. It concerns the fixed order of the verbal stem and the personal verb suffix, analysed by Gabelentz (and many contemporary linguists) as grammatical predicate and grammatical subject. This order is the usual one in many Indo-European languages and those of other families, such as Hungarian: *tud-om* (I know). It indicates an original tendency of these suffixed grammatical subjects to function as psychological predicates, with the verbal stems as psychological subjects. This phenomenon is, according to Gabelentz, related to an alleged inclination attributed to the 'infancy of language' to draw the listener's attention first to conspicuous phenomena and sensations, indicated by the verbal stem, and to mention their cause, indicated by the grammatical subject, the suffix, only afterwards (Gabelentz, 1869, p. 382). This explanation incidentally reveals that for Gabelentz the participant scheme remains in some way connected with the grammatical subject and predicate.

In the article's final sentence, Gabelentz again stresses its tentative character. His aim was 'hinting rather than elaborating'.

14 Only one example of such a restriction is mentioned in passing: verb-preposing is impossible in some languages which also lack 'surrogate' constructions (e.g. Manchu, Mongolian) (Gabelentz, 1869, p. 381). Gabelentz also observes that the fewer the morphological means of a language are to express grammatical relations, the more fixed the constituent order is (ibid., p. 378).

3.2 Widening, deepening and modifying: 'Weiteres zur vergleichenden Syntax. Wort- und Satzstellung' (1875)

Elaboration is offered in the 1875 sequel in two parts. This text is more than eight times longer than the first article, which has only nine pages. The length is mainly due to the wealth of illustrative language data, the majority taken from non-Indo-European languages, so that many examples, translations and explanations were necessary. For example, Gabelentz extensively discusses languages such as Japanese, in which psychological subjects and predicates are marked by specific particles.

Gabelentz begins by recognizing that the earlier article was 'too briefly and poorly conceived' (*zu kurz und knapp gefasst*) and that it has given rise to misunderstandings and objections. He hopes to present corrections and further elucidations.

One objection is rigorously rejected, namely that constituents owe their position to accentuation. According to this *Betonungstheorie* (accentuation theory), emphasized elements prefer an initial sentence position. Gabelentz denies this. Emphasized/stressed elements occur in all positions, irrespective of constituent order. Unlike accentuation, the order of constituents is determined not by rhetorical emphasis, but by their function of psychological subject or psychological predicate (Gabelentz, 1875, pp. 135-137).[15]

These concepts, however, differ from their 1869 predecessors. There is now not a static division, but a dynamic one. During the presentation of a sentence, 'every new constituent further determines the preceding ones; it functions, in other words, as a predicate to them, while they function as its subject' (ibid., p. 137).

By means of various transpositions of the German sentence *Eine Fabrik verfertigt Hüte aus Papier* (A factory manufactures hats from paper), Gabelentz explains how, in all possible orders, a newly added element further completes the preceding totality. Gabelentz emphasizes that this process of successive additions of predicates exclusively applies to the listener. He combines every new constituent he hears with the totality heard before into a new unity. Every constituent provides a new aspect of the picture he receives, makes it clearer and more determinate. The speaker, however, has the complete picture at his disposal before he starts speaking. The

15 By stressing the functional meaning of constituent order, Gabelentz strengthens his argument, presented earlier in the article, against the view, still current in his day, that constituent order belongs to rhetoric. For Gabelentz, it belongs to syntax. Accentuation, on the contrary, belongs to rhetoric.

order of presentation of its constituents is, as far as it is not restricted by grammatical rules, a question of free choice, comparable to the choice of a painter in favour of one element of a scene as its general theme, and the other elements as further clarifying additions.[16]

The new dynamic (*fortschreitend*) idea of psychological subject and predicate is, unlike its predecessor, not simply presented as a law without exceptions. On the one hand, Gabelentz stresses its general-psychological plausibility; on the other hand, its applicability to German sentences is no guarantee for universality. Gabelentz warns against *a priori* speculation and emphasizes the need for further empirical support from a wide variety of languages (ibid., pp. 139, 337).

The new view of psychological subject and predicate is in fact not continued in the rest of the article. In all subsequent examples, only one boundary between psychological subject and psychological predicate is discussed. Generally, this boundary is established after the first constituent. When other boundary locations are mentioned, as in the German example *Dort spricht N. mit X* (lit. 'There talks N. with X'), these are not successive boundaries during the hearing process, but alternative single boundaries, corresponding to different divisions of the sentence into psychological subject and predicate.

Above, Gabelentz's view of the speaker was described as freely choosing constituent order, as far as it is not restricted by grammatical rules. Indeed, unlike the earlier article, its sequel amply discusses fixed constituent order in a broad variety of languages. For German, this results in an impressive new theory of constituent order, in which, for declarative sentences, three syntactic positions can be distinguished.[17]

Fixed constituent order is now explicitly dealt with as an exception to the general law of psychological subject and predicate. 'Such rules concerning the position of certain grammatical sentence elements restrict the freedom to make every sentence element we want psychological subject or predicate; therefore, they are exceptions to the basic rule' (ibid., p. 336). But they do not refute the basic rule; they only tell us that another force may reduce the force of the law. In the same way, says Gabelentz, the law of gravitation is not refuted by the fact that 'my desk stands quietly on the floor instead of rushing to the centre of the earth' (ibid., p. 336).

16 In this context and elsewhere, Gabelentz occasionally applies the term *Thema* (theme) to indicate the psychological subject.

17 Several linguists have evaluated this constituent order theory as far ahead of his time, cf. Knobloch (1988, pp. 337-338). According to Noordegraaf (1997), Gabelentz anticipates modern ideas about strict V2 in German main clauses.

Language use (*Sprachgebrauch*) is now invoked as the main explanatory force behind grammatical order restrictions. They result from a process of conventionalization of frequently used sequences. This process is explained in two ways. Firstly, there is a semantic frequency effect: the content of constituents may facilitate their appearance in specific positions. Other positions, theoretically possible as well, are forgotten, so that what started as a motivated regularity becomes a grammatical rule. Secondly, there is the standardizing effect of socializing. In principle, every individual person develops his own linguistic preferences in situations where choice is possible. But people group together and imitate each other's preferences, especially those of people in authority. Along these lines, fashions and conventions arise, resulting in rules (ibid., p. 160).

The *Völkerpsychologie* approach of fixed constituent order, still hinted at in the first article, is not mentioned at all in its sequel, let alone thematized. The adjective *völkerpsychologisch* occurs once, and in the very broad sense of 'synchronic linguistics', as opposed to 'genealogical linguistics' (ibid., p. 130). But ethnopsychological explanations of fixed constituent order are not entirely absent. One section of the article discusses the general theme 'subjectivity in language'. In this context, languages with obligatory sentence-initial verbal elements are dealt with. In line with the 'infancy of language' idea in the first article, this is explained by assuming a naive, egocentric attitude: direct sensations (the grammatical predicate) are immediately introduced as psychological subject; information about their cause (the grammatical subject) follows afterwards. The idea that this is a symptom of low intellect is, however, rejected, because such languages are spoken by peoples of both high and low cognitive development. The determining factor is, according to Gabelentz, a difference in vivacity of sensations between northern and southern peoples (ibid., pp. 310-311).

Another example appears when Gabelentz discusses the Latin preference for the verb-final order, which contrasts with the general Latin freedom of constituent order. He refers to 'the peculiarly pedantic character of the ancient Roman' (*der eigenthümlich pedantische Sinn des alten Römers*) (ibid., p. 312) which causes his inclination to overregulate everything, also in language. But in order to explain the restrictions actually adopted in Latin, Gabelentz has to provide a further explanation. In the case of the final verb, an appeal to parenthesis is made. This means, as in the first article: secondary predication by modifying preposed elements, which lose their autonomous predicative character and become infixed elements of one complex constituent, representing one complex concept (*Gesammtbegriff*) (ibid., p. 163). This idea is now applied, not only to preposed attributive noun modifiers, but also to objects and adjuncts preceding the verb.

Parenthesis, as free choice and in conventionalized fixed constituent order, is extensively discussed, and especially its variation with secondary predication in French attributive adjectives.[18] In the end, parenthesis is even regarded as a separate word order law, alongside the law of the forward-moving psychological subject. According to Gabelentz, 'the two laws, the law of the forward-moving psychological subject and the law of the infixed additional predicate interweave with each other continuously, like warp and weft' (ibid., p. 157).[19]

With respect to the law of psychological subject and predicate, we observed that Gabelentz regards fixed constituent order as an exception but not as a refutation. He claims that exceptions may even strengthen the law, because in a variety of languages, they allow for exceptions themselves. These second-grade exceptions are consistent with the basic law, which means 'the more weighty support' (*um so schwererwiegende Argumente*) for this law (ibid., p. 336). Although 'exceptions to exceptions' are not discussed in the first article, Gabelentz now claims that they triggered his very first ideas about psychological subject and predicate (ibid., p. 140).[20] Second-grade exceptions play an important role in Gabelentz's Chinese grammar, to which I now turn.

4 Applying the theory to Chinese. Discussion with Misteli

In the final section of 'Weiteres zur vergleichenden Syntax', Gabelentz mentions his initial plan to devote a section to Chinese. He rejected it because explaining Chinese constituent order would require too much background information for readers who do not know the language. This is due in particular to the numerous second-grade exceptions, by which the

18 The semantic difference, partially conventionalized, between parenthesis and secondary predication (roughly: inherent vs. incidental property) is demonstrated through many French examples. The most remarkable pair (from a present-day perspective) – *un savant professeur* (a learned professor) vs. *une femme savante* (a learned woman) – according to Gabelentz, in the former case a usual property, in the latter case a rare property (Gabelentz, 1875, p. 164).

19 Original: 'Zwei Gesetze, das von dem fortschreitenden psychologischen Subjekte und das von den infigirten, beiläufigen Prädikaten kreuzen einander fortwährend, wie Aufzug und Einschlag.'

20 This autobiographical statement is repeated in Gabelentz (1887, p. 102). He speaks about his discovery of exceptions to fixed word orders and his search for their underlying force, which, according to Gabelentz, must be mighty because it can defeat one of the strongest powers: habit.

Chinese language 'likes to break the ties of its word order laws' (*die Banden seiner Wortfolgegesetze zu sprengen liebt*) (Gabelentz, 1875, p. 335).

In Gabelentz's *Chinesische Grammatik*, psychological subjects often appear as 'exceptions to exceptions'. The fixed order of Chinese constituents may be reversed by preposing elements that normally cannot be used in sentence-initial position. This may occur for various reasons; the most important one is the psychological subject function of the preposed element (Gabelentz, 1881, p. 432). In this case, the element has to occupy an 'absolute' position outside the nexus of the sentence and it has to be repeated by a pronoun within the nexus, as in French constructions of the type *Votre frère, j'ai ses nouvelles* (Your brother, I heard from him) (ibid., pp. 114-115, 432).

Special attention is paid to the usual Chinese narrative style of sentence-initial temporal, local and causal adverbials, in this sequence. The grammatical subject, generally the first constituent, follows after these elements. According to Gabelentz, this order reflects a psychological subject and predicate order, characteristic for Chinese. The adverbials constitute a series of psychological subjects (ibid., p. 150). In this position, adverbials are single words. As psychological predicates in final position, however, they take the shape of prepositional phrases.

Psychological subject and predicate are also dealt with in other contexts, for example when particles are discussed whose function consists of signalling these elements (ibid., p. 250).

Misteli's (1887) lengthy review of Gabelentz's Chinese grammar is rather critical in general, but his overall attitude to Gabelentz's information-structural ideas is positive. For the first time, use is made of Gabelentz's terminological innovation to create a clear distinction of three types of subject and predicate: 'grammatical' (formally defined), 'logical' (semantically defined according to the participant scheme) and 'psychological' (semantically defined according to the speech-act scheme). For example, in *Es schwimmt dort ein Cypressenboot* (lit. 'It swims there a cypress boat'), *es* is grammatical subject, *schwimmt* psychological subject, and *Cypressenboot* logical subject and also psychological predicate (Misteli, 1887, p. 60).

However, Misteli also argues against Gabelentz's terminology. Focusing on Gabelentz's frequent appeal to the concept of 'psychological subject' in the case of 'exceptions to exceptions', Misteli claims that Gabelentz's term is actually synonymous with Berthold Delbrück's (1842–1922) term *okkasionelle Stellung* (occasional position), which also applies to exceptionally preposed sentence elements in absolute position (Misteli refers to Delbrücks 1878 *Die altindisiche Wortfolge*). Similarly, Gabelentz's psychological predicate should correspond to what Delbrück calls *Schleppe* (train), an additional element after the formal sentence end. Misteli prefers Delbrück's 'sober and clear' terms to Gabelentz's less clear terminology: its appeal to psychology does not add anything substantial, according to Misteli. He also rejects Gabelentz's explanation of the order of adverbials in Chinese narrative style: 'psychological subject is only a stronger term for initial, occasional position, as long as Gabelentz fails to indicate exactly why temporal and local adjuncts deserve this preferred role in China' (ibid., p. 63).[21] Misteli exhibits a general mistrust of the value of psychology for grammar: psychology cannot explain what grammar itself cannot explain (ibid., p. 65).

Misteli also rejects Gabelentz's view that accentuation is unrelated to word order and irrelevant to psychological subject and predicate. He claims that both accentuation and word order serve the same aim. A psychological subject can be preposed or be accentuated or both. According to Gabelentz, *Heute hab ich Kopfweh* (Today I have a headache) has *heute* as its psychological subject, irrespective of accentuation. For Misteli, there are three ways to present *heute* as psychological subject: *Heute habe ich Kopfweh* (without prominent accentuation), *Ich habe* ***heute*** *Kopfweh* and ***Heute*** *habe ich Kopfweh* (prominently accentuated elements in bold) (ibid., p. 62).[22]

In Gabelentz's reaction to Misteli, the latter criticism is answered by a new demonstration of the independence of constituent order and accentuation. Any constituent can be stressed, irrespective of its position. A constituent is not stressed because it is the psychological subject, but because it is contrasted with alternatives.

As to the issue of terminological adequacy, both Misteli's identification of psychological subject and predicate with Delbrück's *okkasionelle Stellung* and *Schleppe*, and his general criticism of psychological involvement in grammar are rigorously rejected. Although the phenomena covered by Delbrück's terms are psychological subjects and predicates respectively, the reverse is not true. Not all psychological subjects have an irregular initial position and only very few psychological predicates are trains, irregularly postposed elements.

Arguing against Misteli's negative view of the role of psychology in grammar, Gabelentz emphasizes that he deals with grammatical phenomena with a real psychological cause. Gabelentz asks what would be gained if

21 Original: 'Psychologisches Subject ist nur ein vertiefter Ausdruck für: Stellung zu Anfange, okkasionelle Stellung – so lange als der Vf. nicht anzugeben vermag, warum gerade Zeit- und Ortsbestimmungen in China einer solchen Bevorzugung sich erfreuen [...]'.

22 Misteli was not the only nineteenth-century linguist who assumed that the sentence element bearing the main accent is the psychological subject. Paul was the first to criticize confusions on this issue (cf. Elffers, 1991, p. 278).

he had only talked about earlier and later sentence elements, and kept his psychological subjects and predicates to himself. He would merely have indicated positions of sentence elements, something that everyone can observe, but the essence, the function (*das Wesen, die Funktion*), which he was after, would then have remained hidden (Gabelentz, 1887, p. 106).

A psychological explanation of the Chinese trend to prepose temporal and local adverbials can be given in good conscience, according to Gabelentz: Chinese are 'historians par excellence' and their historic style is purely chronological. Narratives start with temporal and local settings, in which the successive occurrences are presented subsequently (ibid., pp. 105-106). This style is reflected in Chinese sentences.

5 Integrating the theory in *Die Sprachwissenschaft*

The fourth and most extensive part of Gabelentz's handbook *Die Sprachwissenschaft*, entitled 'Die allgemeine Sprachwissenschaft' (General Linguistics), is about language competence in the most general sense: its mental basis as well as the scope of its manifestations. The third chapter 'Inhalt und Form der Rede' (Content and form of speech; translated in Chapter 6 of this volume) contains sections on universal language phenomena. The section 'Die Wortstellung' (Word order) has as its subtitle 'Psychological subject and predicate', with a footnote reference to the three earlier articles discussed above. A separate section (the next one) is devoted to accentuation (*Betonung*); the subjects are, in accordance with Gabelentz's standpoint, strictly separated.

A new element in the discussion of psychological subject and predicate is the language-genetic perspective, which dominates more than half of the section. This perspective reflects Gabelentz's idea of the foundational task of General Linguistics. As a child of his time, Gabelentz looks at foundations in a historical-genetic way.

Psychological subject and predicate are introduced as very early sentence elements, belonging to the 'prehistory of human speech'. They constitute 'the preliminary stage of syntax' (Gabelentz, 2016 [1891], p. 388). Gabelentz presents their rise and development in detail. Like many of his contemporaries, he amply appeals to the alleged parallelism between ontogenesis and phylogenesis: general characteristics of child language are attributed to primitive man.

According to Gabelentz, primitive one-word sentences were gradually concatenated, initially without any logical relationship between them. Later on, these utterances could become a unity of complementary elements, with a proto-subject-predicate relation, first as separate mental occurrences, later as related elements of one mental event. So two exclamations 'There!... Deer!' can become one exclamation 'There deer!' (ibid., p. 388).

Just as in child language, the relation between the elements of utterances of primitive man is still largely indeterminate. But they certainly relate to each other as subject and predicate, in the 'dynamic' variety: 'What is heard relates to what is expected as a subject to a predicate' (ibid., p. 389). Gabelentz compares the process with a telegraph with two paper rolls: a growing roll with inscriptions and the blank paper slip that has to be written on.

The difference between the speaker and the listener is again emphasized: only the speaker knows beforehand both what is already written on one roll and what will appear on the blank slip. But, unlike in the 1875 article, the speaker is now assumed to imitate his own previous thought process in the order of presenting sentence elements to the listener. So their psychical processes are similar after all (ibid., p. 389).

Gabelentz claims that psychological subject and predicate are very old, embryonic grammatical categories. For a long time, they remained 'logically undetermined' (*logisch unbestimmt*), because they did not yet take account of the diversity of logical relations (ibid., p. 393).

The view of psychological subject and predicate as categories inherent in human speech supports the idea of their universality. Indeed, Gabelentz appeals to this argument: 'If my deduction is correct, this category must be present in all languages' (*Wenn anders meine Deduction richtig war, so muss diese Kategorie in allen Sprachen gegenwartig sein*) (ibid., p. 393). At the same time, he continues to emphasize the necessity for empirical support (*inductiven Beweis*) (ibid., p. 370).

Research on empirical data must proceed carefully, because constituent order phenomena may differ widely among different languages. To illustrate this, Gabelentz briefly discusses themes earlier dealt with, such as fixed constituent order and 'exceptions to exceptions'.[23] The examples are partially

23 In this context Gabelentz repeats his warning against interaction with contextual factors. Simple autonomous statements, such as proverbs, are the best data; cf. his discussion of the sentences *Mit Speck fängt man Mäuse* (lit. 'With bacon catches one mice') and *Mäuse fängt man mit Speck* (lit. 'Mice catches one with bacon'). Given the meaning of the proverb, the first sentence is a statement about means to catch foolish people. The second is about foolish people (Gabelentz, 2016 [1891], p. 390). Another complicating factor, next to contextual factors, is euphony, which causes rhetorical and poetical influences on constituent order. This factor was not discussed earlier (but see note 10) and is introduced only in the second 1901 edition of *Die Sprachwissenschaft*. Apart from some examples, there are no further substantial differences between the 1891 version and the 1901 version of the *Wortstellung* section.

identical to those in earlier work, but German examples are more prominent now, in accordance with the didactic aim of the book.[24]

A remarkable feature of the section is a partial return to *Völkerpsychologie*. Although the term is not used, the way in which Gabelentz now speaks about fixed constituent order reminds us of his 1869 article. Fixed order is regarded as a signal of a conventional 'march line' (*Marschroute*) of thought elements. Moreover, such 'national thought conventions' are seen as manifestations of national spirit (*Volksgeist*) (ibid., p. 392).

This return to *Völkerpsychologie* is in line with the prominent role of *Sprachwürderung* (language evaluation) in *Die Sprachwissenschaft*. In this programme, language-*Volksgeist* relations are the central theme. The chapter 'Sprachwürderung' is by far the largest of the 'General Linguistics' part of *Die Sprachwissenschaft*.[25] In this chapter, we meet the psychological subject and predicate once again, when Gabelentz discusses languages with a sentence-initial finite verb. The 'southern' sensitivity for new impressions, connected with this order, is now expanded to a host of alleged psychological characteristics, varying from positive (e.g. studiousness) to negative (e.g. greed), of the speakers of the Malay and Semitic languages in question (ibid., pp. 436-37). Discussing the thought patterns leading to these language-*Volksgeist* connections would be addressing a totally new subject (but see McElvenny, 2017a).

6 Psychological subject and predicate in context

In the previous sections, we followed Gabelentz's information-structural ideas, from their first tentative formulations until their integration into General Linguistics as the most fundamental syntactic categories. Some conclusions can be drawn from this overview.

Not surprisingly, Gabelentz's anticipation of modern theories of information structure is amply confirmed, if only by his many examples of information-structural sentence analyses in terms of psychological subject and predicate. Moreover, his description of the content of 'psychological subject' and 'psychological predicate', his exclusive focus on constituent

24 According to Gabelentz, examples must be 'close to' the reader in order to be instructive, which means that they are preferably 'derived from the mother tongue'. This area lies beyond Gabelentz's main professional interests, but he proudly trusts his native speaker intuitions (Gabelentz, 2016 [1891], p. ii; cf. also Elffers, 2012, p. 60).

25 The 'Sprachwürderung' chapter (90 pages) constitutes half of the General Linguistics part, which consists of eight chapters.

order, and his disregard of accentuation classify him as an investigator of the 'aboutness' aspect of information structure.

All these formulations are, however, somewhat misleading, in that they do not take into account the intellectual context of Gabelentz's research. We observed time and again that Gabelentz's central problem was constituent order, not information structure. Our overview confirms the idea that Gabelentz's information-structural ideas arose as an important but nevertheless partial solution to this problem, alongside other solutions (parenthesis, fixed order). Similarly, his disregard of accentuation was not due to an explicit choice within the area of information structure, but to his focus on word order, together with his conviction that accentuation is unrelated to word order and belongs to a different area.[26]

The centrality of order problems also explains Gabelentz's discussion of phenomena that do not belong to information structure, such as fixed constituent order in German declarative sentences. The order of nouns and their attributive modifiers is another example. This constituent-internal order has nothing to do with sentential aspects such as 'aboutness', or 'discourse anchoring'. However, the above-mentioned general nineteenth-century trend of dealing with problems in a genetic way narrows the distance to some degree: attribution was regarded as – originally – identical with predication.[27] In the same way, Gabelentz's application of psychological subject and predicate to a morphological phenomenon, the order of stem and suffix in the finite verb, becomes more understandable in light of the contemporary analysis of these elements as – originally – grammatical predicate and subject.

Apart from the class of phenomena discussed, our overview shows that there are several aspects of Gabelentz's theoretical views which reflect the nineteenth-century linguistic climate and make them differ considerably from present-day theories of information structure. These points have to be reckoned with when we consider Gabelentz as a pioneer of information structure. In the next subsections, three aspects will be dealt with.

26 In Gabelentz (1887 and 2016 [1891]), accentuation is no longer regarded as purely rhetorical, but as related to contextual 'alternatives'. This view also underlies recent 'discourse anchoring' definitions of 'focus' (cf. Krifka, 2007, p.18). Also Gabelentz's (1875, p. 148) discourse-oriented discussion of various accentuations of the conditional auxiliary sentence *Spricht X mit Y* (lit. 'talks X with Y') bears witness of some awareness of information-structural aspects of accentuation. But Gabelentz did not elaborate these isolated observations into a comprehensive theory.

27 Gabelentz (1869, p. 383) reduces the difference between attributive and predicative adjectives to their position: behind the article or behind the copula. According to Paul (²1886 [1880]), the attribute is 'nothing but a degraded predicate' (*nichts anderes als ein degradiertes Prädikat*).

6.1 Völkerpsychologie

In his 1869 article, Gabelentz claimed that the value of his task was primarily *völkerpsychologisch* (ethnopsychological). Indeed, we observed him as an ethnopsychologist, focusing, unlike present-day investigators of information structure, on fixed or preferred constituent orders and connecting them with preferred types of psychological subjects of predicates in language communities with certain characteristics. On the other hand, such examples are scarce. Gabelentz's main focus is on variable constituent orders, which are related to individual speakers' free choices of psychological subject and predicate.

A recurrent theme is the interaction of fixed order and free choice, the 'exceptions to exceptions'. Fixed constituent orders can be broken by the force of the underlying law of psychological subject and predicate. The result is often a preposed constituent in absolute position, functioning as psychological subject. This phenomenon is so prominent in Gabelentz's texts that Misteli mistook it for the normal situation for the psychological subject and equated it with Delbrück's *okkasionnelle Stellung*.

In Chinese narrative style, the exceptions exhibit a regularity themselves: preposed temporal and local adverbials are a common phenomenon. Given the ethnopsychological perspective, relations between this regularity and characteristics of the Chinese people seem a natural theme to explore. However, Gabelentz deals with this theme only in reaction to Misteli's challenge.

We cannot avoid the conclusion that Gabelentz was a wavering *Völkerpsychologe*. This conclusion tallies with a general trend in Gabelentz's work. On the one hand, *Völkerpsychologie* was very dear to him. It was narrowly connected to his lifelong attempts to penetrate deep into exotic languages, including the *Weltanschauung* they embody (cf., for example, Gabelentz, 1887, p. 94). On the other hand, he was very well aware of the pitfalls of this research programme and he abhorred the dubious and prejudiced argumentation patterns in many of his colleagues' ethnopsychological works (cf. Elffers, 2012, pp. 66-67).

6.2 Mental processes

Modern information structure is a discourse-oriented area of linguistics, which implies that the speaker, the listener and their communicative intentions are taken into account. But their internal thought processes are left out of the picture. In nineteenth-century linguistics, the situation was different. According to the then-current psychologism, sentences were regarded as reflecting thought processes: successions of representations and their associative and apperceptive relations. Linguistic meaning was equated with these intra-psychical events. All information-structural terms used by Gabelentz and his contemporaries allegedly referred to entities and occurrences in the speaker's and the listener's mind during thought processes.

Prague structuralist theorists of information structure regarded their predecessors' theories as defective because of this psychologism. Vachek claims that 'Mathesius's merit lies in the fact that by replacing the psychologistic terms with those of functional linguistics, he made the whole conception really workable [...]' (Vachek, 1966, pp. 89-90). Gabelentz is explicitly mentioned as a source of psychologism: 'G. von der Gabelentz, for instance, speaks of the difference between the "psychological subject" and the "psychological predicate". These [...] psychologistic terms proved methodologically barren and operationally intractable. By defining the difference in functional terms, Mathesius made it both highly operative and methodologically fruitful' (ibid., p. 111).

However, this view neglects the distinction between ideas belonging to linguistic metatheory (all aspects of philosophy of linguistics) and ideas belonging to linguistic theory as developed in actual linguistic research practice. Actual linguistic research may be largely independent of the ontological and epistemological status attributed to the objects of the discipline in successive paradigms. Gabelentz's information-structural ideas might be – and have been – useful for linguists (among them Prague linguists) who rejected the nineteenth-century psychologistic metatheory.

Confusion of metatheory and research practice also explains the misunderstanding between Misteli and Gabelentz about psychology and grammar. Misteli, while not a genuine anti-psychologist, claims that psychological terminology in grammar is an empty and mystifying addition.[28] Gabelentz replies that cancelling these terms would come down to mere describing, not explaining order phenomena. However, Delbrück's non-psychological terminology, as explained by Misteli, refers to order phenomena that are functionally explained: the content of the preposed elements in question is

28 Remarkably, to Misteli's sentence 'Let us not fool ourselves about the value of grammar in psychology' (*Täusche man sich doch nicht über den Wert der Psychologie innerhalb der Grammatik*) (Misteli, 1887, p. 63) a footnote is added in which Techmer, the editor of the journal, replies that the value of psychology in general for grammar is here underestimated. Evidently, the relation between psychology and linguistics was a sore point in those days (and it still is).

emphasized (*Sinnbetonung*) (Misteli, 1887, p. 61). Gabelentz misunderstands this. According to his metatheory, meanings are psychological entities and have to be described as such. Removing psychological terminology is removing function and meaning entirely, so that only observable forms remain.

Despite the necessity of distinguishing metatheory and practice, they are sometimes intermingled. In Gabelentz's theory, this is the case in the 'dynamic' version of psychological subject and predicate. The idea is rooted in purely psychological speculation about what is going on in the listener's mind when he hears a sequence of sentence elements. Linguistically, there is no support for the idea of a cumulative 'aboutness' relation. A major problem is that all sentences should have the same, predictable series of information structures. Different partitions of a sentence corresponding to different 'aboutness' relations, as in Gabelentz's example *Dort spricht X mit N* (lit. 'There talks X with N'), are actually impossible.

We observed already that examples in which Gabelentz applies his information structural concepts are virtually never analysed in terms of the dynamic view. Sentences are attributed one or more binary structures. The dynamic view remains a dead letter.[29]

6.3 Subject and predicate

Information structure is today regarded as an autonomous level of description, separate from the grammatical level, to which the concepts 'subject' and 'predicate' belong. Unlike these two-level views, most early information-structural ideas are formulated in terms of 'subject' and 'predicate', which are, at the same time, vindicated as grammatical concepts. This difference evokes the question of how the relation between grammar and information structure *avant la lettre* was conceived. A terminological distinction was made between grammatical and logical/psychological subject and predicate, but what does this mean? The latter pair of concepts retains the semantic features of the earlier unredoubled pair and receives new formal features (features of constituent order, in Gabelentz's case). The former pair retains the formal features of the earlier unredoubled concepts, but seems to stay behind without its semantic features, an unlikely situation for a grammatical category in those days. Also, the very choice in favour of a terminological

29 Gabelentz's 'dynamic' view actually anticipates some twentieth-century non-information-structural semantic theories which are based upon the linear sequence of elements and its alleged interpretive effects on the listener. Bolinger's (^{2}1965 [1952]) theory of linear modification is a clear example.

subdivision instead of unrelated terms, as in Weil's two-level theory, suggests that the original speech-act meaning of the subject-predicate relation is still relevant to the grammatical subject and predicate in some way, which precludes a strict separation of levels.[30]

To get a grip on this situation, we shall look again at the genetic perspective of nineteenth-century language research. This perspective implied that, despite language change, the original situation remains relevant to the present-day situation. A typical nineteenth-century idea about the origin of grammatical categories was that grammatical structure started with the speech-act scheme of subject and predicate. Gradually, these elements of the primitive sentence were enriched, formally with suffixes, semantically with a variety of logical relations. But the speech-act scheme remained relevant. This resulted, in our terms, in a mapping of the participant scheme on the speech-act scheme and a restriction of the freedom to apply the speech-act scheme in any way we like.

Early theorists of information structure assume that the speaker's need for freedom allowed him sometimes to break the restrictions caused by the coinciding functions, and apply the speech-act scheme freely, irrespective of the participant roles to which it is normally connected. This is what occurs when grammatical and psychological subject and predicate do not coincide (cf. Elffers, 1991, Chapter 10, for a more comprehensive discussion of this line of reasoning). So in the case of occasional redoubling, a psychological mechanism ('need for freedom') causes a temporary backgrounding of the speech-act meaning of the grammatical subject and predicate, not its disappearance. The original one-level situation is actually vindicated, albeit in a more complex way.

Of the early students of information structure, Becker (^{2}1842 [1836]) and Paul (^{2}1886 [1880]) present this explanatory narrative most explicitly. Gabelentz describes it only partially. In *Die Sprachwissenschaft* we observed the idea of speech-act subject and predicate as original syntactic categories and of their gradual enrichment with logical relations. Although the rest of the narrative is not mentioned explicitly, the clear assumption of one basic subject-predicate structure in language origin and in child language classifies Gabelentz among the early one-level information structure investigators. His emphasis on the status of psychological subject and predicate as syntactic categories supports this conclusion.

30 For Steinthal and for some philosophers (cf. section 7), the situation is different, because their information-structural concepts are incorporated in a separate discipline: (psychological) logic.

7 Reception of Gabelentz's information-structural ideas

We observed earlier that Gabelentz, as a theorist of information structure, had no genuine precursors. He had no genuine successors either. No linguist continued his project of analysing constituent order in terms of one-level information-structural concepts. In the first post-Gabelentz period, constituent order was hardly discussed. The idea of one-level information structure – that is, subject-predicate redoubling – was, however, adopted by several others, but mainly in the 'discourse anchoring' variety. Wegener (1885) and Paul ([2]1886 [1880]) both redoubled subject and predicate along these lines. Wegener's concepts 'exposition' (a term he prefers to 'logical subject') and 'logical predicate', as well as Paul's 'psychological subject' and 'psychological predicate' (terms borrowed from Gabelentz) are formally defined in terms of accentuation, and semantically in terms of contextually determined choices between alternatives. Word order is not a defining feature: both elements can occupy initial and final positions.

Wegener's and Paul's examples resemble those presented in Steinthal (1855; cf. Section 2 above), but Steinthal is not referred to as a precursor. Wegener nowhere refers to other scholars, so we do not know whether Steinthal's examples were known to him.[31] Paul refers to Gabelentz and to Wegener and criticizes them both.[32]

Subject-predicate redoubling became an idea – especially as a result of Paul's well-known book – that many linguists were more or less acquainted with. However, only very few of them, as well as some psychologically oriented philosophers, actually adopted it, and always in predominantly discourse anchoring varieties.[33]

Redoubling itself soon became controversial. Marty's (1897) and especially Wundt's ([4]1922 [1900]) criticisms were influential. Both scholars claim that redoubling is caused by confusion about what they regard as the 'real' meaning of 'subject' and 'predicate'. This 'real' meaning is assumed always

31 Only in the general introductory section of Wegener (1885) are a few linguists mentioned in a positive way, among them Steinthal.

32 Paul equates his concept of 'psychological subject' with Gabelentz's and thus rejects Gabelentz's view that psychological subjects are always sentence-initial. He also rejects the view, unjustifiably attributed to Wegener, that psychological predicates are usually sentence-initial.

33 Cf. Elffers (1991, section 9.3). The philosopher Fritz Mauthner (1849–1923) is an exception: his view of subject and predicate is very similar to Gabelentz's 'dynamic' view. Cf. his instructive formula for the successive subject-predicate relations between constituents a–e: '{[(a + b) + c] + d} + e' (Mauthner, [3]1913 [1902], p. 218).

to apply to the grammatical subject and predicate, so redoubling is rejected.[34] Wundt suggests that the phenomena captured by psychological subject and predicate have to be described by means of the term *dominierende Vorstellung* (dominant representation). This two-level view, the first after Weil, was actually rather confusing, but it inspired several linguists to develop their own two-level concepts and/or terms.[35]

During the first decades of the twentieth century, subject and predicate were gradually redefined in terms of (pre-)structuralist constituent theory, which reduced the importance of discussions about their exact semantic content. So redoubling was discouraged, and, moreover, the genetic narrative connected with redoubling was now precluded by the assumption of a strict separation of synchrony and diachrony.[36] Otto Jespersen (1860–1943) may be the last scholar who, despite redefining subject and predicate in terms of 'primary' and 'secondary' of a 'nexus', amply discusses the semantics of subject and predicate, including questions of redoubling. Gabelentz's observations about discourse effects of word order are praised, but, according to Jespersen, 'the analogy between this and the subject-predicate relation is far too loose for the same name to be applied to both' (Jespersen, [10]1968 [1924], p. 147).[37]

While grammarians lost interest in subject-predicate redoubling, the simultaneous rise of synchronic linguistics stimulated a renewed interest in word order. The main impetus came from Prague structuralism. On the margins of linguistics, the rise and development of the area of 'stylistics' also involved a focus on word order (especially in French). Both approaches appeal to a separate level of information-structural concepts, and both

34 Marty characterizes the subject-predicate relation in terms of *Zuerkennung* (attribution). Wundt adopts the Aristotelian view of subject as *das Zugrundeliegende* (the underlying element) and predicate as *Aussage* (statement).

35 Terms may be misleading in this phase. Ludwig Sütterlin (1863–1934), for example, defends the one-level view, including a detailed version of the genetic subject-predicate narrative (cf. section 6.3), but he eventually prefers the terms 'prepared representation' (*vorbereitete Vorstellung*) and 'new representation' (*neue Vorstellung*) to 'psychological subject' and 'psychological predicate', in order to avoid confusion with the grammatical counterparts (Sütterlin [4]1918 [1907], p. 300). Conversely, the 'stylistic' scholars (section 7) adopt a two-level view but retain the terms 'psychological subject' and 'psychological predicate'.

36 However, Albert Sechehaye (1870–1943) reconstructs the narrative in terms of the Saussurean distinction between *langue* and *parole*. Grammatical and psychological subject and predicate differ from each other when *parole* breaks the restrictions of *langue*. His redoubling is of the discourse anchoring type (Sechehaye, 1926). His view is partially continued in the work of Charles Bally (1865–1947).

37 Jespersen regards other redoubling proposals as equally confusing. His own definition of the subject-predicate relation is based upon class-inclusion.

draw on earlier one-level theorists such as Gabelentz. But the bird's eye perspective of the received view was soon established, and 'aboutness' and 'discourse anchoring' were frequently mixed.[38]

The best tribute to Gabelentz's insights can actually be found in more recent information-structural theories, which explicitly take the difference between 'aboutness' and 'discourse anchoring' into account. Halliday (1967), Chafe (1976) and Krifka (2007) are examples of this approach. These authors apply different concepts to different aspects of information structure. Halliday's 'theme' and 'rheme', and Krifka's 'topic' and 'comment' especially are, apart from their two-level framework, very similar to Gabelentz's psychological subject and predicate. His strict separation of word order and accentuation is one of the numerous ideas in which he was ahead of his time.

Works cited

Adjémian, Christian. 1978. 'Theme, rheme and word order. From Weil to present-day theories'. *Historiographia Linguistica* 5, pp. 253-273.

Becker, Karl F. ²1842 [1836]. *Ausführliche deutsche Grammatik als Kommentar der Schulgrammatik*. Frankfurt am Main: G.F. Kettembeil.

Bolinger, Dwight. ²1965 [1952]. 'Linear modification'. In: *Forms of English, Accent, Morpheme, Order*, pp. 279-307. Cambridge, Mass.: Harvard University Press.

Chafe, Wallace L. 1976. 'Givenness, Contrastiveness, Definiteness, Subjects, Topics, and Point of view'. In: *Subject and Topic*, Charles N. Li (ed.), pp. 25-56. New York.: Academic Press.

Delbrück, Berthold. 1878. *Die altindische Wortfolge*. Halle: Verlag der Buchhandlung des Waisenhauses.

Elffers, Els. 1991. *The historiography of grammatical concepts. 19th and 20th-century changes in the subject-predicate conception and the problem of their historical reconstruction*. Amsterdam: Rodopi.

Elffers, Els. 2012. 'The rise of general linguistics as an academic discipline: Georg von der Gabelentz as a co-founder'. In: *The making of the humanities. Vol. II. From early modern to modern disciplines*, Rens Bod, Jaap Maat & Thijs Weststeijn (eds.), pp. 55-72. Amsterdam: Amsterdam University Press.

Firbas, Jan. 1974. 'Some aspects of the Czechoslovak approach to problems of functional sentence perspective'. In: *Papers on Functional sentence perspective*, František Daneš (ed.), pp. 11-37. The Hague: Mouton.

38 Cf. Keysper (1975) for the mixed approach of Prague structuralists. Stylistics is discussed in Adjémian (1978) and Elffers (1991, section 11.3).

Gabelentz, Georg von der. 1869. 'Ideen zu einer vergleichenden Syntax. Wort- und Satzstellung'. *Zeitschrift für Völkerpsychologie und Sprachwissenschaft* 6, pp. 376384.

Gabelentz, Georg von der. 1875. 'Weiteres zur vergleichenden Syntax. Wort- und Satzstellung'. *Zeitschrift für Völkerpsychologie und Sprachwissenschaft* 8, pp. 129165, 300338.

Gabelentz, Georg von der. 1881. *Chinesische Grammatik*. Leipzig: Weigel.

Gabelentz, Georg von der. 1887. 'Zur Chinesischen Sprache und zur allgemeinen Grammatik'. *Internationale Zeitschrift für allgemeine Sprachwissenschaft* 3, pp. 92-109.

Gabelentz, Georg von der. 2016 [1891]. *Die Sprachwissenschaft, ihre Aufgaben, Methoden und bisherigen Ergebnisse*, Manfred Ringmacher & James McElvenny (eds.). Berlin: Language Science Press.

Halliday, Michael A.K. 1967. *Intonation and Grammar in British English*. The Hague: Mouton.

Heusinger, Klaus von. 1999. *Intonation and Information Structure*. Konstanz: University of Konstanz.

Jespersen, Otto. 1924. *The philosophy of grammar*. London: Allen & Unwin.

Keysper, Cornelia E. 1985. *Information Structure. With examples from Russian, English and Dutch*. Amsterdam: Rodopi.

Knobloch, Clemens. 1988. *Geschichte der psychologischen Sprachauffassung in Deutschland von 1850 bis 1920*. Tübingen: Niemeyer.

Krifka, Manfred J. 2007. 'Basic Notions of Information Structure', *Interdisciplinary Studies on Information Structure* 6, pp. 13-55.

Marty, Anton. 1897. 'Über die Scheidung von grammatischem, logischem und psychologischem Subject resp. Prädicat'. *Archiv für systematische Philosophy* 3, pp. 175190, 294333.

Mathesius, Vilém. 1964 [1928]. 'On Linguistic Characterology with illustrations from Modern English'. In: *A Prague School Reader in Linguistics*, Josef Vachek (ed.), pp. 59-67. Bloomington: Indiana University Press.

Mauthner, Fritz. ^{3}1913 [1902]. *Beiträge zu einer Kritik der Sprache*, vol. 3, *Zur Grammatik und Logik*. Stuttgart: Cotta'sche Buchhandlung Nachfolger.

McElvenny, James. 2017a. 'Grammar, typology and the Humboldtian tradition in the work of Georg von der Gabelentz'. *Language and History* 60:1, pp. 1-20.

McElvenny, James. 2017b. 'Georg von der Gabelentz'. *Oxford Research Encyclopedia of Linguistics*. http://linguistics.oxfordre.com/view/10.1093/acrefore/9780199384655.001.0001/acrefore-9780199384655-e-379, DOI: 10.1093/acrefore/9780199384655.013.379, accessed 2 August 2018.

Misteli, Franz. 1887. 'Studien über die Chinesische Sprache'. *Internationale Zeitschrift für allgemeine Sprachwissenschaft* 3, pp. 2791.

Noordegraaf, Jan. 1997. 'Gabelentz, Paardekooper en de eenzinsdeelproef'. In: *Voorlopig verleden. Taalkundige plaatsbepalingen, 1797-1960*, Jan Noordegraf (ed.), pp. 220-221. Münster: Nodus.

Paul, Hermann. [2]1886 [1880]. *Prinzipien der Sprachgeschichte*. Halle an der Saale: Niemeyer.

Sechehaye, Albert. 1926. *Essai sur la structure logique de la phrase*. Paris: Champion.

Sgall, Petr. 2003. 'From functional sentence perspective to topic-focus articulation'. In: *Language and Function: to the memory of Jan Firbas*, Josef Haldký (ed.), pp. 279-288. Amsterdam: Benjamins.

Spieghel, Hendrik Laurensz. 1962 [1585]. *Ruygh-bewerp vande redenkaveling*. In: *Twe-spraack : Ruygh Bewerp : Kort begrip : Rederijckkunst*, J.H. Caron Willem (ed.), pp. 65-158. Groningen: Wolters.

Steinthal, H. 1855. *Grammatik, Logik und Psychologie, ihre Prinzipien und ihre Verhältnis zueinander.* Berlin: Dümmler.

Sütterlin, Ludwig. [4]1918 [1907]. *Die Deutsche Sprache der Gegenwart*. Leipzig: Voigtländer.

Vachek, Josef. 1966. *The linguistic school of Prague. An introduction to its theory and practice*. Bloomington: Indiana University Press.

Wegener, Philipp. 1885. *Untersuchungen ueber die Grundfragen des Sprachlebens*. Halle an der Saale: Niemeyer.

Weil, Henri. [3]1879 [1844]. *De l'ordre des mots dans les langues anciennes comparées aux langues modernes*. Paris: Vieweg.

Wundt, Wilhelm. [4]1922 [1900]. *Völkerpsychologie. Eine Untersuchung der Entwicklungsgesetze von Sprache, Mythus und Sitte*, Zweiter Band, Zweiter Teil, *Die Sprache*. Leipzig: Kröner.

About the author

Els Elffers is a retired lecturer in linguistics at the University of Amsterdam. She has researched and published extensively on the history and foundations of linguistics, in particular the subfields of semantics and pragmatics, and the relations between linguistics and psychology.
e.h.c.elffers@uva.nl

4 The Basque-Berber connection of Georg von der Gabelentz

Bernhard Hurch and Kathrin Purgay

Abstract
Discussions of Gabelentz's *œuvre* tend to omit the fact that he attempted in some of his later writings to demonstrate a genealogical connection between the Basque and Berber languages. Based on an examination of Gabelentz's texts, unpublished notes and correspondence, as well as several contemporary reviews, this chapter examines Gabelentz's proposal and what it shows about his theoretical views vis-à-vis historical-comparative linguistics and his place in the linguistic community of the time. It shows how Gabelentz's critique of the prevailing historical-comparative approach led him to abandon all established methods and draw wildly implausible conclusions. Even the most vehement critics of the Neogrammarians, such as Hugo Schuchardt (1842–1927), looked on uncomprehendingly at the turn Gabelentz had taken.

Keywords: Basque, Berber, historical-comparative linguistics, Afro-Asiatic languages, Hugo Schuchardt, Neogrammarians

1 Introduction[1]

In the last years of his life, Georg von der Gabelentz dealt intensively with the question of a possible genetic relationship between Basque and Berber.

1 We would like to thank Leopold von der Gabelentz for granting us permission, on behalf of the Gabelentz family, to consult the papers and documents stored in the Thüringisches Staatsarchiv in Altenburg. We are grateful to Doris Schilling of this Archive for her kind help. For providing information to us and for fruitful discussion, we wish to thank Frans Plank, Manfred Ringmacher and Frank Zimmer. Last but not least, we owe a special debt of thanks to James McElvenny for his critical reading of the article.

McElvenny, James (ed.). *Georg von der Gabelentz and the Science of Language*. Taylor & Francis Group, 2019
DOI: 10.5117/9789462986244/CH04

This resulted in two publications. The first is his lecture at the meeting of the Philosophisch-historische Classe der Königlich Preussischen Akademie der Wissenschaften (Philosophical-Historical Class of the Royal Prussian Academy of Sciences) on 22 June 1893, which was immediately published in the proceedings of the Academy (Gabelentz, 1893); the second is available in book form, published posthumously one year after his death by his nephew Albrecht von der Schulenburg under the title *Die Verwandtschaft des Baskischen mit den Berbersprachen Nord-Africas* (The genetic relationship between Basque and the North African Berber languages) (Gabelentz, 1894). These two writings were largely ignored at the time and have hardly received any attention in later Gabelentz scholarship, even though the arguments for the Basque-Berber relationship are also taken up in the revised second edition of *Die Sprachwissenschaft* (Gabelentz, 2016 [1901]),[2] where Gabelentz expounds the implications that, in his opinion, necessarily result for historical linguistic theory.

The different publication dates for the Academy lecture and Basque-Berber book are somewhat deceptive in that there cannot have been much time between the composition of the two works, given that Gabelentz passed away only five months after delivering the lecture. He himself announces in the Academy version a 'more detailed text' (*ausführlichere Schrift*) (Gabelentz, 1893, p. 608) and, in a letter of 22 July 1893 to his sister Clementine von Münchhausen (printed in Münchhausen, 2013 [1913], pp. 138-139), he writes somewhat more extensively in this regard:

> You've never kept me waiting so long for a letter when I sent you something new from me, and I think it was a good thing this time, even if it's just a report on research results and the precursor of an entire book. [...] In the world of my colleagues, the matter seems at first to be perplexing, partly even shocking,[3] like a heretical attack against the dogma of the infallible sound laws. I can't stand such axioms, which are based on a limited field

2 The first edition of Gabelentz's *Die Sprachwissenschaft* appeared in 1891 and the second posthumous edition in 1901. Page numbers provided in references in this chapter are to the 2016 critical edition (Gabelentz, 2016 [1891/1901]). The critical edition clearly indicates whether a passage is found in only the first or second edition or is common to both of them. For added clarity, when reference is made in this chapter specifically to the first edition, the citation is given as 'Gabelentz, 2016 [1891]'. When reference is made specifically to the second edition, the citation is 'Gabelentz, 2016 [1901]'.

3 Behind these lines stands an untranslatable pun, which builds on the assonance of German *verdutzen* 'to perplex' with the quantifier *Dutzend* 'dozen', as well as *schockieren* 'to shock' with the quantifier *Schock* '60 pieces, heap'.

> of experience but even there still leave behind a lot of open questions, and if the second edition of my *Sprachwissenschaft* comes about, you will read more about it.[4]

Since the proceedings appeared only on 29 June 1893 and the written critiques, to which we will return in detail later, had not yet been published, Gabelentz can presumably only be referring in his letter to verbal objections given after the lecture.[5] The fact that in the later 'more detailed text' (Gabelentz, 1894) he still does not address the published objections – such as those Schuchardt makes in his review (Schuchardt, 1893), as discussed below – suggests that this text was written in parallel with the lecture.[6] In this last phase of life, Gabelentz was occupied with other publications (as recorded in Gimm, 2013b, p. 116) and was already in very poor health

4 The passages that Gabelentz refers to in this letter can be found on pp. 179-180, 198-202, 205-207, 293 and 307-308 of the second edition (Gabelentz, 2016 [1901]). Original text of Gabelentz's letter: 'Du hast mich noch nie so lange auf einen Brief warten lassen, wenn ich Dir etwas Neues aus meiner Feder geschickt habe, und ich denke, diesmal ist es etwas recht Gutes gewesen, wenn es auch nur ein Bericht über Forschungsergebnisse und der Vorläufer eines ganzen Buches sein soll. [...] In der fachgenössischen Welt scheint die Sache zunächst verdutzend, zum Teil, da ein Schock mehr ist als ein Dutzend, shocking gewirkt zu haben, als ein ketzerischer Angriff gegen das Dogma von den unverbrüchlichen Lautgesetzen. Solche Heischesätze, die sich auf ein beschränktes Erfahrungsgebiet stützen, und selbst da noch vieles unerklärt lassen, kann ich in der Seele nicht leiden, und kommt es zur zweiten Auflage meiner Sprachwissenschaft, so wirst Du davon noch mehr lesen.'

5 In the file with the signature II-V, 157, of the Berlin-Brandenburgische Akademie der Wissenschaften, p. 97 reads, 'After the reading and approval of the minutes of the session of 8 June, Mr von der Gabelentz held a lecture "on Basque and Berber", whose inclusion in the proceedings is approved.' (*Nach der Verlesung und Genehmigung des Protokolls der Sitzung v. 8. Juni hielt Hr. von der Gabelentz einen Vortrag ,über Baskisch und Berberisch', dessen Aufnahme in die Sitzungsberichte genehmigt wird.*) Present at the session were: Heinrich Kiepert (geography, cartography), Albrecht Weber (Indology), Adolf Kirchhoff (classical philology, classical studies), Eduard Zeller (philosophy, history), Johannes Vahlen (classical philology), Alexander Conze (archaeology), Adolf Tobler (Romance studies), Wilhelm Wattenbach (history, auxiliary sciences of history), Hermann Diels (classical philology), Alfred Pernice (law, Roman law), Heinrich Brunner (law), Johannes Schmidt (comparative linguistics), Otto Hirschfeld (ancient history, archaeology), Eduard Sachau (oriental studies), Gustav von Schmoller (political science), Wilhelm Dilthey (philosophy), Ernst Dümmler (medieval history), Adolf von Harnack (history of Christianity, religious history, history of the academy).

6 In the posthumous book (Gabelentz, 1894), there are only two references in reaction to Schuchardt's criticisms. The first one, on p. 4, is only very selective and does not mention his Academy lecture at all but simply indicates that Schuchardt had drawn his attention in the *Literaturblatt* to a certain Basque word form, *alhargunt(a)*. In the second one, on p. 136, Gabelentz notes in a footnote that Schuchardt derives Basque *zamari* 'horse' (*Pferd*) from *sagmarius*, which, however, seems unlikely to him.

and general condition (Münchhausen, 2013 [1913]). In our opinion, it is also conceivable that the book manuscript was begun earlier and that the Academy lecture and its publication should only be understood as an excerpt or snapshot of work in progress, or a short version of the longer text.

The manuscript on which the book is based has been lost. There is no trace of it in the extant *Nachlass* and remaining library collections of Georg and his father Hans Conon von der Gabelentz, which are kept in the Thüringisches Staatsarchiv (Thuringian State Archive) in Altenburg. It has not been possible to locate Albrecht von der Schulenburg's own *Nachlass*, in which the manuscript could lie.[7]

A small comment from the publisher accompanies the publication:

> † Professor von der Gabelentz himself described the above work as **'his greatest achievement'**; it will cause the greatest sensation among all linguists and ethnographers, all the more since up to now the Basque language, **the most mysterious language**, has not been the subject of any publication in which its position within the world of languages and peoples has really been determined. – The ready-to-print manuscript was edited by the nephew of the deceased, Dr. phil. Albrecht Graf von der Schulenburg. [emphasis in original][8]

7 The various branches of the family were unfortunately unable or not prepared to provide oral information concerning papers and left subsequent written enquiries unanswered. We have contacted both the Münchhausen branch of the Gabelentz family and entered into oral and written contact with the Schulenburg branch in Nordsteimke. There seem to be post-World War II remnants of a family library in Nordsteimke, but also here it was not possible to get information on its inventory.
In the Thüringisches Staatsarchiv, however, apart from some reviews of the book, there is also a file with the complete correspondence between the publisher Richard Sattler of Braunschweig, Albrecht von der Schulenburg and Georg's widow, Gertrud (née von Oldershausen, widowed by von Adelebsen). This correspondence is extremely unpleasant and deals almost exclusively with the publisher's unpaid financial claims against the family. Books from the publisher Richard Sattler are only listed in German libraries between 1890 and 1905, after which a few books from Sattler were published in Leipzig. Linguistic literature is virtually non-existent in the publishing programme except for Gabelentz (1894). A publisher's archive, in which the manuscript could possibly be located, could not be found.

8 Original: 'Der † Professor von der Gabelentz hat seine obige Arbeit selbst als **‚seine grösste Leistung'** [sic] bezeichnet; dieselbe wird bei allen Sprachforschern und Ethnographen das grösste Aufsehen erregen, umsomehr, als über das Baskische, **dieses seither grössten Sprachräthsels**, bislang keine Publication erschien, welche die Stellung desselben innerhalb der Sprachen- und Völkerwelt auch nur einigermassen richtig näher bestimmt hätte. – Die Herausgabe des druckfertigen Manuscriptes übernahm der Neffe des Verstorbenen, Dr. phil. Albrecht Graf von der Schulenburg.' [emphasis in original]

There are various views on the character of the manuscript. The description above of the manuscript as 'ready to print' is most certainly not true. The first part of the text is much too fragmentary, while the accompanying dictionary section is incomplete and uncommented. At the beginning of the preface, Schulenburg admits that the 'work would certainly have been completed and improved in many parts if the deceased had been able to put the finishing touches on it' (*Arbeit gewiss noch in vielen Teilen vervollständigt und verbessert worden wäre, wenn der Verstorbene die letzte Hand an sie hätte legen dürfen*). Gabelentz's student Wilhelm Grube (1905, p. 554), for his part, described the manuscript as a 'work that was by no means ready to print' (*keineswegs druckreife Arbeit*). In doing so, he may have merely been trying to shield his teacher from criticism, but there is much in favour of his position. There is no discernible development in the argumentation between the Academy lecture and the book; indeed, the Academy lecture has a more explicitly theoretical approach that is completely missing in the book version. In our opinion, it would be most reasonable to emphasize the complementary nature of the two works, as Gabelentz (1893, p. 608) himself suggests: 'It is not the place here to share all my material; this must be done in a more detailed text' (*Es ist hier nicht der Ort, mein ganzes Material mitzutheilen; das muss in einer ausführlicheren Schrift geschehen*). This is also supported by the fact that the tables of the Academy lecture, which are not easily comprehensible in their short presentation, find their equivalent in the lists of the book version. For example, the numbers given in the *u:i*-alternations on p. 14 of the book correlate exactly with table II on p. 598 of the essay. This means that these are different ways of presenting the same facts and that the tabular form is due to the concise presentation of the lecture, while the individual language examples are given in the book.

The two publications have provoked almost exclusively the most negative reactions, from simple rejection to even complete incomprehension at how a linguist of the stature and genius of Georg von der Gabelentz could produce such dubious work. But only some of these reactions contain substantive arguments, and the detailed criticism is often used as an excuse for ignoring the theoretical considerations deliberately foregrounded by Gabelentz. However, it is advisable to discuss the work in detail first.

2 Gabelentz (1893)

From the beginning of his Academy lecture, Gabelentz leaves no doubt that the genetic relationship between Basque and Berber is beyond question for him. He sees his task therefore as simply a matter of illustrating this relationship and developing new methodological principles of historical-reconstructive linguistics to support it. Although Gabelentz emphasizes the importance of this second aspect in many parts of his work, it is often overlooked in later assessments. His final sentence (1893, p. 613) is quite explicit:

> In addition, such individual questions are by far less important than the knowledge we have acquired, firstly, that Basque is a Hamitic language related to the Berber family and, secondly, that the Hamitic languages present completely new data to the study of sound change, whose evaluation requires new points of view, whose exploitation demands a new method.[9]

The material he uses in his representation of the languages does not go beyond lexical comparisons. This is sufficient for the purposes of his argument, but Gabelentz is walking on thin ice and seems to be aware of it.

He wangles his way out of explaining the lack of grammatical correspondence between Basque and Berber. Gender, which is the only example he mentions, is an inappropriate object of comparison as it appears very marginally and is present only in some Basque dialects (e.g. Gipuzkoan) as a semantic but not a grammatical category, and it also designates exclusively biological sex. There is no nominal category of gender in Basque. Gabelentz gives examples of non-compatibilities, but in a sense he throws the baby out with the bathwater when he adds:

> Faith in the constancy of outer and inner language form is an achievement to which our science clings in the most tenacious way, and the facts that could shake it are new acquisitions themselves and little known, as they lie in the area of Indochinese and Melanesian. (Gabelentz, 1893, p. 594)[10]

9 Original: 'Auch sind solche Einzelfragen bei weitem weniger wichtig, als die gewonnene Erkenntniss, erstens, dass das Baskische eine hamitische, der Berberfamilie verwandte Sprache ist, und zweitens, dass die hamitischen Sprachen der lautgeschichtlichen Forschung ganz neue Bilder vorführen, deren Beurtheilung neue Gesichtspunkte, deren Verwerthung eine neue Methode erfordert.'

10 Original: 'Der Glaube an die Beständigkeit der äusseren und inneren Sprachform gehört zu den Errungenschaften, an denen unsere Wissenschaft am zähesten festhält, und die Thatsachen,

This brief statement may be a sideswipe to many of the Neogrammarians, who merely concerned themselves with sound change, but in this context, at the beginning of the 1890s, it is no longer true as a general critique. Moreover, it is not easy to avoid the fact that, by assuming a common genealogy between Basque and Berber, one would have to propose a scenario, for example, for the development of the Basque auxiliary system, of ergativity, and other grammatical characteristics.[11] In addition, as Schuchardt (1893, p. 337) correctly points out, it is not necessary to look as far as East Asia or Oceania to call into question belief in the constancy of inner and outer language form. Indications of common grammatical characteristics in related languages are – regardless of whether inner and outer language form is constant or not – always desirable and advantageous for a genealogical argument.

In his own *Sprachwissenschaft*, Gabelentz (2016 [1891]) postulates that historically oriented language comparisons should be based on older or the oldest attested language stages. Especially if

> [...] genetically related languages [are] so dissimilar to each other that their relationship is not obvious at first sight, then this relationship is a distant one; that is, a very long time has passed since their former unity. Consequently, one always has to go back to the oldest recognizable sounds and meanings of words and formatives when comparing languages. (Gabelentz, 2016 [1891], p. 164)[12]

Although this remains his first principle in offering a proof of linguistic relationship (cf. ibid., p. 164), he does not follow it in his Basque-Berber comparisons, especially because of the absence of usable evidence. In this case, he argues, it is allowed, for the time being, to select 'the seemingly best preserved language of the family, especially if it is sufficiently studied' (*die scheinbar besterhaltene Sprache des Stammes, zumal wenn sie leidlich erforscht ist*) – there is still time for corrections and explanations afterwards (ibid., pp. 184-185).

die ihn erschüttern könnten, sind ihrerseits neuer Erwerb und wenig bekannt, da sie auf indochinesischem und melanesischem Gebiete liegen.'

11 Gabelentz (1894) explained his approach in such a way that it was first of all sufficient to find enough plausible similarities which were neither due to coincidence nor to borrowing; it is only then that we have to seek the explanations for cases of doubt (Gabelentz, 1894, p. 3). See also the discussion of Gabelentz (1894) in section 3.

12 Original: '[...] verwandte Sprachen einander so unähnlich [sind], dass ihre Verwandtschaft nicht ohne Weiteres in die Augen fällt, so ist diese Verwandtschaft eine entferntere, also seit der vormaligen Einheit eine sehr lange Zeit verstrichen. Daraus folgt, dass man bei der Vergleichung immer auf die ältesten erkennbaren Lautformen und Bedeutungen der Wörter und Formative zurückzugehen hat.'

Gabelentz is well aware of the problems that the sometimes enormous differences in form between the assumed correspondences cause in his lexical comparisons. Countless examples can be cited – and indeed many were cited in critiques – in which the correspondence is anything but immediately understandable and demands a great deal of imagination on the part of the reader; here, for example, Basque *borroka* 'fight' (*Kampf*) to Kabyle *amerzi* or Basque *erraz* 'easy' (*leicht*) to Tuareg *elluq* 'poor' (*arm*) etc. However, Gabelentz does not see this as a weakness of his proposal for a linguistic relationship, but rather as an attempt to forge a new kind of methodology, to which, interestingly enough, the literature on Gabelentz does not respond. In many places he showed himself to be an open and vehement critic of the Neogrammarians,[13] and accordingly he is more than sceptical about the concept of sound laws. He only allows sound laws to operate in a restricted way and, if at all, only in recent phases of language development, in so-called 'more regular times' (*geordneteren Zeiten*). In his eyes 'constant sound substitutions [are] the product of regular sound change' (*constante Lautvertretungen [sind] das Erzeugniss gesetzmässiger Lautverschiebung*) (Gabelentz, 1893, p. 602). As there are hardly any such substitutions between Basque and Berber, he transposes their genetic relations into a 'prior period of uncertain articulation' (*Vorperiode unsicherer Artikulation*) (ibid., p. 602), for which he uses expressions such as 'wildness' (*Verwilderung*), 'mixing' (*Vermischung*), and 'blurring' (*Verwischung*) of the sound system, 'confusion' (*Wirrsal*), etc. The sound correspondences of the Basque dialects alone

> [...] give a picture of phonetic wildness which, to my knowledge, has hardly any equal in the world of languages. Only in a few cases, which he [van Eys] correctly recognized and which I do not need to repeat here, are there fixed laws of sound correspondences between the dialects, alongside which are the strangest variations sometimes in one dialect and sometimes in the other. (ibid., p. 596)[14]

As a first step, Gabelentz seeks to create an inventory of sound correspondences in the Basque dialects. He then makes a similar attempt for the Kabylian

13 In the literature there are some excellent works on this subject, such as Plank (1991) and various publications by Ringmacher (e.g. 2011). Cf. passages of Gabelentz quoted here, such as the above letter to his sister Clementine.

14 Original: '[...] geben ein Bild lautlicher Verwilderung, das meines Wissens in der Sprachenwelt kaum Seinesgleichen hat. Nur in wenigen Fällen, die er [van Eys] richtig erkannt hat, und die ich hier nicht zu wiederholen brauche, zeigen sich zwischen den Dialekten feste Lautvertretungsgesetze, daneben bald in dem einen, bald in dem anderen die wunderlichsten Varianten.'

language and uses the same procedure in the Basque-Kabyle comparison. Finally, he describes (ibid., pp. 604-606) six types of sound variation and alternation: free variation in the articulatory space, in the perceptive space, through contact and mixing, through fortitive processes, through evolutionary disambiguations, and finally fixation of irregular to regular alternations. This results in a line of development that he describes as follows:

> […] because the result is the same both for later linguistic history and for the method to be applied here: during the time of chaos there was no room for fixed sound laws; it is only a later steadier development that could stand a regular sound shift instead of sound confusion and distortion. This is the thinner, younger layer in our case. Underneath it, however, lies the massive layer of debris, which is still recognizable everywhere by its enormous consequences. (ibid., p. 607)[15]

We will leave open to what extent Gabelentz's theory may anticipate aspects of the concept of allophones in later phoneme theories. This is an interesting point in itself but not relevant in this context. The almost unlimited range of variations that Gabelentz sketches on the synchronic plane opens the door to all conceivable lexical correspondences from a historical perspective: these no longer need to be demonstrated because in fact almost every single sound correspondence seems possible.[16] Even if Gabelentz adds quite suggestively that there are three possibilities for dealing with the problem – negating it, ignoring it or facing it (ibid., p. 607) – and insinuates, of course, that he is the only one facing it, his position nevertheless demands a few fundamental observations.[17]

15 Original: '[…] denn das Ergebnis für die weitere Sprachgeschichte und für die hier anzuwendende Methode der Forschung ist in beiden Fällen das gleiche: während der Zeit des Chaos war für feste Lautgesetze kein Platz; erst eine spätere ruhigere Entwicklung konnte an Stelle der Lautverwirrung und -verzerrung eine geordnete Lautverschiebung vertragen. Das ist in unserem Falle die dünnere, jüngere Schicht. Darunter aber lagert, noch überall an gewaltigen Nachwirkungen erkennbar, jene mächtige Schicht durcheinander geworfenen Gerölles.'

16 Gabelentz (2016 [1901], p. 307) states: 'It seems to me that where the roots of related languages resemble each other to some extent in sound and meaning, one should assume an inexplicable sound change, the influence of an unrecognized sound law, or uncertain articulation, rather than an inexplicable new creation' (*Mir scheint, wo sich die Wurzeln verwandter Sprachen einigermaßen in Klang und Sinn ähneln, sollte man lieber einen unerklärlichen Lautwandel, das Walten eines noch unerkannten Lautgesetzes oder unsichere Articulation, als eine noch unerklärliche Neuschöpfung vermuthen*).

17 In the same spirit, however, Gabelentz repeatedly provides surprisingly strong judgements about languages and establishes connections between anthropological forms of life or living conditions and types of language. For example, in his inaugural speech to the Preußische

Critical fieldwork on unwritten languages admits to a higher degree of variation than can be found in the comparatively confined space of the written languages of the Indo-European world. However, this does not in any way mean that centrifugal variation in earlier times would have led to open chaos and that order is only a product of more recent epochs, because this would deny from the outset the possibilities of linguistic reconstruction based on regular change. But this is exactly what Gabelentz does: where he cannot discern the system, he theorizes disorder. In his far-reaching work *Fonética Histórica Vasca* (Michelena, 1961), Luis Michelena, the doyen of twentieth century Basque studies, demonstrated systematic principles precisely where Gabelentz saw nothing but chaos and debris; that is, in Basque dialectology and therefore in the historical reconstruction of the Basque language.[18] That this work was only possible 70 years later is not due to any methodological developments, but rather to the quality of the sources.

3 Gabelentz (1894)

This book consists basically of two parts, which are distinguished from one another in their design and layout.[19] There are, in fact, two frontispieces: the first with the title *Die Verwandtschaft des Baskischen mit den Berbersprachen Nord-Africas* (The genetic relationship between Basque and the North African Berber languages), which precedes the first general part of pages 1 to 91, and then the *Wörterbuch zur Vergleichung des Baskischen mit den*

Akademie der Wissenschaften (Prussian Academy of Sciences), he states (Gabelentz, 1890, p. 785): 'A mixed Anglo-Chinese people, however high its level of civilization, would only make its miserable Pidgin-English a suitable medium of its spiritual life after the work of many generations. And if the Cherokee were to reach their goal with their cultural aspirations, we would one day have a people of farmers and citizens who spoke the language of a hunting people. Who knows if there is not already such a people living now – at the Bay of Biscay' (*Ein anglo-chinesisches Mischvolk, so hoch seine Gesittung sein möchte, würde erst nach der Arbeit vieler Geschlechter sein elendes Pitchen-English zu einer tauglichen Trägerin seines Geisteslebens gestalten. Und sollten doch noch die Tscheroki mit ihren Culturbestrebungen an's Ziel gelangen, so hätten wir dereinst ein Volk von Ackerbauern und Bürgern, das die Sprache eines Jägervolkes redete. Wer weiss, ob nicht schon jetzt ein solches lebt, – am biskaischen Meerbusen*).

18 It is an interesting fact that Michelena does not deal with Gabelentz at any point in his entire work, which is now available in the 15 volumes of the *Obras completas*. This cannot be explained by ignorance or lack of interest in the history of linguistics, nor even by a blindness to linguistic typology. Michelena must have deliberately refused to write more about Gabelentz's Basque excursions than what had already been said in earlier discussions.

19 The copy of the book which is available in the Graz university library carries the library stamp of Hugo Schuchardt and is provided with some critical handwritten notes made by him.

Berbersprachen: Kabylisch, Tuareg, Γadamsi, Šilha, mit Ägyptisch und Koptisch und mit anderen hamitischen Sprachen (Dictionary for comparing Basque with the Berber languages: Kabyle, Tuareg, Ghadamès, Shilha, with Egyptian, Coptic, and other Hamitic languages) from pages 93 to 285. This second part consists exclusively of a comparative dictionary, which is presented without commentary. Gabelentz used its framework – the lexical entries organized according to semantic groups – for the first time in Gabelentz & Meyer (1882) and it reappears in the same form in the *Handbuch zur Aufnahme fremder Sprachen* (Handbook for recording foreign languages) (Gabelentz 1892a).[20]

Gabelentz himself worked on the handbook for both Basque and the Berber reference languages: research in the Gabelentz collection of the Thüringisches Staatsarchiv in Altenburg has revealed the following copies of the *Handbuch*, all of which have been filled in by Gabelentz in his own handwriting, although not all have been filled in completely:

– the most complete copy is the Basque version; it is discussed in detail in Hurch (2009; 2011);[21]
– for 'the language of the Kabyle people' (*die Sprache der Kabylen*), Gabelentz also compiled a relatively complete copy, dated 1893;
– for 'Tuareg together with Šilha and Γadamsi' (*Tuareg nebst Šilha und Γadamsi*) (with no handwritten year) there is a copy filled in by him only partially;
– another copy (again without indication of the year) for 'the language of the Egyptians and Copts' (*die Sprache der Aegypter und Kopten*) is only very sparsely filled in.

It can be assumed that all these *Handbuch* versions originated in the same period, between 1892 and 1893, and served the same end, namely to be a tool for the comparison of Basque and Berber.

The *Handbuch* copies do not explicitly list their sources. As far as the Basque version is concerned, there is no doubt about the fact that the most important source by far was the dictionary of van Eys (1873), although its dialectologically oriented comparative representation of the Basque dialects is not reproduced by Gabelentz. Nevertheless, the Basque version has a special status, in that he first writes the Basque entries into the *Handbuch*, followed by the forms to

20 The differences between these versions are limited to individual entries in the later versions, which do not receive their own numbers but which are differentiated by letters. There is a critical (linguistic and social) study of the *Handbuch* in Kürschner (2014). Cf. also Hurch (2011). We only know a few copies which were actually used for the *recording of foreign languages*. They are enumerated in Hurch (2011), with acknowledgements to Manfred Ringmacher.

21 The existence of this copy was brought to our attention by Manfred Ringmacher.

be compared from Kabyle and Tuareg. This suggests that the Basque volume was the chronologically oldest part. No sources are mentioned in the Berber handbooks, either. In the essay, Gabelentz (1893) explicitly mentions Newman (1882; 1887), Hanoteau (1858; 1860) and Ben Sedira (1887), but the orthography used by Gabelentz for the language samples and the glossary does not allow any conclusions to be drawn about the origin of the entries.

In all these versions of the *Handbuch*, the focus is not on the recording of the languages, but rather on the editing of existing sources for comparative purposes.[22] The template for this task is one that, in Gabelentz's opinion, had already proven successful in another part of the world. Here he obviously hoped that he would be able to discover and establish a linguistic relationship of his own by means of a tried and tested method. However, his justification remains extremely poor and methodologically questionable.

The print version of the comparative dictionary from 1894 is not just the sum of the earlier *Handbuch* versions mentioned above. It follows the same lexicographic pattern in all details and even takes over the numbering of the earlier versions, except that a few entries have been omitted, such as no. 180, *Perlenmuschel* (pearl oyster), which plays no role in either the Basque or the Berber cultural sphere, but also such entries as no. 203, *Bräutigam* (bridegroom). Instead, some numbers that have a few more entries are assigned an additional subdivision with a, b, c, etc. The languages are presented in columns in the order Basque / Kabyle / Tuareg / Shilha / Ghadamès[23] / Egyptian / Coptic. The glossing language is German, the first reference language is Basque. Since this part of the dictionary is largely devoid of commentary, and since some entries are not filled in at all – a few are filled in only for Basque but not for the other languages – it can be assumed that Gabelentz does indeed regard the forms he cites as forms that are related not only semantically but also in their form. He has therefore not included in this final version those forms from the different versions of his *Handbuch* that he could not connect to Basque, although he considers them as having equivalents in different Berber languages as they are Hamitic. On the other hand, the dictionary also contains entries and senses that he has not listed in the handbooks. Some words are marked by Gabelentz with either an exclamation or question mark; we can assume that this is supposed to indicate whether the relationship is certain or questionable – however, Gabelentz employs all of them equally in his sound comparisons.

22 It is not known how it came about that Gabelentz was commissioned to write this *Handbuch* for the Kolonialabtheilung des Auswärtigen Amtes (Colonial Department of the Foreign Office).

23 The spelling of this language alternates between initial *Gh*- and *Γ*- in Gabelentz (1894).

The first part of Gabelentz (1894) has to be subdivided once more, because only a short initial section of a few pages deals with the question of possible grammatical correspondences. He lists a number of typologically relevant characteristics that have no equivalents between Basque and Berber (order of grammatical elements, ergative, presence of individual categories, etc.). Solving these questions, he says, 'is the task of the internal history of a language, whose work can only begin once the genetic relationship is proven' (*ist die Sache der inneren Sprachgeschichte, deren Arbeit erst beginnen kann, wenn die Stammverwandtschaft erwiesen ist*) (Gabelentz, 1894, p. 3). The various statements that follow about gender, diminutives, causatives, a Basque prefix *ma-* that is not defined further, case endings, and plural suffixes indicate a lack of detailed knowledge of Basque grammar, and Gabelentz makes no attempt to formulate any serious claims.[24] This short section is so fragmentary that one can hardly imagine that the author could have considered it to be ready to print.

The vast majority of this first section, namely pages 9-91, gives lists of words again, arranged according to sound variations, sound alternations, sound correspondences, starting with Basque dialectal alternations, then, if available, internal Berber ones, which are finally put in relation to the Basque ones. According to Gabelentz, they show, on the one hand, the correspondences but, on the other hand, they often simply illustrate the chaotic state of these correspondences, which in fact do not seem conducive to formulating sound laws at either an initial glance or even after a more detailed examination (such examples as Basque *belar* to Kab. *amlageγ*, Basque *belarri* to Kab. *amezzug*, Basque *badarik* to Shil. *meqqar*, Basque *undar* to Kab. *anegger* could be multiplied indefinitely). These comparisons also remain largely uncommented from a phonological point of view, although sometimes there are brief comments on the lack of plausibility of certain correspondences. It seems that merely postulating these correspondences is enough to make them real for Gabelentz.

The book ends with four tables of sound correspondences, Table II, 'sound change within the Kabyle language' (*Lautwandel innerhalb des Kabylischen*), and Table IV 'sound change in the Basque language' (*Lautwandel im Baskischen*), which are identical with Tables III and I of the Academy

24 It is therefore not clear to what extent Gabelentz would still have relied on the similarity of formatives, or whether he would have considered them to be deceptive evidence, since 'affixes consist mostly of few, light-weight speech sounds' (*Affixe bestehen ja meist aus wenigen, leichtwiegenden Lauten*), where 'chance [has] easy play' (*Zufall [hat] leichtes Spiel*) (Gabelentz, 2016 [1891], p. 161). It is not entirely clear why they, of all sounds, should show regularity in all the phonetic chaos.

essay.[25] Table I, which is missing in Gabelentz (1893), gives a summary of the Basque-Berber sound correspondences according to the same pattern.[26] Table III is an evaluation of the consonant permutations according to Campión (1884, p. 116), which deviates in some points from Gabelentz's own results, although he fails to mention this in any way.[27]

4 Gabelentz (1901)

The intervention of the editor Schulenburg in the second edition of *Die Sprachwissenschaft* is often referred to in the literature; the various changes are highlighted by Ringmacher and McElvenny in their critical edition (Gabelentz, 2016). Among these additions in Schulenburg's second edition is a new and even clearer formulation of Gabelentz's views relating to the putative Basque-Berber relationship, which was probably introduced by Gabelentz himself. In particular, Gabelentz (2016 [1901], pp. 314-316) corresponds to the explanations of 1893 – this 'theoretical' part, as stated above, is missing in the 1894 version – but provides further clarifications. In keeping with the general character of *Die Sprachwissenschaft*, the presentation is now more accessible to a general audience and embellished to a certain extent, but the aim of integrating this account into the model of general linguistics is nevertheless explicitly in the foreground.[28] Gabelentz also emphasizes the 'wildness' (*Verwilderung*) as well as the 'uncertain contours of the sound images' (*unsicheren Umrisse der Lautbilder*) and dedicates himself more specifically to the implications for general linguistics.

> Confronted with such confusion, however, the tried and tested method of phonetic language comparison fails. [...] Another may try to forge new tools to deal with the new subject; and anyone who wants to extend linguistics as far as human languages reach must dare to try.

25 Minimal differences between the Basque tables suggest inaccurate proof-reading rather than separate evaluation of the data.

26 Table V (Gabelentz, 1893) also shows a much more confusing comparison of the two languages, but here several consonants are grouped together into whole categories and the figures of the two tables do not always agree completely.

27 Perhaps Gabelentz refers here to Campión, whom he quotes nowhere else as a source, since the latter had also categorized the sound correspondences according to the frequency of their occurrence in 'normal, deviant and sporadic' (*normales*, *anormales* and *esporádicas*), while van Eys (1873, pp. xliii-xlv) only contains a pure enumeration of the possibilities.

28 We will leave aside here the renewed tabular representations of sound correspondences and their only slightly altered interpretations.

> New tools, that means new categories. And in our case this is incredibly difficult: we want to draw with a mop, paint [a picture] with a paint roller, add an 'or something like that' to every sound and every word we write down. In Basque, a small hill is called: *muru, murru, mora, murko, burko, morroko, mulko, mulho, muillo, mulzo, muno, munho*, or something like that; [...] There are no sound shifts or substitutions, but sound shufflings and blendings, no sound laws but sound possibilities, each of which could be easily demonstrated by a sufficient number of other examples. And that is what matters: the facts must be absolutely compelling before we can sacrifice the most proven rules of research to them. But if we close our eyes to compelling facts we sacrifice even more, for we sacrifice knowledge to our anxiousness or arrogance. (Gabelentz, 2016 [1901], pp. 315-316)[29]

Gabelentz calls for new categories to describe these supposedly uncertain relations, but he does not explain what they might be. It is clear that he rejects sound laws as a mechanism for explaining the correspondences he mentions. But even here he fails due to a lack of knowledge of Basque, namely the highly productive method of diminution through suffixation and/or palatalization, as well as simple orthographic traditions. This ignorance leads him to accept unexplained correspondences where there is an easily perceived system.

5 Contemporary criticism

It did not take long for a critical examination of the two writings to appear. As early as the beginning of September 1893 – only about two months after the academic lecture – a devastating assessment by Hugo Schuchardt was

29 Original: 'Einem solchen Wirrsale gegenüber versagt freilich die alterprobte Methode der phonetischen Sprachvergleichung ihren Dienst. [...] Ein Anderer mag versuchen, sich neue Werkzeuge zu schmieden, womit er den neuen Stoff bearbeitet; und wer die Sprachwissenschaft soweit erstrecken will, wie menschliche Sprachen reichen, der muss den Versuch wagen. „Neue Werkzeuge", das heisst neue Kategorien. Und das ist in unserem Falle entsetzlich schwierig: man möchte mit dem Wischer zeichnen, mit dem Vertreiberpinsel malen, zu jedem Laute, jedem Worte, das man niederschreibt, möchte man beifügen: „oder so ähnlich". Auf Baskisch heisst ein kleiner Hügel: muru, murru, mora, murko, burko, morroko, mulko, mulho, muillo, mulzo, muno, munho, oder so ähnlich; [...] Es handelt sich hier nicht um Lautverschiebungen oder Lautvertretungen, sondern um Lautverwischungen und -vermischungen, nicht um Lautgesetze, sondern um Lautmöglichkeiten, deren jede leicht an einer genügenden Zahl anderer Beispiele nachzuweisen wäre. Und darauf kommt es allerdings an: die Thatsachen müssen geradezu zwingend sein, ehe man ihnen die erprobtesten Regeln der Forschung opfern mag. Aber wer vor zwingenden Thatsachen die Augen verschliesst, opfert noch weit mehr, denn er opfert seiner Ängstlichkeit oder Rechthaberei eine Erkenntnis.'

published in the *Literaturblatt für germanische und romanische Philologie*. His so-called 'exhibitions' (*Ausstellungen*) address practically all aspects of the work: Gabelentz does not know the relevant literature; his lack of knowledge of the Basque language and grammar led to erroneous morphological analyses and assumptions; he does not confront his sources critically enough; he makes for the most part unfounded assertions about formal and semantic correspondences that are not at all clear; he is not sufficiently familiar with the history of the Romance and Basque languages and postulates wrong etymological connections or, as the case may be, wrong paths of borrowing; he often does not recognize Arabic loan words whose integration took different routes on the Iberian peninsula from in the rest of Latin Europe; he confines himself to word correspondences and avoids necessary questions which a grammatical comparison would require. In short, Schuchardt shows that Gabelentz's work has serious defects and contains beginner's mistakes in practically all respects, and since the book version of 1894 did not bring about any changes in the points mentioned above, critics of the latter in turn refer explicitly to Schuchardt's detailed criticism (e.g. Stumme, 1895; Unamuno, 1895; Vinson, 1901–1902). The reproaches are of course also addressed to the editor of the posthumous volume, Albrecht von der Schulenburg, who should have taken note of the earlier criticism and who had 'not given over to the flames' (*nicht den Flammen überantwortet*) such an obviously unfinished manuscript, which is apparently only able to 'attach a stain hard to remove' (*einen schwer tilgbaren Fleck anzuhängen*) to the good name of Gabelentz (Gustav Meyer, 1895, p. 785). The unanimous conclusion is that Schulenburg should not have published the manuscript in this form, and only did so out of a false piety towards his uncle (loc. cit.).[30]

30 Thus the Orientalist and linguist H[ans] St[umm]e (1895) in the *Literarisches Centralblatt* and the Balkan specialist and Indo-European scholar Gustav Meyer (1895) in the *Berliner Philologische Wochenschrift*. The latter discusses Gabelentz's volume together with Topolovšek's (1894) *Die basko-slavische Spracheinheit*, which, from a linguistic point of view, is the worst company in which one could be placed. In the file in the Altenburg archive mentioned in note 7 above, there is a letter in which the publisher Sattler sends these two reviews to Schulenburg, accompanied by the accusation of having deceived him before printing.
Friedrich Müller (1895) also links Gabelentz and Topolovšek because of the lack of grammatical correlations. Passages from Schuchardt and Müller are translated into Spanish by Aranzadi (1902) without further commentary.
Again and again the critics reproach Gabelentz for making, in these two works, the same methodological mistake he criticized in Bopp's discussion of the putative relationship between Indo-European and Austronesian languages (Gabelentz, 2016 [1891], pp. 152, 163, 280): he sacrificed the grammatical comparison to superficial phonetic similarities.

One serious objection Schuchardt raises is that Gabelentz claimed he was breaking new ground with his proposal, while in reality he had not taken note of the existing literature. In particular, Schuchardt refers to the works of Louis Gèze (1883–1885)[31] and Claudio Giacomino (1892).[32] In those years, however, the problem of Basque-Berber was apparently the subject of publications in various countries and from different perspectives. Brinton (1894, p. 43) mentions in his brief review in *Science* that 'so far back as 1876 Dr. Tubino [Tubino, 1876], of Madrid, in his "Aborigenes Ibericos," compared the two idioms for the same purpose'. The Iberian perspective was also taken up by Taylor (1893, p. 77), but using arguments based on physical anthropology: Broca's measurements of the skull had shown a similarity between the southern Basques and the Iberians, while the northern Basques belonged to an independent type. Basque as a language had thus spread from the northern to the southern Basque country, which created the contact situation with Iberian (and therefore presumably with Berber). However, it seems doubtful to us that Taylor correctly understood Gabelentz's text in this essay.

It would be all too easy to demonstrate further errors on the part of Gabelentz – methodological, historical, linguistic, in his understanding of Basque and Berber and in his treatment of the sources – apart from those already discussed.[33] But this has already been done in the literature. What

31 Louis Gèze was a specialist of Basque studies who wrote a Souletin grammar with a famous dictionary in 1873. In a letter from Gabelentz to Schuchardt, which follows Schuchardt (1893), Gabelentz asks for more detailed information on the works on Basque and Berber Schuchardt mentioned in his review. Schuchardt's answer is not preserved, but apparently he sent the aforementioned text by Gèze to his colleague in Berlin. This can be inferred from a letter from Schulenburg, which is kept in Schuchardt's *Nachlass* in Graz (available electronically at: http://schuchardt.uni-graz.at/id/letter/2780, accessed 2 August 2018). Schuchardt had apparently tried to get this publication back after the death of Gabelentz, but it no longer appears in Schuchardt's library holdings.

32 Claudio Giacomino was an Italian grammar school teacher and linguist. He later succeeded Ascoli, who had supported him for many years. With regard to the present topic, he wrote two studies: a shorter work from 1892, which was more focused on Old Egyptian and which could only be the base of the criticism in question, and a later, much longer one, which was published in 1895. Schuchardt could in fact also have known the latter, because Giacomino had submitted it in 1890 for the *Premio Reale* of the *Accademia dei Lincei* (without success). In 1896, however, he received the award for this work (the award was shared equally with Pio Rajna). Although Schuchardt was not a member of the Commission, he was probably informed as a member of the Academy; the manuscript is now in the archives of the *Lincei* in Rome.

33 The latter criticism was not mentioned in the reviews. But there are quite a lot of passages where Gabelentz did not handle his primary Basque source (van Eys, 1873) properly. For example, he neither looked at the alternation width of *m* already shown by van Eys in his foreword (1873, p. xliv) – or at least he did not take note of it in his presentation (Gabelentz, 1883, p. 597) – nor did

is surprising is that his mistakes are partly those of a beginner, which one would not have expected of Georg von der Gabelentz. Rather, it is important for us to find out why Gabelentz was so enthusiastic about an undertaking for which he simply lacked the specialist competence.

Vinson (1901–1902) continues his commentaries on new publications in Basque studies, which he started in his *Essai* (1891) and the *Additions* (1898), two standard works that are still used today. It is within this context that Vinson (1901–1902, pp. 145-146) deals with Gabelentz and criticizes his bold assumptions about lexical correspondences from a formal and semantic point of view. But he also finds that the book has missed several opportunities to establish grammatical correlations. In his article *Les Aryens*, Vinson (1904a, p. 183), who was known for his polemical directness, is merciless with the researchers on the Basque-Berber approach:

> [...] he [de Michelis] sees, for example, the relation between Basque and Berber, Coptic and Egyptian as established, due to the works of Gèze, Gabelentz and Giacomino; and yet nothing is more open to dispute, and among the works on Basque of these past years there is none more insufficient than those of Gèze, none more absurd than those of Gabelentz and none more far-fetched than those of Giacomino: risky comparisons and adventurous etymologies have never proven anything.[34]

He also comments on these theories, which lack methodological coherence, in a similar way in Vinson (1904b) in the *Journal Asiatique*. Miguel de Unamuno (1895) takes a gentler approach, which is, however, still just as damning: on the one hand, the criticism of Schulenburg's premature

he take into consideration van Eys's Romance etymologies. A correspondence between Tuareg *tablelt* and Basque *berun* 'lead' (*Blei*), also questioned by Basset (1896, p. 90) because of the strong dissimilarity, would find a much simpler and more plausible explanation in the Romance stem *plumb-*, especially if one takes into account the simplification of Latin-Romance onset clusters by epenthetic vowels and the change from *m* to *n* in final position, which Gabelentz himself emphasizes very prominently, on the second page of his Academy essay, as being characteristic. It is incomprehensible why he decided against this extremely obvious etymology, which was also suggested by van Eys (1873, p. 64), and preferred a barely tenable alternative. He does not give any reasons himself.

34 Original: '[...] il [de Michelis] regarde, par exemple, comme établie, grâce aux travaux de MM. Gèze, Gabelentz et Giacomino, la parenté du basque et du berbère, du copte, de l'égyptien; or rien n'est plus contestable, et parmi les travaux dont le basque a été l'objet depuis ces dernières années, il n'en est pas de plus insuffisants que ceux de M. Gèze, de plus absurdes que ceux de M. Gabelentz et de plus fantaisistes que ceux de M. Giacomino: des rapprochements hasardés et des étymologies aventureuses n'ont jamais rien prouvé.'

publishing activity is cutting, but, like Schuchardt (1893), he focuses on concrete word forms and their erroneous or questionable semantic and grammatical analyses. Unamuno, however, sharply criticizes Gabelentz's sources: he questions the reliability of van Eys's *Grammaire comparée* and especially of his *Dictionnaire*, claiming that they are too philological and therefore linguistically rather useless. But Unamuno is also much more conciliatory than Vinson, as he leaves open the possibility that some progress could be achieved in this domain with more precise work and more reliable knowledge. He also admits that Gabelentz could have obtained other results if he had used his materials more carefully.

It remains open whether this positive assessment is correct or not. Because of his early death, Gabelentz could not participate in the critical discussion of his publication of 1893. There is only one letter – that is, the letter to Schuchardt dated 5 September 1893 –[35] in which Gabelentz responds to the criticisms in the review. His answer is more defensive than understanding. He admits being unaware of the earlier works (Gèze and Giacomino) as well as the fact that Schuchardt himself had published on Basque,[36] but points out that his colleagues in Berlin – he obviously means those present at his Academy lecture – did not direct him to the missing literature. However, Gabelentz remains unwilling to admit methodological failings in other regards: he only wants to accept words of Romance origin 'for the smallest part' (*zum kleinsten Theile*); that is, if they are of Indo-European origin themselves. He appeals to folk etymology to justify this approach: words can be transferred to a new etymon by means of a similar sound structure. By doing so he tries to get around the criticism that he often takes erroneous analyses of morpheme as a basis. He does not react to other criticisms (e.g. that Arabic loan words may have entered Basque via another route, such as Spanish).

Schuchardt expresses his astonishment at Gabelentz's unwillingness to see reason less than a week after the death of Gabelentz in a letter to Otto

35 This letter is printed in Hurch (2011, pp. 251-252) and is available electronically at http://schuchardt.uni-graz.at/id/letter/2981, accessed 2 August 2018.

36 This statement is not very credible. Schuchardt (1892) discussed Giacomino's study in the *Literaturblatt* of Leipzig. Gabelentz himself did not publish anything there (Gabelentz sent almost all reviews to the *Literarisches Centralblatt*), but it can be assumed that he received the *Literaturblatt*, because he was reviewed there in 1892 by Otto Behagel (pp. 257-58); Schuchardt's above-mentioned discussion of Giacomino can be found there on pp. 426-430. Schuchardt's (1888) review of Gerland, which is indeed listed in the card catalogue in Poschwitz (see Appendix 2), also appeared in the *Literaturblatt* (it remains an open question as to when it was added to the bibliography).

Jespersen of 16 December 1893,[37] in the context of a discussion on '*anmeldelser* (review essays – for I now write mere reviews only in exceptional cases) and *smasafhandlinger*' (anmeldelser (*Rezensionsabhandlungen – denn blosse Rezensionen schreibe ich jetzt nur ausnahmsweise*) *und* smasafhandlinger) (Schuchardt):

> Take a work such as that of Gabelentz on linguistics [*Die Sprachwissenschaft*]; how stimulating that is! and all the problems are only *touched on*, many of them barely mentioned. I deplore the death of this researcher all the more because I inevitably subjected his last paper on *Berber* and *Basque* to a very negative review. He wrote me a very kind letter afterwards, but I was surprised that my arguments and proofs did not impress him much at all. I had very much been looking forward to an amicable discussion of one and the other of the thousand points touched on in his book, and now he can't even reply to what I've just said against him in a *smaaafhandling* [short essay].[38]

Contemporary reactions from specialists on Berber languages are no less vehement. In an appendix to his article, Basset (1896, pp. 90-91), after a few general remarks about the subject of Gabelentz's study, examines his treatment of the names for metals as representative of Gabelentz's approach. First of all, Basset criticizes Gabelentz for limiting himself to too few dialects/languages of Berber and the fact that his sources (Newman, 1882) are 'absolutely incomplete and often faulty' (*absolument incomplet et souvent fautif*) and continues:

> But the same is true for the dialects: the author [Gabelentz] lacks thorough knowledge [of them]. He undertakes his comparison of Basque not only with *Arabic* (!) words but also with *French* (!!) words borrowed into Kabyle.

37 The whole letter is available electronically at http://schuchardt.uni-graz.at/id/letter/144, accessed 2 August 2018.

38 Original: 'Nehmen Sie ein Werk wie das von Gabelentz über die Sprachwissenschaft; welche ungeheure Fülle von Anregungen! und alle Probleme sind nur *gestreift*, ja viele kaum erwähnt. Ich beklage den Tod dieses Forschers um so mehr als ich seine letzte Schrift über *Berberisch* und *Baskisch* nothgedrungen einer sehr absprechenden Beurtheilung unterzog. Er schrieb mir darauf einen sehr liebenswürdigen Brief, aus dem ich aber doch mit einiger Verwunderung ersah dass meine Argumente und Nachweise keinen besondern Eindruck auf ihn gemacht hatten. Wie sehr hatte ich mich darauf gefreut, einen und den anderen von den tausend Punkten die in seinem Buche berührt werden, freundschaftlich mit ihm zu diskutiren, und nun kann er mir nicht einmal darauf mehr erwidern was ich eben in einer smaaafhandling gegen ihn vorgebracht habe.'

> In this way, Basque *burdi* 'carriage' is compared to Kabyle *ϑabruedt* (an inexact form for *ϑabruet = thabrouet'*, from French *brouette* 'wheelbarrow'!). [...] It is, I believe, useless to insist on the value of results obtained in this way.[39]

Subsequently Basset deals in detail with Gabelentz (1894, pp. 116-117); that is, the designations for metals. Basset (1899, p. 43) notes tersely a few years later: 'The attempts of Gabelentz to connect Basque to Berber do not deserve to be dwelt upon. The author's knowledge – and I only talk about Berber – does not permit him to undertake such work' (*Les tentatives de M. von der Gabelentz pour rapprocher le basque du berbère ne méritent pas qu'on s'y arrête. Les connaissances de l'auteur, je ne parle que du berbère, ne lui permettent pas d'entreprendre un pareil travail* [...]).[40]

6 On Gabelentz's sources

6.1 Gabelentz's sources and knowledge of Basque

The study of Basque is a rather marginal topic in Gabelentz's research up until the last years of his life and he did not leave a great deal of written evidence behind of his engagement with this topic. Even in his *Sprachwissenschaft* (Gabelentz, 2016 [1891]), Basque hardly receives any mention. It is questionable how deeply he familiarized himself with the Basque language and existing research on it. It is therefore all the more astonishing that in his final works – although he could not have known that they would be his final works – Basque occupies such a central place, and was dealt with so incompetently. There are no extant independent manuscripts or publications on Basque by Gabelentz and only two reviews out of 235 that he wrote in

39 Original: 'Mais même pour ce qui concerne ces dialectes, une connaissance approfondie manquait à l'auteur; aussi fait-il porter sa comparaison du basque, non pas seulement sur des mots *arabes* (!), mais même sur des mots *français* (!!) passés en Kabyle; c'est ainsi que le basque *burdi*, voiture, est comparé au kabyle *ϑabruedt* (forme inexacte pour *ϑabruet = thabrouet'*, du français *brouette !*). [...] Il est, je crois, inutile d'insister sur la valeur des résultats ainsi obtenus.'

40 Gabelentz's theories on the Hamitic (Austroasiatic) character of Basque seldom appear in recent literature on the subject. Only Mukarovsky (1963–1964) attempted to resume the discussion, although with incorrect premises regarding the literature; his approach has remained without further resonance. At the time, Schuchardt (1907; 1913) tried to establish the North African link of Basque via Iberian, but his attempts were questioned by progress at the time in reading of Iberian and Celtiberian texts. Finally, Zyhlarz (1932) dealt with this subject, once more in detail and negatively. See also Michelena (1964, pp. 171-175) for discussion.

the course of his life are devoted to Basque works: Gabelentz (1883b; 1885) discuss W. van Eys's *Outlines of Basque Grammar* and Arno Grimm's *Über die baskische Sprache und Sprachforschung*. In 1888 he also wrote a review of the 1886 *Glosario* by Eguilaz y Yanguas. The latter, however, is little more than a brief discussion and does not go into the details of Basque research at all, even though it is precisely the *Glosario* that could and should be relevant for his two later publications. It is indeed surprising that it does not reappear in Gabelentz's work on Basque and Berber.[41]

The language library of Poschwitz is only partially preserved and the remnants are kept in the Thüringisches Staatsarchiv in Altenburg. We know little about how many volumes were originally in the library; current descriptions are largely based on guesses or indirect and less concrete anecdotal evidence, much of it stemming from an oral tradition within the Gabelentz family.[42] Various catalogue lists and, above all, the card catalogue itself provide information on those works on Basque which might have been available to Gabelentz.[43] Appendix 2 to this chapter lists the various relevant inventories. However, Gabelentz used only a small portion of these works for his Basque-Berber writings and materials.[44] In his two publications he mentions several times that his most important and almost exclusive source was the Basque-French dictionary and comparative grammar of Willem van

41 However, in his library in Poschwitz, which is described in detail below, Gabelentz actually lists Eguilaz y Yanguas among the Basque books and writings relevant for Basque studies.

42 Emig (2013, p. 312) summarizes this 'state of knowledge'.

43 The card catalogue in Altenburg is kept in a few meticulous hands with the same ink. Since the library grew over decades, and also based on the type of index cards used, it is reasonable to assume that the now existing catalogue was not kept continuously, but was (newly) created at a later date. However, there is no doubt about its being original and in our opinion it is not clear why Vogel (2013, p. 200) speaks of the 'few, still remaining card index boxes' (*wenigen, noch erhaltenen Karteikästen*). The completeness of the alphabet in the nominal catalogue does not suggest that there is anything missing; the subject catalogue also includes, for example, those works on Basque which can be assumed to have been available to Gabelentz for his research, and Gabelentz does not mention any works which are not also represented in the file boxes. In addition, in the old photograph of his study (ibid., p. 198) there are 8 boxes, which correspond in appearance and extent exactly to the ones available today.
In the appendix to this chapter, comparisons have also been made between the card boxes and handwritten book lists referring to Basque and Berber, which have been preserved in Gabelentz's *Nachlass* in the Thüringisches Staatsarchiv in Altenburg; these do not indicate either that the card index boxes are now only roughly incomplete.

44 These lists hold some surprises. Gabelentz mentions, for example, Humboldt (1821), but not Humboldt's contribution to *Mithridates* (1817). As other contemporaries (such as A.F. Pott) state, *Mithridates* was still the main reference work on the Basque language in the German-speaking area in the second half of the century. Gabelentz's interest in the Vasco-Iberian or Celtiberian discourse of Humboldt (1821) does not go beyond the mere mentioning of it, however.

Eys (1873; 1879), but he also refers to the works of Louis-Lucien Bonaparte (without pointing to any concrete passage). Campión's *Gramática de los cuatro dialectos literarios* (1884) appears only in the appendix to the book of 1894, in connection with the above-mentioned table of the 'permutations of consonants' (*permutaciones de consonantes*).

A recurring mistake on the part of Gabelentz shows how slight his knowledge of Basque really was: he does not consider the internal structures of Basque and Berber words in his comparison. This means that he simply compares word structures without worrying about the question whether a (supposedly) corresponding sound belongs to the stem or to an affix. Such a systematic methodological error is surprising and, being committed by the author of *Die Sprachwissenschaft*, demands to be understood rather than simply rebuked. Gabelentz responds somewhat unconvincingly to criticism of this approach by saying that in folk etymology such examples are explained without being decomposed into their constituents.[45] Basque scholars of the Basque language, however, only comment on Gabelentz's obvious lack of understanding of the language, language history and grammar. Gabelentz's Basque studies were the product of a rather peculiar approach: there were two literary overviews of the Basque language in German, an older one (Humboldt, 1817 [1816]) and a relatively recent and more extensive one (Pott, 1887), which Gabelentz does not seem to have considered either.

6.2 Gabelentz's sources and knowledge of Berber

On 16 January 1885, Gabelentz wrote to his sister that he had studied Kabyle 'more or less during the past years' (*innerhalb der letzten Jahre mehr oder minder*) (Münchhausen, 2013 [1913], p. 121). His activities date back to 1882 at least, when his entry 'Kabyle (linguistic)' (*Kabylen* (*sprachlich*)) appears in the *Allgemeine Encyclopädie der Wissenschaften und Künste* (General Encyclopaedia of the sciences and arts) of Ersch and Gruber.[46] It is a short

45 An example for illustration: *et(h)orki* 'clan, family' (*Sippe, Familie*), according to Gabelentz, is therefore more likely to belong to Berber than to be a completely regular derivation of Basque *etor* 'come' (*kommen*) and the nominal suffix *ki(n)*. Such Basque words would thus have their origin in the Berber *ϑerga* or *δerga* and the phonetic similarity would have caused their ascription to another etymon (that of *etorri*). Gabelentz does not give any justification for this from a scientific point of view, which opens the door to methodical arbitrariness.

46 Between 1882 and 1889, Gabelentz wrote a total of 26 articles beginning with the letter K and three beginning with L for the *Allgemeine Encyclopädie*. Primarily, these are language portraits (e.g. 'Kamilaroi language' [*Kamilaroi Sprache*], 'Khamti', 'Kuki', 'Lepcha') and contributions on East Asian literature and philosophy (e.g. 'Laozi' [*Lao-tse*]), but Gabelentz also contributed a short biography of Julius Klaproth.

and as such simplified representation of the most important grammatical phenomena, based on Adolphe Hanoteau's grammar of Kabyle (1858). Gabelentz also cites two dictionaries, Delaporte (1844)[47] and 'Creuzat' (i.e. Creusat, 1873), which are not mentioned in his later works. Here already, he characterizes the sounds as fluctuating and speaks of uncertain articulation.

Later Gabelentz reviewed Newman's *Lybian* and *Kabail vocabulary* (Gabelentz, 1883a; 1889), which would be important sources for his later comparison to the Basque language.[48] In between, he confines himself to the proof the genetic relationship between Hamitic and Semitic languages by first citing similar grammatical constructions and only secondarily giving lexical examples in *Die Sprachwissenschaft* (Gabelentz, 2016 [1891], p. 169).

As already mentioned, it is not possible to clearly determine his actual sources on the basis of the spellings he uses for examples. The works he explicitly cites are either in French or English and not only reproduce individual phonemes by different letters, but also partly diverge in the vowel systems used. In his article in the *Encyclopädie* Gabelentz still follows Hanoteau (1858) to a great extent, but for his later studies he adapts the spelling to his German-speaking audience. He does not say so at any point, but he may have used Lepsius's standard alphabet for *Tamašeq* (Lepsius, 1863, pp. 205-206) as his starting point.

7 An assessment

Proving that Basque is a 'Hamitic language, related to the Berber family' (*hamitische, der Berberfamilie verwandte Sprache*) and developing proposals for sound correspondences in prehistoric times that contradict the regularity of sound change are perhaps the two main goals of Gabelentz's final project. This work has, however, been completely disregarded by later research on Gabelentz. Ignoring this aspect of his work is consistent with the fact that it does not fit with the idealized picture of Gabelentz that is cultivated in the history of linguistics. But ignoring this work and leaving the relevant texts (Gabelentz, 1893; 1894) out of anthologies such as Ezawa, Hundsnurscher & Vogel (2013) eschews scholarly responsibility.[49] In this ideal conception,

47 The dictionary was compiled by J.D. Delaporte, E. de Nully, Ch. Brosselard and Sidi Ahmed ben el Hadj Ali under the direction of Amédée Jaubert. It is therefore known as Jaubert (1844) in the literature.

48 Basset's critique (1896) was discussed here earlier.

49 With the exception of the list of his publications, which does mention the two works, this statement applies to all essays assembled in the anthology.

the master's works end with the publication of *Die Sprachwissenschaft* in 1891, after which he seems to have devoted himself only to the preparation of a second edition of this book.[50]

After almost 125 years it is still difficult to assess the place of these works. Gabelentz (1894) only partially corresponds to the picture the history of linguistics has made of him. It is true that there are similarities, a same basic tendency and cross-references between the posthumous book and Gabelentz (1893), the text he published himself. Substantial parts are not formulated in words, however, but mainly contain uncommented material. This suggests that it was by no means a 'ready-to-print manuscript' (*druckfertiges Manuskript*), as the accompanying text states. Gabelentz refers in various places to a longer work on the subject, but it is unlikely that he would have wanted to see it published in this form. The editor, Schulenburg, was perhaps reckless and lacking in sufficient respect in simply putting Gabelentz's notes into print.[51] Ringmacher and McElvenny (Gabelentz, 2016) have already complained when comparing the two editions of *Die Sprachwissenschaft* that Schulenburg intervened in a not insignificant and often incompetent way. In the present case, we do not assume that Schulenburg interfered textually with the posthumous book version of 1894, because it contains hardly any continuous running text.[52] Nevertheless, the fact remains that Gabelentz published the Academy lecture of 1893, in which he puts forward his bold theories and questionable linguistic correspondences, and that he reacted quite uncomprehendingly to the criticism he faced before his death. He used unreliable sources for all languages involved and did not acquire sufficient linguistic knowledge of the languages, available sources, grammars, and lexical works to start an undertaking of this kind. However, such an enormous misjudgement can of course occur in exceptional situations – and perhaps this was the case for Gabelentz, who found himself in a difficult period both in terms of his family life and his health (see Vogel & McElvenny, this volume). In those decades, other linguists also proposed bold theories about genetic relationships, and the Basque language has

50 Gabelentz's Basque-Berber theory was still praised by Münchhausen (2013 [1913], pp. 135, 138); she does not seem to have noticed its dubious nature at the time of writing and publishing of her memoires.

51 This circumstance is also regrettable because Gabelentz appreciated his nephew very much and 'was pleased by him' ([*hat s*]*eine Freude an ihm*) (letter of 30 April 1891, printed in Münchhausen, 2013 [1913], p. 126). He also gave a positive review (1892b) of Schulenburg's grammar of the language of Murray Island (1891).

52 For this, he was clearly lacking in professional competence. However, textual interventions would hardly be detectable today.

always stimulated not only scientific activity but also – and above all – the imagination of many researchers. But it is hardly excusable in the history of science and of linguistics that *this Gabelentz* is simply ignored and that the secondary literature ends with *Die Sprachwissenschaft*.

8 Annexes

8.1 Appendix 1: materials for the time in Berlin

Klaus Kaden (1993)[53] has written a very well-researched article based on the files concerning the appointment of Georg von der Gabelentz to the Friedrich-Wilhelms-Universität in Berlin. Understanding Gabelentz's last years in Berlin and his activities there provides a key to understanding his Basque-Berber works. For this, Kaden evaluates the documents that lie in the archives in Berlin, in particular in those of the present-day Humboldt-Universität and the Berlin-Brandenburgische Akademie der Wissenschaften (Berlin-Brandenburg Academy of Sciences), as well as today's Geheimes Staatsarchiv Preußischer Kulturbesitz (Secret State Archive of the Prussian Cultural Heritage Foundation). Gabelentz's chief mentors in the Berlin faculty were the Indo-Europeanist Johannes Schmidt[54] and the geographer Ferdinand von Richthofen.[55] Schmidt was responsible for assessing and recommending Gabelentz to the faculty on the basis of his linguistic work, while Richthofen wrote a detailed letter of recommendation to the ministry, supported by some faculty members and the dean.[56] At that time, Gabelentz was already a member of the Academy of Berlin, succeeding Wilhelm Schott. In 1889 he also succeeded Schott as professor at the university, but his position was upgraded from Schott's extraordinary professorship to a full professorship, a step taken to expand Sinology as a subject in Berlin and thus respond to new scientific and social perspectives. But in these years Gabelentz's own interests shifted: in teaching (see Kaden,

53 This essay is now available in an easily accessible form in Ezawa & Vogel (2013, pp. 271-288).

54 A few years earlier, it was precisely Johannes Schmidt, at that time professor of comparative and Indo-European linguistics in Graz, who had led Schuchardt's call to Graz, with great ambition and, ultimately, success for the faculty. See the relevant documents in the electronic *Hugo Schuchardt Archiv* (http://schuchardt.uni-graz.at/home, accessed 2 August 2018) or in the publication of the correspondence Schmidt-Schuchardt contained in the archive.

55 Ferdinand Frhr. von Richthofen (1833–1905) is considered the founder of geomorphology and gave lectures in Berlin from 1886.

56 Thankfully, all these documents were edited by Kaden (2013 [1993]).

2013 [1993], p. 287, for the list of lectures held by Gabelentz in Berlin), Sinology and East Asia-related lectures clearly dominate, but his publications are increasingly devoted to general linguistic and typological topics, as well as to the field of Basque and Berber. In 1891 Gabelentz published a – or perhaps *the* – magnum opus, namely *Die Sprachwissenschaft*, in 1892 the aforementioned *Handbuch*, and in 1893, the year of his death, he devoted himself to the Basque-Berber studies. These are altogether very extensive works on non-Sinological topics. The Berlin faculty did not seem to be very pleased by these developments.

Below are two documents, which originate from Schuchardt's *Nachlass*, who was visibly affected by the death of Gabelentz. Both of them are letters from Adolf Tobler, a Romance scholar and linguist as well as a member of the Academy, who therefore took part in Gabelentz's appointment to Berlin and who certainly sympathized with him. Unfortunately, Schuchardt's letters have not been preserved in Tobler's papers.[57]

Lib. no. 11715
Berlin, 18 Dec. 1893.
Dear colleague,
I regret to say that I am not able to provide you with any further information about the life and nature of my late colleague v. d. Gabelentz. As far as I am aware, he was not in close contact with any of his colleagues, except for example with the geographer v. Richthofen, whom he knew from Leipzig, to whom he was probably drawn by their shared interest in China, and perhaps also by their shared social status [i.e. they were both aristocrats], which is less important among men, but which sometimes comes into consideration for the social intercourse of those who are married. G. was probably not of unsociable nature, he also liked to talk in a lively way about non-specialist subjects, but he remained quite lonely here. In the beginning, he was unmarried, several times on leave for a longer period of time, which he spent on a small estate near Altenburg, where his books had remained. When he remarried about a year and a half ago (a young aristocratic widow, who seemed to me to be very kind), he probably stayed in his four walls even more than before. He leaves behind some children of his first marriage, which I believe have never been here,

57 The two letters are kept in Schuchardt's Nachlass in the special collections of the Graz university library. They can be accessed electronically at http://schuchardt.uni-graz.at/id/letter/4046, accessed 2 August 2018, and http://schuchardt.uni-graz.at/id/letter/4045, accessed 2 August 2018.

a child of his second wife, & she expects the birth of a second, I am told.[58] The huge, seemingly strong man suffered from kidney stones; he died of nephritis accompanied by pneumonia. His academic effectiveness was minimal, not only as far as it concerned Chinese, Manchu etc. but also general linguistics. I do not know why. He stuttered quite a lot, which, in connection with his being quite cross-eyed, did not make conversation with him pleasant;[59] but in Leipzig he is said to have been very successful as a lecturer, as it is called. It seemed to me as if he was not really able, even with a bit of willpower, to make the effort to connect with his audience. But I don't know whether this was the cause or effect of the low participation on the part of the students. I met him every 14 days during the winter in a company of 15 men of scholarship,[60] where – every time a lecture was given – I heard a lot of stimulating things and spoke myself with joy when it was my turn. Here too, he did not understand the appropriate way to do so; one did not get the impression of the certainty that comes from methodical work, but rather that of a certain inconstant amateurism. He was certainly a man of great talent & varied knowledge; but it seems to me that he lacked a thorough education. You know that he had actually studied law & had been working in government administration for a long time (in Alsace); the weakness of his work may be explained by the fact that he took up his father's studies only later, I believe, and entirely on his own initiative & that he perhaps started teaching a little too early in a discipline where we work largely unmonitored. A great deal of effort was made to get him to Berlin back then, but in general there was a certain disappointment & I should not be surprised if, after this experience, it would be a while before the deceased's chair is reoccupied. For Chinese we have the proficient extraordinarius,[61] to whom G. himself gladly referred those students he was not inclined to teach. That's all I can tell you about G. What I have read about him in local newspapers is the useless bilge to which we poor devils regularly fall victim when we die. Richthofen (Geh.

58 Gabelentz's widow, however, suffered a miscarriage after his death. The two sons Albrecht and Wolf were born from his first marriage, Hanns-Conon from his second marriage (cf. Münchhausen, 2013 [1913], esp. p. 140; Vogel & McElvenny, this volume).

59 Gabelentz's sister Clementine writes that since childhood his 'left eye would tend to leer outwards' (*mit dem linken Auge nach außen zu schielen*) when he felt attacked, and that he would stutter 'when he had the impression that someone listened to him without sympathy' (*wenn er den Eindruck hatte, daß ihm ohne Wohlwollen zugehört wurde*) (Münchhausen, 2013 [1913], p. 90).

60 This is the *Berliner Mittwochsgesellschaft*, which was founded on 19 January 1863 and was limited to sixteen persons. A series of lectures was held in turn and recorded.

61 Carl Arendt (1838–1902).

Rat Freiherr Ferdinand von Richth., Kurfürstenstr. 117) would most likely be able to send you more detailed information.[62]
I thank you very much for your kind regards on the occasion of my silver wedding anniversary.
Yours devotedly
A. Tobler

Berlin d. 18. Dez. 1893.
Verehrter Herr Kollege,
irgend welche genauere Auskunft über Leben & Wesen meines verstorbenen Amtsgenossen v. d. Gabelentz Ihnen zu geben oder zu verschaffen bin ich zu meinem Bedauern nicht in der Lage. Er hat, so weit ich weiß, mit keinem seiner Kollegen in vertrauterem Verkehr gestanden außer etwa mit dem Geographen v. Richthofen, den er von Leipzig her kannte, mit dem ihn wohl auch das beiderseitige Interesse für China zusammenbrachte, vielleicht auch die Standesgemeinschaft, die unter Männern zwar weniger zu bedeuten pflegt, für den geselligen Verkehr Verheirateter jedoch manchmal in Betracht kommt. G. war wohl nicht ungeselliger Natur, sprach gern & lebendig auch über andere als Fachgegenstände; aber er ist hier doch ziemlich einsam geblieben. Anfangs war er noch unverheiratet, mehrfach auch auf längere Zeit beurlaubt, die er auf einem kleinen Besitztum bei Altenburg, wo seine Bücher geblieben waren, verbrachte. Als er dann vor etwa 1½ Jahren sich wieder verheiratete (mit einer, wie mir schien, liebenswürdigen jungen Wittwe von Adel), blieb er wohl noch mehr als zuvor in seinen vier Wänden. Er hinterläßt außer mehreren Kindern erster Ehe, die glaub ich nie hier gewesen sind, ein Kind seiner zweiten Frau, & diese erwartet, wie ich höre, die Geburt eines zweiten. Der riesengroße, anscheinend kräftige Mann litt an Steinbeschwerden, gestorben ist er an Nieren- verbunden mit Lungenentzündung. Seine akademische Wirksamkeit war ganz gering & zwar nicht allein, soweit sie Chinesisch, Mandschuh u. dgl. galt, sondern auch wenn sie allg. Sprachwissenschaft zum Gegenstand hatte. Woran das gelegen haben mag, weiß ich nicht. Er stotterte allerdings ziemlich stark, was in Verbindung mit sehr auffälligem Schielen die Unterhaltung mit ihm nicht eben angenehm machte; aber in Leipzig soll er trotzdem als Dozent, was man so nennt, großen Erfolg gehabt haben. Mir kam es so vor, als ob er sich nicht hinlänglich dazu verstehen könne, sich um eine ernstliche Förderung seiner Zuhörer auch mit einiger Selbstüberwindung zu bemühen. Aber ob das Ursache oder Wirkung der geringen Teilnahme von Seite der Studenten

62 There is no answer of Richthofen in Schuchardt's *Nachlass* in the Graz university library.

war, weiß ich nicht. Ich bin den Winter über alle 14 Tage mit ihm in einer Gesellschaft von 15 Männern des Gelehrtenstandes zusammengekommen, wo jedesmal ein Vortrag gehalten wurde, ich viel Anregendes gehört & mit Freude auch selbst gesprochen habe, wenn an mir die Reihe war. Auch da hat er die angemessene Art nicht zu treffen verstanden; den Eindruck der Sicherheit, die ein methodisches Arbeiten giebt, bekam man nicht, viel eher den eines gewissen unsteten Dilettantismus. Er war gewiß ein Mann von großem Talent & mannigfaltigem Wissen; aber mir scheint, es habe ihm eine gründliche Schulung gefehlt. Sie wissen, daß er eigentlich die Rechte studiert hat & auf lange Zeit (im Elsaß) in der Verwaltung thätig gewesen ist; daß er erst später, wie ich glaube, ganz auf eigene Faust die Studien seines Vaters aufgenommen hat & vielleicht etwas zu früh auf einem Gebiete lehrend aufgetreten ist, wo man unter spärlicher Kontrole arbeitet, mag die Gebrechen seiner Thätigkeit erklären. Man hat seiner Zeit große Anstrengungen gemacht, um ihn für Berlin zu gewinnen; aber allgemein ist doch eine gewisse Enttäuschung gewesen, & es sollte mich nicht wundern, wenn nach der gemachten Erfahrung man sich längere Zeit dazu nähme den Lehrstuhl des Verstorbenen wieder zu besetzen. Für Chinesisch haben wir den tüchtigen Extraordinarius, an den G. selbst die Studenten gerne wies, die er zu unterrichten keine Neigung empfand. Das ist, was ich Ihnen über G. zu sagen weiß. Was ich in hiesigen Zeitungen über ihn gelesen habe, waren die ganz nichtsnützigen Elaborate, denen wir arme Teufel, wenn wir sterben, regelmäßig verfallen. Zu genaueren Daten würde Ihnen wohl am ehesten Richthofen (Geh. Rat Freiherr Ferdinand v. Richth., Kurfürstenstr. 117) verhelfen können.
Für Ihren freundlichen Gruß zu meiner silb. Hochzeit sage ich Ihnen herzlichen Dank.
Ihr ganz ergebener
A. Tobler.

Lib. no. 11716
Berlin, 30 March 1894
Dear colleague,
it is very surprising to me that Richthofen has not sent you an answer; usually he is not one of those who give reason to complain of impoliteness. When I see him again, I will remind him of the matter. Fortunately, I can tell you myself what will happen to Gab.'s library. Gabelentz has disposed by will that it will remain in Goßnitz,[63] where it had always been, until

63 Tobler should have written *Poschwitz*.

> the son of his second marriage, who is still quite young, comes of age. If he should not turn to linguistic studies, it should be sold, & namely Gabelentz's nephew & student, the Count Schulenburg, should have the first right to acquire it for 250,000 Mark. Well, I hear, Sch. already wants to take it over under the condition that it will have to be returned to the young Gabelentz if and when he claims it. […]
>
> Berlin, d. 30. Mz 1894
> Verehrter Herr Kollege,
> es ist mir sehr überraschend, daß Richthofen Sie ohne Antwort gelassen hat; er ist sonst nicht von denen, die Anlaß zu Klagen über Unhöflichkeit geben. Wenn ich ihn wieder sehe, will ich ihn an die Sache erinnern. Die Frage nach dem Schicksale von Gab.s Bibliothek kann ich zum Glück selbst beantworten. Gabelentz hat letztwillig bestimmt, daß sie in Goßnitz, wo sie bisher immer sich befunden hat, auch weiter bleibe, bis sein noch ganz junger Sohn zweiter Ehe volljährig sei. Sollte dieser sich nicht linguistischen Studien zuwenden, so soll sie verkauft werden, & zwar soll Gabelentz' Neffe & Schüler der Graf Schulenburg das erste Recht haben sie für 250000 M. zu erwerben. Nun, höre ich, soll Sch. sie jetzt schon übernehmen wollen mit dem Vorbehalt, daß sie seiner Zeit dem jungen Gabelentz zurückzugeben sei, wofern er Anspruch darauf erhebe. […]

These two letters draw – certainly without any ulterior motives – a picture that contrasts with the consistently optimistic assessment that is widespread in biographies of Gabelentz (see, e.g., Gimm, 2013a, in particular pp. 58-70). It is only for this reason that they are introduced here without further comment. Some of the criticisms that emerge in the contemporary and later Basque-Berber discussions can easily be understood in light of Tobler's statements here.

8.2 Appendix 2: On the Basque and Berber collections in Poschwitz

8.2.1 About Basque

As mentioned above, there is a lack of clarity as to the extent of the holdings of the language library in Poschwitz.[64] In all the library inventories that have come to us the Basque collections are of such a size that they can easily be surveyed. In the following we deal with all three sources that are available.

64 See 6.1 and note 43 above.

Together with the assessment of his Basque studies, this results in an overall picture which seems to reflect what was available to Gabelentz for his work.

First, a bibliographic catalogue of Basque literature in Gabelentz's own handwriting is reproduced. In Gimm (2013b, p. 118), this is referred to as the 'catalogue of the Poschwitz Library' (*Katalog der Poschwitzer Bibliothek*) under no. 333 on the list of publications.[65] It is less extensive than the card catalogue of the Poschwitz language library, whose cards are then reproduced below. The third source is a single keyword index card for Basque (from the same catalogue). We assume that, on the whole, this represents the complete range of Basque literature that was available in Poschwitz.

A.) From the handwritten list of Gabelentz (Thüringisches Staatsarchiv in Altenburg, no. 1061):[66]

IV., *Baskische Sprache*
M. de Larramendi El impossible vencido. Arte de la lengua Bas- 3153
congada. Salamanca 1729. 8º.
F. Lécluse Grammaire basque. Toulouse & Bayonne 1826. 8º. 3148
Abbadie & *Chaho* Etudes grammaticales sur la langue Euskarienne. 3150
Paris 1836. 8º.
F.J. de Lardizaval Gramatica Vascongada. S. Sebastian 1856. 4º.
Yrizar y Moya de l'Eusquere & de ses Erderes, ou de la langue 3143-45
basque. Paris 1841-45. IV. 8º.
F. Lécluse Dissertation sur la langue basque. Toulouse 1826. 8º. 3142
Le même Sermon sur la montagne, en Grec et en Basque, précédé 3149
du Paradigme de la conjugaison basque. Toulouse 1831. 8º.
Inchauspe le verbe basque. Paris 1858. 4º.
S.H. Blanc Grammaire de la langue basque. Lyon 1854. 12º. 3157
W.G. [sic] *van Eys* Outlines of Basque Grammar. Lond. 1883. 8º. 3152
K. Hannemann Prolegomena zur baskischen oder kantabrischen 3156
Sprache. Leipzig 1884, 8º.

65 Unfortunately, this entry has several mistakes: firstly, Gimm assigned the number 333 twice, secondly it is a *manu*script but not a *type*script. In addition, this entry is questionable on the whole, since the document indicated is not a publication, but simply an ordered list of books in handwritten form left behind by Gabelentz. The exact date of its creation cannot be determined. In any case, it was laid out in such a way that Gabelentz could always add new literature, which he did indeed do.

66 The entries Hannemann, Grimm and Gabelentz are written in *Kurrent*, all others in Latin script. The numbers added on the side and in another ink and writing appear to refer to a specific location.

A. Grimm Die baskische Sprache und Sprachforschung. Allgemeiner Theil. Breslau 1884, 8º. 3155
A. Chaho Dictionnaire basque, français, espagnol et latin. Bayonne 1856.s. 4º. 3166
M. de Larramendi Diccionario trilingue, Castellano, Bascuence y Latin. Nueva ed. p. Pio de Zuazua. San Sebastian 1854. 4º. 3163
G.v.d. Gabelentz Baskisch und Berberisch. Berlin. 1893, 8º.
W.J. van Eys Grammaire comparée des dialectes basques. Paris 1879, 8º. 3162

B.) In the card catalogue of the Poschwitz language library there are additional index cards to the following Basque writings (only works not mentioned under A above are listed):

Astarloa, Pablo Pedro. 1883. Discursos filosóficos sobre la lengua primitiva ó Gramática y análisis razonada de la euskara ó bascuence. Bilbao: Pedro Velasco.
Campión, Arturo. 1884. Gramática de los cuatro dialectos literarios de la lengua euskara. Toulouse [sic]: Eusebio Lopez.
Chaho, Augustin. 1856. La guerre des alphabets. Règles d'orthographe euskarienne, adoptées pour la publication du dictionnaire basque, français, espagnol et latin. Bayonne: P. Lespés.
D.A.P.I.P. [Pascual Iturriaga, Agustín]. 1842. Dialogos basco-castellanos. Para las escuelas de primeras letras de Guipúzcoa. Hernani.
Eys, Willem Jan van. 1873. Diccionnaire basque-français. Paris – London: Maisonneuve – Williams u. Norgate.
Humboldt, Wilhelm von. 1821. Prüfung der Untersuchungen über die Urbewohner Hispaniens vermittelst der Vaskischen Sprache. Berlin: Dümmler.
Mahn, C.A.F. 1857. Denkmaeler der baskischen Sprache. Berlin: Dümmler.
Pott, August Friedrich. 1875. Über vaskische Familiennamen. Zur Erinnerung an den glücklichen Schluß des durch Otto Böhtlingk und Rudolph Roth 1852 begonnenen und 1875 vollendeten Sanskrit-Wörterbuchs. Detmold: Meyersche Hofbuchhandlung.
Schuchardt, Hugo. 1888. Besprechung von Gerland, Georg, die Basken und die Iberer. Literaturblatt für germanische und romanische Philologie 9: 225-234.
Schuchardt, Hugo. 1893. Baskische Studien I. Über die Entstehung der Beziehungen des baskischen Zeitwortes. Wien: Tempsky (Denkschriften der Kais. Akademie der Wissenschaften).

C). Single index card for the keyword *Baskisch* (Basque) (also part of the card catalogue of the Poschwitz language library, Thüringisches Staatsarchiv in Altenburg)

Schlagwort: Baskisch (Versch. alte Spr. III c)	
Siehe: 1. Gesner, C… 'Mithridates'	LIb,1
2. 'Orient u. occident. *Sprachmeister…*'	L.Ib./8.
3. Hervás, L. 'Catálogo de las lenguas…'	L.Ib./9c+d
Einzelne Stämme: Elgua = Alava, Eldue = Guipuzcoa, Elgriva = Eliberri = Biscainisch	
4.) Rüdiger, J.C.C. 'Grundriss … Sprache'	LIb15
5.) Erro y Azpiroz, J.B.d. 'Alfabeto de la lengua…'	L.IV.d.5./1.
Andre Namen: Euskarisch, Guipuzcoa, Vaskisch, Biscai(-y) nisch	
6.) C.; A.J. 'Censura crítica del Alfabeto…'	L.IVd.5/2.
7.) Schnakenburg, J.F. 'Tableau … Patois de la France…'	L.IV.d.8/25.
8.) Vallancey, C. 'An Essay … of the Irish Language…'	L.IV.e.5/1.
Fortsetz. s. Einz. Stämme. Labourdin; Nieder-Navarra; Ober-Navarra; Souletin.	
9.) Eguilas y Yanguas., L. Glossario etymológico [sic]	4LIVd5/4
10.) Verschiedene Bibelteile	½ B.II.a/1-7.
11.) Catechismus	½ B.II.a/8
12.) Gebetbuch.	½ B.II.a/9.
13.) Oihenhart [sic], A: 'Proverbes basques; poésies basques.'	E.V.b.1./1.
14.) Oriental. Archiv: Winkler: 'Die mongol. Völk… u. d. Basken.'	4K.XXVI.a./3,4.

This card is written in two different hands, neither of which is Gabelentz's. It contains entries that are of no or of only very peripheral interest for Basque, and also lists works that are rarely found in the German-speaking world, like Oihenart's collection of proverbs. However, Gabelentz definitely did not use it for his studies.[67] It is also strange that the last work mentioned on this card was published about twenty years after Gabelentz's death.[68]

67 In contrast to Humboldt, who worked on them in great detail; for example, in his contribution to *Mithridates* of 1816–1817, but also on dozens of other unpublished papers.

68 And also nearly ten years after Schulenburg's death.

8.2.2 *About Berber*[69]

A.) The catalogue of the language library contains among other things the following index cards for Kabyle:

Ben Sedira, Belkassem. 1887. Cours de langue kabyle. Grammaire et versions. Alger: Adolphe Jourdan.

Hanoteau, Adolphe. 1858. Essai de grammaire kabyle renfermant les principes du langage parlé par les populations du versant nord du Jurjura et spécialement par les Igaouaouen ou Zouaoua suivi de notes et d'une notice sur quelques inscriptions en caractères dits Tifinar' et en langue Tamacher't. Alger – Constantine – Paris: Bastide – Bastide et Amavet – Challamel & Duprat.

Newman, Francis William. 1887. Kabail Vocabulary. Supplemented by Aid of a New Source. London: Trübner & Co.

Sierakowski, Adam. 1871. Das Schaūï. Ein Beitrag zur berberischen Sprachen- und Völkerkunde. Dresden: Kraszewski.

B.) Card for the keyword *Hamitische Sprachen, allgemein* (Hamitic languages, general)

Newman, Francis William. 1882. Lybian [sic] Vocabulary. An Essay towards Reproducing the Ancient Numidian Language out of Four Modern Tongues.

C.) Card with the keyword *Berberisch* (Berber)

Schlagwort: Berberisch (Lybisch, Hamitisch XVI.c.1.)	
Siehe: 1.) Adelung, J.Chr. 'Mithridates'	L.I.b./2c.
2.) Cust, R. 'The ... of Africa'	L.I.b./36a.
3.) Duret, Cl. 'Thrésor de l'histoire des langues ...'	L.Ib./6.
4.) Orient. u. Occident. Sprachmeister	L I b/8
5.) Rüdiger, J.C.C. 'Grundriss ... Sprache'	L I b/15
6.) *Lybische* Sprachen	
7.) Sierakowski, Graf A. 'Das Schaūï'	L.XVI.c.2/4.
8.) 12 Chapitres de. S. Luc. (London 1833)	B.XIV b.1/1.
Dialecte: 1.) Marocco-B. 2.) Sahára. 3.) Algeria. 4.) Tunisia-B.	

69 The following three sources are part of the card catalogue of the Gabelentz language library in the Thüringisches Staatsarchiv in Altenburg.

D.) Card with the keyword *Kabylisch* (*Lybisch, Hamitisch*) (Kabyle (Libyan, Hamitic)) (only works not mentioned under A.) to C.) above are listed)

Lepsius, Richard. 1880. Nubische Grammatik mit einer Einleitung über die Völker und Sprachen Afrika's.
Pharaon, Joanny. 1835. Les Cabiles et Boudgie.

Works cited

Aranzadi, Telesforo de. 1902. 'El supuesto parentesco del euskera y el berberisco. [trad. de F. Müller y H. Schuchardt]'. *Euskal-Erria* XLVI, pp. 38-40.

Basset, René. 1896. 'Les noms des métaux et des couleurs en berbère'. *Mémoires de la Société de Linguistique de Paris* IX, pp. 58-91.

Basset, René. 1899. 'Rapport sur les études berbères et haoussa'. *Actes du Onzième Congrès International des Orientalistes. Paris 1897. Cinquième, sixième et septième sections*, pp. 39-70.

Behagel, Otto. 1892. Review of Georg von der Gabelentz, *Die Sprachwissenschaft. Literaturblatt für germanische und romanische Philologie* 13, pp. 257-258.

Ben Sedira, Belkassem. 1887. *Cours de langue kabyle. Grammaire et versions.* Alger: Adolphe Jourdan.

Brinton, Daniel G. 1894. 'Basque and Berber' *Science* 23 (26 January 1894), p. 43. (Current Notes on Anthropology No. XXXVIII).

Campión, Arturo. 1884. *Gramática de los cuatro dialectos literarios de la lengua euskara.* Tolosa: Eusebio Lopez.

Creusat, Jean-Baptiste. 1873. *Essai de dictionnaire français-kabyle* (*Zouaoua*) *précédé des éléments de cette langue.* Alger: Jourdan.

Eguilaz y Yanguas, Leopoldo de. 1886. *Glosario etimológico de las palabras españolas* (*castellanas, catalanas, gallegas, mallorquinas, portuguesas, valencianas y bascongadas*) *de orígen oriental* (*árabe, hebreo, malayo, persa y turco*). Granada: La Lealtad.

Emig, Joachim. 'Das Familienarchiv v.d. Gabelentz im Thüringischen Staatsarchiv Altenburg'. In: Ezawa & Vogel (2013), pp. 311-321.

Eys, Willem J. van. 1873. *Dictionnaire basque – français.* Paris – London: Maisonneuve – Williams & Norgate.

Eys, Willem J. van. 1879. *Grammaire comparée des dialectes basques.* Paris – London – Amsterdam: Maisonneuve – Williams & Norgate – Frederik Muller.

Eys, Willem J. van. 1883. *Outlines of Basque Grammar.* London: Trübner & Co.

Ezawa, Kennosuke & Annemete von Vogel (eds.). 2013. *Georg von der Gabelentz. Ein biographisches Lesebuch.* Tübingen: Narr Verlag.

Ezawa, Kennosuke, Franz Hundsnurscher & Annemete von Vogel (eds.). 2014. *Beiträge zur Gabelentz-Forschung*. Tübingen: Narr Verlag.

Gabelentz, Georg von der. 1881. 'Sur la possibilité de prouver l'existence d'une affinité généalogique entre les langues dites indochinoises'. In: *Atti IV Congresso Internazionale Orientalisti tenuto in Firenze nel settembre 1878*, vol. II, pp. 283-293. Firenze: Successori Le Monnier.

Gabelentz, Georg von der. 1882. 'Kabylen (sprachlich)'. In: *Allgemeine Encyklopädie der Wissenschaften und Künste*, Johann Samuel Ersch & Johann Gottfried Gruber (eds.), 2. Sect. 32. Theil, pp. 28-29. Leipzig: Brockhaus.

Gabelentz, Georg von der. 1883a. Review of Francis William Newman, *Libyan vocabulary. An essay towards reproducing the ancient numidian language, out of four modern tongues. Literarisches Centralblatt*, p. 330.

Gabelentz, Georg von der. 1883b. Review of Willem van Eys, *Outlines of Basque Grammar. Literarisches Centralblatt*, pp. 1314-1315.

Gabelentz, Georg von der. 1885. Review of Arno Grimm, *Über die Baskische Sprache und Sprachforschung. Literarisches Centralblatt*, pp. 24-25.

Gabelentz, Georg von der. 1888. Review of Leopoldo de Eguilaz y Yanguas, *Glosario etimológico* etc. *Literarisches Centralblatt*, p. 924.

Gabelentz, Georg von der. 1889. Review of Francis William Newman, *Kabail vocabulary. Supplemented by aid of a new source. Literarisches Centralblatt*, pp. 824-825.

Gabelentz, Georg von der. 1890. 'Antrittsrede [zur Aufnahme in die Berliner Akademie]'. *Sitzungsberichte der königlich preussischen Akademie der Wissenschaften* II, pp. 782-785.

Gabelentz, Georg von der. 2016 [^{1}1891, ^{2}1901]. *Die Sprachwissenschaft, ihre Aufgaben, Methoden und bisherigen Ergebnisse*. First edition: Leipzig: T.O. Weigel Nachfolger. Second edition: Albrecht von der Schulenburg (ed.). Leipzig: Tauchnitz. Critical edition: Manfred Ringmacher & James McElvenny (eds.). Berlin: Language Science Press.

Gabelentz, Georg von der. 1892a. *Handbuch zur Aufnahme fremder Sprachen. Im Auftrage der Kolonial-Abtheilung des Auswärtigen Amts*. Berlin: Mittler.

Gabelentz, Georg von der. 1892b. Review of Albrecht von der Schulenburg, *Grammatik, Vocabularium u. Sprachproben der Sprache von Murray Island. Literarisches Centralblatt*, pp. 410-411.

Gabelentz, Georg von der. 1893. 'Baskisch und Berberisch'. *Sitzungsberichte der königlich preussischen Akademie der Wissenschaften zu Berlin. Phil.-hist. Cl. vom 22. Juni 1893*, pp. 593-613.

Gabelentz, Georg von der. 1894. *Die Verwandtschaft des Baskischen mit den Berbersprachen Nord-Africas nachgewiesen von ___*, Albrecht C. Graf von der Schulenburg (ed.). Braunschweig: Sattler.

Gabelentz, Georg von der. *Katalog der Poschwitzer Bibliothek*, started by H. Conon von der Gabelentz, unprinted manuscript, Thüringisches Staatsarchiv Altenburg, Legat Gabelentz, Nr. 1061 (s.d.).

Gabelentz, Georg von der & Adolf Bernhard Meyer. 1882. 'Beiträge zur Kenntnis der melanesischen, mikronesischen und papuanischen Sprachen. Ein erster Nachtrag zu Hans Conon's von der Gabelentz Werke "Die melanesischen Sprachen"'. *Abhandlungen der philologisch-historischen Classe der Königlich Sächsischen Gesellschaft der Wissenschaften* VIII, pp. 373-490.

Gèze, Louis. 1873. *Éléments de grammaire basque, dialecte souletin, suivis d'un vocabulaire basque-français & français-basque*. Bayonne: Veuve Lamaignère.

Gèze, Louis. 1883–1885. 'De quelques rapports entre les langues berbères et basques'. *Mémoires de la Société archéologique du Midi de la France*, Seconde Série, Tome XIII, pp. 30-36.

Giacomino, Claudio. 1892. 'Delle relazioni tra il Basco e l'Antico Egizio'. *Rendiconti*, Serie II, Vol. XXV, Adunanza del 14 luglio 1892, pp. 1063-1077.

Giacomino, Claudio. 1895. 'Delle relazioni tra il Basco e l'Egizio'. In: *Supplementi periodici all'Archivio Glottologico Italiano* II, pp. 15-96 [original manuscript, Archivio dell'Accademia die Lincei, 1890].

Gimm, Martin. 2013a. *Georg von der Gabelentz zum Gedenken. Materialien zu Leben und Werk*. Wiesbaden: Harrassowitz.

Gimm, Martin. 2013b. 'Schriftenverzeichnis (in chronologischer Folge nach Erscheinungsjahren)'. In: Gimm (2013a), pp. 79-118.

Grimm, Arno. 1884. *Ueber die baskische Sprache und Sprachforschung. Allgemeiner Teil*. Breslau: Hirt.

Grube, Wilhelm. 1905. 'Gabelentz: Hans Georg Conon von der'. In: *Allgemeine Deutsche Biographie*, vol. 50. (Harkort – v. Kalchberg), Historische Commission bei der Königlichen Akademie der Wissenschaften (ed.), pp. 548-555. Leipzig: Duncker & Humblot.

Hanoteau, Adolphe. 1858. *Essai de grammaire kabyle renfermant les principes du langage parlé par les populations du versant nord du Jurjura et spécialement par les Igaouaouen ou Zouaoua suivi de notes et d'une notice sur quelques inscriptions en caractères dits Tifinar' et en langue Tamacher't*. Alger – Constantine – Paris: Bastide – Bastide et Amavet – Challamel & Duprat.

Hanoteau, Adolphe. 1860. *Essai de grammaire de la langue Tamachek', renfermant les principes du langage parlé par les Imouchar' ou Touareg, des conversations en Tamashek', des fac-simile d'écriture en caractères Tifinar', et une carte indiquant les parties de l'Algérie où la langue berbère est encore en usage*. Paris: Imprimerie impériale.

Humboldt, Wilhelm von. 1817 [1816]. 'Berichtigungen und Zusätze zum ersten Abschnitte des zweyten Bandes des Mithridates über die Cantabrische oder

Baskische Sprache'. In: *Mithridates oder allgemeine Sprachenkunde*, J.C. Adelung & J.S. Vater (eds.), vol. 4, pp. 275-360. Berlin: Voss.

Humboldt, Wilhelm von. 1821. *Prüfung der Untersuchungen über die Urbewohner Hispaniens vermittelst der Vaskischen Sprache*. Berlin: Dümmler.

Hurch, Bernhard. 2009. '"Emakume-bahitzea" eta lege fonetikoak. Georg von der Gabelentzen *Hizkuntza arrotzak jasotzeko esuliburua*-ren ingurukoak'. In: *Beñat Oihartzabali Gorazarre – Festschrift for Bernard Oyharçabal*, Ricardo Etxepare, Ricardo Gómez & Joseba Lakarra (eds.), *ASJU – International Journal of Basque Linguistics* XLIII, 1-2, pp. 503-516.

Hurch, Bernhard. 2011. 'Über "Weiberraub" und Lautgesetze. Anmerkungen zu Georg von der Gabelentz' *Handbuch zur Aufnahme fremder Sprachen* in baskischer Version'. *Beiträge zur Geschichte der Sprachwissenschaft*, 21, pp. 239-262.

Jaubert, Amédée et al. 1844. *Dictionnaire français-berbère*. Paris: Imprimerie royale.

Kaden, Klaus. 1993. 'Die Berufung Georg von der Gabelentz' an die Berliner Universität'. In: *Sinologische Traditionen im Spiegel neuer Forschungen*, Ralf Moritz (ed.), pp. 57-90. Leipzig: Leipziger Universitätsverlag [repr. in Ezawa & Vogel (2013), pp. 271-288].

Kürschner, Wilfried. 2014. 'Georg von der Gabelentz' "Handbuch zur Aufnahme fremder Sprachen" (1892) – Entstehung, Ziele, Arbeitsweise. Wirkung'. In: Ezawa, Hundsnurscher & Vogel (2014), pp. 239-259.

Lepsius, Carl Richard. ^{2}1863. *Standard Alphabet for Reducing Unwritten Languages and Foreign Graphic Systems to a Uniform Orthography in European Letters*. London – Berlin: Williams & Norgate – Hertz.

Meyer, Gustav. 1895. Review of Johann Topolovšek, *Die basko-slavische Spracheinheit* und Georg von der Gabelentz, *Die Verwandtschaft des Baskischen* etc. *Berliner Philologische Wochenschrift* 15 (15 June 1895), pp. 783-785.

Michelena, Luis. 1961. *Fonética Histórica Vasca*. San Sebastián: Publicaciones del Seminario 'Julio de Urquijo' de la Excma. Diputación Foral de Guipuzcoa. [repr. in Luis Michelena. 2011. *Obras completas*, J.A. Lakarra and I.R. Arzalluz (eds.), VI: *Fonética Historica Vasca* (Anejos del ASJU, 59). San Sebastián – Donostia / Vitoria – Gasteiz: UPV – EHU].

Michelena, Luis. 1964. *Sobre el pasado de la lengua vasca*. San Sebastián: Auñamendi. (Colección Auñamendi, 36). [repr. in Luis Michelena. 2011. *Obras completas*, J.A. Lakarra and I.R. Arzalluz (eds.), V: *Historia y geografía de la lengua vasca* (Anejos del ASJU, 58), pp. 1-115. San Sebastián – Donostia / Vitoria – Gasteiz: UPV – EHU].

Mukarovsky, Hans G. 1963–1964. 'Baskisch und Berberisch'. *Wiener Zeitschrift für die Kunde des Morgenlandes* 59/60, pp. 52-94.

Müller, Friedrich. 1895. 'Die neuesten Arbeiten über das Baskische'. *Globus* 68, p. 14.

Münchhausen, Clementine von. 2013 [1913]. 'H. Georg v. d. Gabelentz. Biographie und Charakteristik. zusammengestellt von ___'. In: Ezawa & Vogel (2013), pp. 85-171.

Newman, Francis William. 1882. *Libyan Vocabulary. An Essay towards Reproducing the Ancient Numidian Language out of Four Modern Tongues*. London: Trübner & Co.

Newman, Francis William. 1887. *Kabail Vocabulary. Supplemented by Aid of a New Source*. London: Trübner & Co.

Plank, Frans. 1991. 'Hypology, Typology: the Gabelentz Puzzle'. *Folia Linguistica* 25, pp. 421-458.

Pott, August Friedrich. 1887. *Zur Literatur der Sprachenkunde Europas*. Leipzig: Barth. (*Internationale Zeitschrift für allgemeine Sprachwissenschaft*, Supplement 1).

Ringmacher, Manfred. 2011. 'Der Humboldtianismus von Georg von der Gabelentz'. In: *Miscellanea Linguistica. Arbeiten zur Sprachwissenschaft*, Wilfried Kürschner (ed.), pp. 321-334. Frankfurt: Peter Lang.

Schuchardt, Hugo. 1888. Review of Georg Gerland, *Die Basken und die Iberer. Literaturblatt für germanische und romanische Philologie* 9, pp. 225-234.

Schuchardt, Hugo. 1892. Review of Cl. Giacomino, *Delle relazioni tra il basco e l'antico egizio. Literaturblatt für germanische und romanische Philologie* 13, pp. 426-430.

Schuchardt, Hugo. 1893. Review of G. von der Gabelenz, *Baskisch und Berberisch. Literaturblatt für germanische und romanische Philologie* 14, pp. 334-338.

Schuchardt, Hugo. 1907. 'Die iberische Deklination'. *Sitzungsberichte der Kaiserlichen Akademie der Wissenschaften in Wien. Phil.-hist. Klasse* 157, pp. 1-90.

Schuchardt, Hugo. 1913. 'Baskisch-hamitische Wortvergleichungen'. *Revista Internacional de Estudios Vascos* 7, pp. 289-340.

Schulenburg, Albrecht von der. 1891. *Grammatik, Vocabularium und Sprachproben der Sprache von Murray Island.* Leipzig: Friedrich.

St[umm]e, H[ans]. 1895. Review of von der Gabelentz, *Die Verwandtschaft des Baskischen* etc. *Literarisches Centralblatt*, p. 581.

Topolovšek, Johann. 1894. *Die basko-slavische Spracheinheit*, Bd.1: *Vergleichende Lautlehre*. Wien: Gerold.

Taylor, Canon Isaac. 1893. 'The Affinities of Basque and Berger [sic]'. *Science* 22 (11 August 1893), p. 77.

Tubino, Francisco M.. 1876. *Los aborígenes ibéricos ó Los berèberes en la península*. Madrid: Secretaría de la Sociedad Antropológica.

Unamuno, Miguel de. 1895. Review of Gabelentz, Georg von der (1894). *Revista crítica de historia y literatura españolas* 1.2, pp. 40-41.

Vinson, Julien. 1891. *Essai d'une bibliographie de la langue Basque*. Paris: Maisonneuve.

Vinson, Julien. 1898. *Essai d'une bibliographie de la langue Basque. Additions et corrections*. Paris: Maisonneuve.

Vinson, Julien. 1901–1902. 'Revue des études basques (1891–1899)'. *L'Année linguistique* I, pp. 135-197.

Vinson, Julien. 1903–1904. 'Les études basques de 1901-1904'. *L'Année linguistique* II, pp. 81-104.

Vinson, Julien. 1904a. 'Les langues indo-européennes. Les Aryens'. *Revue de l'École d'Anthropologie de Paris* XIV, pp. 169-184.

Vinson, Julien. 1904b. Review of E. de Michelis, *L'origine degli Indo-Europei*. *Journal Asiatique*, Dixième Série, Tome II, pp. 543-544.

Vogel, Annemete von. 2013. 'Aus der Gabelentz-Ausstellung 2010 in Berlin'. In: Ezawa & Vogel (2013), pp. 185-222.

Zyhlarz, Ernst. 1932. 'Zur angeblichen Verwandtschaft des Baskischen mit afrikanischen Sprachen'. *Praehistorische Zeitschrift* 23, pp. 69-77.

About the authors

Bernhard Hurch is Professor of General Linguistics at the University of Graz. His areas of expertise include the history of linguistics, history of science, Basque studies, Mesoamerican languages, as well as phonology, morphology and typology.
bernhard.hurch@uni-graz.at

Katrin Purgay is a master's student in General Linguistics at the University of Graz, specializing in the study of language typology and history of linguistics (19th and early 20th century).
katrin.purgay@edu.uni-graz.at

5 Phenomenological aspects of Georg von der Gabelentz's *Die Sprachwissenschaft*

Klaas Willems

Abstract

When Georg von der Gabelentz's (1840–1893) book *Die Sprachwissenschaft* (1891) was published, the historical-comparative paradigm was the dominant perspective in empirical linguistic enquiry, and the then current theory of language was firmly rooted in language psychology. While itself based in this tradition, Gabelentz's *Sprachwissenschaft* nevertheless strikes another chord, which sets it apart from contemporary sources. This chapter argues that the book is particularly noteworthy for its wide-ranging bottom-up approach to linguistic phenomena and its propensity to conceive of language both as object and instrument of enquiry in a sense akin to the 'reflexive' stance that later would become a hallmark of Edmund Husserl's (1859–1938) phenomenology. The 'phenomenological mindset' can in particular be retraced in the way a different philosophy of the language sciences and theory of meaning take shape in *Die Sprachwissenschaft*.

Keywords: phenomenology, philosophy of science, theory of meaning, 19th-century psychology of language

> ... it is necessary to divide the general but at the same time to bring together into a unity that which emerges as specific. To reach such a centre, the nature of language itself gives us a hand.
>
> – Wilhelm von Humboldt, *Über die Verschiedenheit des menschlichen Sprachbaues*, § 13.[1]

1 '... ist es zugleich notwendig, das Allgemeine mehr auseinanderzulegen, und das dann hervortretende Besondere dennoch mehr in Einheit zusammenzuziehen. Eine solche Mitte zu erreichen, bietet die Natur der Sprache selbst die Hand' (Humboldt, 1998 [1836], § 13).

McElvenny, James (ed.). *Georg von der Gabelentz and the Science of Language*.
Taylor & Francis Group, 2019
DOI: 10.5117/9789462986244/CH05

1 Introduction

In this chapter, I examine the philosophy of science and theory of meaning in Georg von der Gabelentz's (1840–1893) treatise on general linguistics, *Die Sprachwissenschaft* (2016 [1891]), and the extent to which they can be regarded as being 'phenomenological' in nature for reasons to be explained below.[2] The term phenomenological, to be sure, does not occur in Gabelentz's book. I use the term to refer to a general philosophical mindset towards consciousness, scientific analysis and language. Authors such as Spiegelberg (1982, p. xxvii), Lembeck (1994, p. 3), Sokolowski (2000, p. 47) and others rightly observe that phenomenology is not, as is sometimes assumed, a clear-cut doctrine or body of knowledge which scholars can either adopt or reject. It is, however, widely accepted that a detailed description of the specific 'phenomenological mindset' or 'attitude' in the modern sense of the term has been provided by Edmund Husserl (1859–1938) from c. 1900 onwards. Obviously, the possibility of any direct influence of Husserl's work on Gabelentz is ruled out in advance, nor is there any evidence for a dependency in the other direction. Gabelentz died in 1893 and Husserl's first major – and, strictly speaking, pre-phenomenological – publication, *Philosophie der Arithmetik* (HUA XII),[3] appeared in 1891, the same year as the first edition of *Die Sprachwissenschaft*. Moreover, I do not claim that Gabelentz's *Sprachwissenschaft* is the only book on general linguistics in the second half of the nineteenth century which features aspects of a phenomenological mindset, but I argue that Gabelentz's book stands out as particularly interesting in this respect. Of course, drawing attention to a number of correspondences and similarities between Gabelentz and Husserl should not make us forget that there are many differences between the two authors, who in their work pursue different aims. This is not surprising, Husserl being a philosopher and Gabelentz a linguist. Husserl's primary concern is creating a new basis for philosophy in which due attention is also paid to language. Gabelentz, in his *Sprachwissenschaft*, aims at providing an overview of the language sciences, their achievements and their diverse areas of application with a special focus on theoretical and methodological

2 All references are to the critical edition of *Die Sprachwissenschaft* edited by Manfred Ringmacher and James McElvenny (Gabelentz, 2016, henceforth abbreviated as *SPRW*). The issues I will discuss draw on the original text written and published by Gabelentz himself in 1891 and not on passages added by his nephew Albrecht von der Schulenburg after Gabelentz's death in the second edition of the book published in 1901.

3 HUA = Husserliana (Collected Works), published since 1950 by The Hague: Martinus Nijhoff, The Hague: Kluwer Academic Publishers and New York: Springer (in that order).

issues. I do not intend to investigate similarities and differences by tracing the positions of both scholars in the history of philosophy and linguistics.[4] Nor can a brief enquiry such as the present one claim to do full justice to the richness of ideas in Husserl's or Gabelentz's work. However, I believe that it is particularly illuminating to confront Gabelentz's and Husserl's views on language precisely because of a number of convergences which are not due to any direct dependency but rooted in a general phenomenological mindset.

The chapter is organized as follows. Section 2 defines two complementary thematic domains of general importance to any phenomenology of language, viz. the philosophy of science and the theory of meaning. I argue that these two thematic domains are closely linked from a phenomenological viewpoint and that they can be used as guidelines to determine relevant phenomenological aspects in works on general linguistics such as Gabelentz's *Sprachwissenschaft*. After briefly introducing aspects of a phenomenological perspective on the philosophy of science in Section 2.1, I present in more detail Husserl's theory of meaning in Section 2.2. Twentieth-century phenomenology of language cannot be restricted to Husserl's work, considering that many scholars have elaborated on Husserl's views in several respects, including Martin Heidegger, Hermann Ammann, Hans Lipps, Hendrik Pos, Maurice Merleau-Ponty, Alfred Schütz, Paul Ricoeur, Hans-Georg Gadamer, Jacques Derrida, Aron Gurwitsch, Jitendra Nath Mohanty, Thomas Luckmann and Robert Sokolowski, among many others. However, Husserl's phenomenological analysis of language, with a particular focus on the constitutive function of meaning in his *Logische Untersuchungen* (1913 [1900–1901]; HUA XVIII–XX), is generally regarded as basic to any endeavour to clarify the relationship between language and consciousness from a phenomenological point of view. Section 3 turns to a close reading of some pertinent passages in Gabelentz's *Sprachwissenschaft* in view of the two aforementioned thematic domains. As I shall document, several of Gabelentz's observations are phenomenological to different degrees. The section focuses on Gabelentz's stance towards psychology as part of linguistic enquiry and his conception of linguistic meaning. Section 4 rounds off the chapter with some general conclusions and with a comment on one of Gabelentz's more puzzling enterprises in the *Sprachwissenschaft*

4 For historical surveys of the phenomenological approach ('Einstellung'), in particular its roots in the work of Franz Brentano and Carl Stumpf (1848–1936), and overviews of Husserl's influence on twentieth-century philosophers, see Spiegelberg (^{3}1982), Rollinger (1999), Moran (2000), among others. Gabelentz's place in the history of linguistics is briefly discussed in Morpurgo Davies (1998, Chapter 10).

which seems to be somewhat neglected in the newly revived Gabelentz scholarship. I have in mind Gabelentz's chapter on the appraisal of languages (*Sprachwürderung*) in the last part of the book (*SPRW*, pp. 409-502).

2 Phenomenology of language

Husserl's general phenomenological enterprise touches on a diverse array of topics including consciousness, logic, corporeality, temporality and language. Phenomenology of language, too, covers a wide spectrum, ranging from issues such as meaning and reference and the 'syntactic' combination of signs (Mohanty, 1977; Bernet, Welton & Zavota, 2005; Mattens, 2008) to communication and social interaction (Luckmann, 1980). Husserl's interest in language cannot be judged in isolation from his general phenomenological programme, which always encompasses the overarching, reflective question of how the phenomenological mode of philosophizing is itself possible. Husserl's pervasive transcendental *Rückfrage* (Husserl, 1939, pp. 47-49; HUA VI, p. 73; HUA XV, p. 614) on the origins in human intentionality of any science and scientific experience is central to the phenomenological enterprise.

The present section revolves around the assumption that two thematic domains are of general importance to the phenomenology of language to the extent that they can be used as guidelines to determine phenomenological aspects in works on general linguistics. These two thematic domains are the philosophy of science and the theory of meaning. I briefly consider the phenomenological view on the philosophy of science first and then elaborate on Husserl's theory of meaning, which is the centrepiece of his phenomenology of language.

2.1 First thematic domain: the philosophy of science

A phenomenology of Husserlian observance claims to be a presuppositionless and rigorous method of philosophical enquiry (Husserl XXV, pp. 2-62, 'Philosophie als strenge Wissenschaft'). It purports to reveal the true nature of 'the things themselves' (*die Sachen selbst*), i.e. the objects of our experience as they present themselves to us. To this end, Husserl proffers an unbiased perspective on the phenomena that are the objects of intentional acts of knowing and understanding. Such a perspective requires, first of all, a critical attitude towards existing accounts in the history of philosophy and science. Husserl's entire phenomenological enterprise rests on the conviction that a consistent theory of consciousness and knowledge has to abstract away from previous theories and assumptions, including potentially invalid assumptions of the 'natural attitude' of everyday life (cf. Moran, 2000, p. 11). Once this has been achieved, one can try to reconstruct the modes of consciousness and knowledge to which human beings are capable, including scientific knowledge. According to Husserl, this is a challenge that calls for a re-evaluation of intuition: that which is originally given in intuition is the necessary and incontrovertible basis of knowledge (HUA III, p. 52). Husserl's point of view entails a conception of philosophy of science that extends not only to the natural sciences, logic and mathematics but also to the humanities. While Husserl is most explicit regarding the natural and formal sciences, his criticism of 'blind' specialism in any science (cf. HUA VI) and his emphasis on the detailed description of that which is intuitively given in consciousness as a basis of scientific analyses is important to a phenomenology of the humanities as well.

According to Husserl, it is possible to determine specific areas of science (*Wissenschaftsgebiete*) on the basis of intuition. As a consequence, the phenomenological perspective runs counter to any deductive-nomological approach to scientific description and explanation (Sinha, 1969, Chapter 6; Lembeck, 1994, pp. 71-72). The basis of phenomenology as a philosophy of science is the rich world of pre-scientific experience, the so-called *Lebenswelt* (HUA VI, pp. 142-146). However, if the world of sciences is a construction on the basis of the 'life world', it is important that the ways the life world is intentionally constituted and intuitively given to consciousness are properly understood. At the same time, engaging in phenomenological research means that one has to be aware that the obvious often hides the most difficult problems ('sich gerade hinter dem "Selbstverständlichen" die schwierigsten Probleme verbergen', HUA XIX/1, p. 350). This is an important aspect in the philosophy of science conceived from a phenomenological perspective for two reasons: first, phenomenology sets out to explore the conditions of knowledge and science in a way that is as unbiased and comprehensive as possible; second, it entails that any analysis of some domain or object has to go hand in hand with a critical questioning of the very conditions of the enquiry itself and its modes of analysis.

2.2 Second thematic domain: the theory of meaning

The question of meaning and consciousness takes pride of place in the phenomenology of language. This is a direct consequence of Husserl's seminal *Logische Untersuchungen* (1913 [1900–1901]) (HUA XVIII–XX). In particular, I. Investigation on expression and meaning (*Ausdruck und Bedeutung*) and

IV. Investigation on independent and dependent meanings and the concept of pure grammar (*Der Unterschied der selbständigen und unselbständigen Bedeutungen und die Idee der reinen Grammatik*) are important in this respect (HUA XIX/1).[5] Husserl argues that linguistic meaning cannot be properly accounted for in terms of the psychological theory of mental representations and associations that held sway in philosophy in the nineteenth and early twentieth centuries (as it did in linguistics, cf. Knobloch, 1984; 1988). On the other hand, linguistic meaning, properly understood, is not simply derivative on logical thought either. Whatever Husserl's indebtedness to logic and logical concepts even in his account of 'expression and meaning' in his *Logische Untersuchungen*, one should keep in mind that he is concerned, in this part of his work, with natural language and its intuitive systematicity, not merely with logical reasoning (Edie, 1977, p. 139).[6]

As the founder of modern phenomenology, Husserl's lifelong interest in natural language is particularly noteworthy. Whereas he devotes a lot of attention to elucidating the possibility of an *a priori* or 'pure' grammar in the early *Logische Untersuchungen* (cf. HUA XIX/I, p. 302), in later years he points out the importance of the historical dimension of natural language (Husserl, 1939 [1938]; 1939; and HUA XIII-XV). In this way, he strikes a balance between accounting for the ideality of linguistic meanings and recognizing their constitutive, intersubjective nature in the concrete consciousness of speakers as members of historical linguistic communities (Mattens, 2008, pp. xi-xvii). According to Husserl, linguistic meaning (*Bedeutung*) should be distinguished from reference by means of signs that have no proper meaning (so-called *Anzeichen* and other 'signals', HUA XIX/1, pp. 31, 37-38). On the one hand, linguistic signs *have* meaning; they do not acquire meaning as a result of their being used. In the terminology of Husserl's *Logische Untersuchungen*, a linguistic sign is uniquely associated with a meaning-intention (*Bedeutungsintention*) which is the basis for diverse uses on particular occasions which in turn give rise to meaning-fulfilments (*Bedeutungserfüllungen*) (HUA XIX/2, p. 44). On the other hand, being

5 The *Logische Untersuchungen* remain the touchstone of Husserl's phenomenology of language but important observations on language can also be found in other texts by Husserl, which moreover shed light on the development of his thought through the years, in particular: *Vorlesungen über Bedeutungslehre* (HUA XXVI), *Zur Phänomenologie der Intersubjektivität* (HUA XIII–XV), *Formale and transzendentale Logik* (HUA XVII), 'Die Frage nach dem Ursprung der Geometrie als intentional-historisches Problem' (Husserl, 1939 [1938]) and *Erfahrung und Urteil* (Husserl, 1939).

6 Cf. the volumes edited by Mohanty (1977) and by Bernet, Welton & Zavota (2005) for some of the criticisms that have been levelled at Husserl's theory of meaning.

conscious of something means that one is intentionally directed towards it. Using a linguistic sign therefore adds to its proper ('ideal') meaning aspects of content which are essential to the interpretation of the sign in discourse although they are not part and parcel of the sign's meaning. Thus there is a fundamental difference between the meaning of, for example, a common noun and its referent which is the object of the sign (*Gegenständlichkeit*) whenever it is used in an act of discourse. For Husserl, a linguistic sign has a unitary meaning (*einheitliche Bedeutung*, HUA XIX/1, pp. 10, 51-57, 83, 102, 317, 326) which is moreover general (*unbestimmt*) to the extent that it is compatible with an array of possible fulfilments (*eine Sphäre möglicher Erfüllung*, HUA XIX/1, pp. 55, 59; cf. HUA XX/2, pp. 238-239) rather than one single fulfilment. This is correlative with the fact that a sign's meaning is an 'essence': knowing such a meaning is an original intuitive knowledge of a general 'idea' (XIX/1, pp. 102-117, cf. Mohanty, 1977, pp. 78-82; and Lembeck, 1994, p. 32, for discussion). Linguistic meanings pertain to a realm of consciousness of their own, i.e. they are not reducible to any other mental or psychological level (Benoist, 2008). A 'meaning is entirely different from the mental images which may accompany an act of thinking' (Moran, 2000, p. 111). This is the basis for the flexibility and (using a modern term) multifunctionality of linguistic signs from a phenomenological point of view.

Although Husserl unambiguously rejects the view nowadays known as 'linguistic relativism' which holds that thinking, including perception and remembering, crucially relies on language (HUA XX/2, p. 20), he does acknowledge the foundational role of language in conscious acts of judgement, surmising, questioning etc. (Welton, 2005 [1973], pp. 92-97). A community of human beings is fundamentally different from one of animals by virtue of its being founded on natural language: 'Die Heimwelt des Menschen [...] ist grundsätzlich von der Sprache her bestimmt' (HUA XV, pp. 224-225). However, phenomenology not only acknowledges the role of language for thought but also seeks to uncover the structures of a pre-linguistic and even pre-logical nature that underlie language and logic alike. It is interesting to note that in so doing, Husserl focuses on the 'syntax' of experience and is more interested in syntactic structures (including their *a priori* combinatorial rules) than in lexical structures of language (De Palma, 2008).

Intimately linked to Husserl's later views on language and intersubjectivity, apart from the 'life world' (*Lebenswelt*), is the concept of 'preliminary knowledge' (*Vorwissen, Vorbekanntheit*, Husserl, 1939, pp. 26-32). While the concept of 'life world' captures the original human experience of an intersubjective, lived world which is very different from the world that is constructed as an 'object' of the natural sciences (Sokolowski, 2000, Chapter 10), 'preliminary knowledge' refers to passive performances of consciousness and includes the fact that every object that is present in an intentional act of consciousness is embedded in a 'horizon' of already constituted knowledge and former experiences (Husserl, 1939, p. 26; HUA XV, p. 225: *sprachlich-apperzeptiver Horizont*; Luckmann, 1980, pp. 31-32). The importance of this aspect of Husserl's phenomenology for linguistics is that language usage is not an activity in isolation. Language is used by speakers in a physical, social, historical and psychological – hence cognitive – context. In other words, language is embedded cultural activity in a broad sense of the term.

To conclude this succinct presentation of some of the hallmarks of Husserlian phenomenology of language, a final observation on terminology is in order. First of all, Husserl does not adopt Frege's (1892) famous distinction between the 'sense' (*Sinn*) of a linguistic sign and its referent (which Frege chose to call, against conventions, *Bedeutung*). However, Husserl's terminology changes over time. In the *Logische Untersuchungen*, the terms *Sinn* and *Bedeutung* are used interchangeably (HUA XIX/1, p. 58). This is important, because for Husserl the intuitive grasp of a meaning (*Bedeutung* or *Sinn*) and the reference to an object on the basis of that meaning are not separated in the way Frege claims (Vandevelde, 2008). Yet Husserl concurs with Frege that meanings are not to be equated with *Vorstellungen* (representations), which according to Frege (1892, p. 29) are of a subjective nature. Later, in the *Ideas* (HUA III, § 124), Husserl uses *Sinn* to designate any sense in general, covering the whole range of intentional acts, whereas *Bedeutung* refers to the specifically linguistic (or 'expressive') meaning of expressions (cf. Benoist, 2008). From a linguistic point of view, this terminological distinction is a felicitous one. In linguistic analyses one should be able to distinguish language-particular meanings (*Bedeutungen*) from the general realm of content, i.e. senses at different levels of intentionality, including (but not restricted to) speech.[7]

In the next section, I clarify the ways in which the two thematic domains can be brought to bear on the phenomenological character of parts of

7 From the point of view of linguistic theory, several of Husserl's other terminological differentiations also deserve to be taken into account, e.g. the distinction between equivocal and universal (*vieldeutig* and *vielwertig*, HUA XIX/1, pp. 53-59), between expressions without meaning and expressions without an object (*bedeutungslos* and *gegenstandslos*, HUA XIX/1, p. 60), between the nonsensical and the absurd (*Unsinn* and *Widersinn*, HUA XIX/1, p. 334), or between lexical and syntactic meaning (HUA XIX/1, p. 333). Due to reasons of space, I cannot elaborate on these aspects in the present chapter. Suffice it to say that they are part and parcel of a comprehensive theory of meaning which Husserl develops with the aim of clarifying the relationship between empirical evidence (*Anschaulichkeit*) and consciousness (*Bewusstsein*) of which general meanings are, if one considers the normal intentional experiences that make up our (inter)subjective experiences, an integral part.

Gabelentz's *Sprachwissenschaft*. I start off with Gabelentz's philosophy of science (Section 3.1) before turning to his theory of meaning (Section 3.2).

3 The scope of Gabelentz's *Sprachwissenschaft*

Although the second half of the nineteenth century saw a general broadening of the scope of the language sciences (Morpurgo Davies, 1998, pp. 296-300), the contents of Gabelentz's *Sprachwissenschaft* strike us as uncommonly diverse and wide-ranging for its time. As a matter of fact, no other book published in the second half of the nineteenth century covers the variety of topics Gabelentz addresses, in this respect even surpassing Hermann Paul's (1846–1921) *Principien der Sprachwissenschaft* (1886 [1880]). Gabelentz writes about the philosophy of language (including the origins of language), history of linguistics, linguistic theory and methodology, language typology and comparative linguistics (with examples taken from a large number of languages and language families). The book further contains occasionally detailed accounts of a plethora of linguistic phenomena relating to historical and synchronic linguistics, language contact, first language acquisition, quantitative linguistics (cf. also Gabelentz, 1894), morphology and word formation, sound symbolism (*das lautsymbolische Gefühl*, cf. Willems, 2016), prosody, writing systems, stylistic issues etc.

3.1 Gabelentz's philosophy of science

Gabelentz frequently says that linguistics has many branches which have to deal with all sorts of linguistic phenomena (*Erscheinungen*, *SPRW*, pp. 3, 35, 41, 45, 50, 82 *et passim*). Not surprisingly, Gabelentz advocates an inductive approach. However, its ultimate aim is a systematic analysis of language which also requires logic and philosophy:

> The logical work of a linguist is mainly inductive. One has to find the laws behind the phenomena; therefore one must collect samples of the phenomena, examine the collected samples, recognize the commonalities, formulate the acquired knowledge in precise and clearly expressed theorems, and finally merge the theorems organically into a well-integrated system of the whole. All of this requires a logically-versed, the latter even a philosophically talented mind. (*SPRW*, p. 50)[8]

8 Original: 'Die logische Arbeit des Sprachforschers ist vorwiegend inductiv. Es gilt den Erscheinungen ihre Gesetze abzulauschen; darum gilt es, die Erscheinungen als Beispiele

Similar to Husserl's warning that the obvious comprises many problems, Gabelentz writes:

> an account is scientific only if it is appropriate, and when the subject matter has its own integral order, then the account must follow this order, otherwise there are unscientific demands put on it. It would seem obvious that things that are so closely related that even the dim-witted eye can see the relation must not be torn apart; but even the obvious requires intellect. (*SPRW*, p. 86)[9]

For Gabelentz, the difficulty of analysing a specific language (*Einzelsprache*) is that the linguist has to be 'entirely subjective and entirely objective at the same time [...] because it is precisely his subjectivity that becomes his object' (*zugleich ganz subjectiv und ganz objectiv* [...], *denn eben seine Subjectivität wird ihm zum Objecte*). Even if the analysis is driven by the study of empirical phenomena, the linguist is no less interested in the question 'what constitutes grammatical knowledge, i.e. how it should be accounted for according to its essential nature' (*worin das grammatische Wissen bestehe, wie es also seinem Wesen gemäss dargestellt sein wolle*) (*SPRW*, p. 87).

At one point, Gabelentz (*SPRW*, p. 143) refers to Paul's *Principien der Sprachgeschichte* (1886 [1880]) and William Dwight Whitney's (1827–1894) *The Life and Growth of Language* (1875). Paul's book in particular had quickly acquired a canonical status among the *Junggrammatiker* (Neogrammarians) at the end of the nineteenth century (Morpurgo Davies, 1998, Chapter 9). But whereas Paul's *Principien* aim at providing a synthesis of the then-current theory and methodology of historical linguistics (which does not mean that the book is only about historical linguistics!), Gabelentz's *Sprachwissenschaft* brings literally every aspect of linguistic enquiry within its purview. In so doing, Gabelentz raises the notion of a linguistic *Prinzipienwissenschaft* to a more general level compared to Paul. Paul's *Principien* are moreover firmly rooted in an all-pervasive psychologistic point of view. By contrast,

zu sammeln, die gesammelten zu sichten, das ihnen Gemeinsame zu erkennen, die erlangte Erkenntniss in scharf und klar ausgesprochenen Lehrsätzen zu formuliren, endlich die Lehrsätze zu einem wohlgefügten Lehrgebäude organisch zu vereinigen. Alles dies verlangt einen logisch gebildeten, das Letzterwähnte sogar einen philosophisch beanlagten Kopf.'

9 Original: 'eine Darstellung ist nur dann wissenschaftlich, wenn sie sachgemäss ist, und wo die Sache ihre Ordnung in sich selbst trägt, da muss die Darstellung dieser Ordnung folgen, sonst thut sie der Sache unwissenschaftlichen Zwang an. Man sollte meinen, es wäre selbstverständlich, dass wenigstens das Nächstverwandte, dessen Zusammenhang auch dem blöden Auge einleuchtet, nicht auseinandergerissen werden darf; aber auch das Selbstverständliche verlangt Verstand.'

although subscribing to the broad lines of *Völkerpsychologie*, it seems that Gabelentz is more cautious when it comes to psychologising explanations of linguistic phenomena, without for that matter ignoring the role and merits of psychology in the development of the language sciences. This particular disposition of Gabelentz's *Sprachwissenschaft* may be considered as a general phenomenological property of the book: it combines a broad descriptive ('bottom-up') focus on linguistic phenomena with a genuinely epistemological interest in issues concerning the linguistic analysis itself (cf. Willems, 2002, p. 56). Gabelentz's book is an important step in paving the way for a linguistic theory guided by genuine principles and methods which are not derived from some non-linguistic discipline, in particular psychology, but from the nature of language and speech as they present themselves to us in all their diversity and complexity. One should not be misled by the fact that Gabelentz uses the terms 'psychology' and 'psychological' profusely in the *Sprachwissenschaft* and other writings, in line with nineteenth-century scholarship. But the designation of these terms had broadened to refer to an ever-increasing range of phenomena, in particular since Brentano (1874) had called his investigation of 'intentional acts' of consciousness '(descriptive) psychology'. The aim of Franz Brentano (1838–1917) was to define psychology as the study of 'psychic phenomena', yet contrary to Wilhelm Wundt's (1832–1920) 'physiological psychology', for example, Brentano attempted to distinguish psychology from the natural sciences in a novel way. He explicitly aimed at overcoming the limits of traditional associationist psychology. It was in this sense that Brentano would become a forerunner of Husserl's phenomenology, who in his later years occasionally conceived his phenomenology as a 'pure psychology' (Spiegelberg, 1982, pp. 27, 34-45; Lembeck 1994, pp. 9-16).[10]

Husserl does not criticize psychology generally but to the extent that methods of psychological explanation are illegitimately used to reduce one science to another (Holenstein, 2005 [1975], p. 18). A similar combination of moderation and anti-reductionism can be found in the *Sprachwissenschaft*. On several occasions, Gabelentz strives to assign a proper place to psychology as part of linguistic research (*SPRW*, pp. 15, 67, 100). For example, Gabelentz calls the articulation of thought and the connection of ideas by means of language ('*etwas* aussprechen') the 'logical' aspect of language, but this aspect is permeated by another one which he calls the 'psychological' aspect of language. The latter concerns the relationship between the speaker and

10 Note that in Gabelentz's *Sprachwissenschaft* there are numerous references to contributions by H. Steinthal (1823–1899), including the latter's philosophy and psychology of language, but no references to W. Wundt or F. Brentano.

the very fact of performing speech acts ('*mich* aussprechen', *SPRW*, p. 86). Gabelentz also emphasizes that psychology is one of the driving forces behind the creation of grammatical categories, which can be at odds with logical reasoning (Gabelentz *SPRW*, pp. 495-499). He devotes an entire section to the 'psychological training' of linguists and carefully explains with a number of examples what the object of such a training could be (*SPRW*, pp. 41-49; the other ones are phonetic, logical and general linguistic training). Finally, Gabelentz conceives of meaning primarily in terms of representations (*Vorstellungen*), albeit in a way that sets him apart from his contemporaries (see Section 4.2 below). One of Gabelentz's most conspicuous and pervasive references to psychology in his treatment of grammar is his distinction between psychological subject and psychological predicate (*psychologisches Subject und Prädicat*).[11] However, Gabelentz is not so much concerned with the psychological, or cognitive, representations of subject and predicate. His aim is rather to clarify the age-old distinction between the theme (or topic) and rheme (or comment) of a sentence with a view to establishing their role in the information structure of different types of sentences in relation to constituent order in various languages. For instance, according to Gabelentz (*SPRW*, p. 391), in the French sentence *Cette lettre, je l'ai lu* the pronoun *je* is the grammatical subject but *cette lettre* is the psychological subject and the rest of the sentence is the psychological predicate. While this account is plausible, other applications of the categories are less conclusive (Harweg, 2014, p. 223; see Elffers, this volume, for ample discussion).

As was pointed out before, Gabelentz conceives of the phenomenon of language in a comprehensive way, unlike most historical linguists of his time. He deems every single aspect of language(s) worthy of scholarly interest and assumes a descriptive and profoundly empirical perspective towards them. Ultimately, the study of language is part of anthropology (*SPRW*, p. 14). Yet Gabelentz is also interested, in every part of the book, in the ways of doing linguistics as a science, its methods and theoretical assumptions, in what he himself calls 'hodegetische Fragen' (*SPRW*, p. i; cf. pp. 79, 317). The latter perspective is subservient to the former. This approach is reminiscent of the famous leitmotif in Husserlian phenomenology, 'to the things themselves' (*zu den Sachen selbst*) (cf. Husserl LU XIX/1, p. 10). This slogan expresses the belief that an adequate scientific perspective (be it in

11 The distinction is introduced in Gabelentz (1869) and further elaborated in Gabelentz (1875) and SPRW (pp. 385-396). It was taken up by contemporaries, e.g. Paul (1886 [1880], p. 100), who is sometimes erroneously credited to be the first linguist who made the distinction. Cf. Elffers, this volume, for a brief historical survey.

philosophy or in a particular branch of science) reveals the true nature of the object under investigation if, and only if, the focus on the object of enquiry is consistently guided by a focus on the mode of enquiry itself. It is therefore a phenomenological corollary that Gabelentz advocates methodological pluralism, given that languages constitute a complex and multifarious object of enquiry: 'There are as many different tasks of science as there are different aspects and different points of view' (*Soviele Seiten, soviele Standpunkte der Betrachtung, soviele Aufgaben der Wissenschaft*) (*SPRW*, p. 8; compare also Gabelentz's remark about the adverse consequences of 'native speaker prejudices' [*muttersprachliche Vorurtheile*] on linguistic analysis, *SPRW*, p. i). At the same time, Gabelentz (*SPRW*, pp. 96, 130, 222, 317 etc.) does not shy away from invoking, time and again, 'die Natur der Sache' (*the nature of the thing*) when dealing with various aspects of language and linguistic research. This, too, is a phenomenological attitude: scientific enquiry ought not be aprioristic but guided by the conditions that the object of enquiry itself imposes on the ideally unbiased perspective of the investigator, which is grounded in intuition. Against this background, the sheer amount of topics that are discussed in Gabelentz's book and which draw on important areas of theoretical and empirical linguistic research, the large number of languages from which examples are presented and the overall breadth of the account may be appreciated as genuine phenomenological properties of the *Sprachwissenschaft*. It might be noted in passing that one of the first attentive readers of Gabelentz in the twentieth century, Louis Hjelmslev (1899–1965), in his first major work (Hjelmslev, 1928), did not fail to stress this combination, i.e. reliance on 'the very nature of language' (*la nature même du langage*) (1928, p. 59; cf. p. 39 with explicit reference to Gabelentz) combined with methodological pluralism, in particular as far as synchronic and diachronic linguistics are concerned (ibid., pp. 59-63).

It is worth pointing out that the reception of Gabelentz's *Sprachwissenschaft* was mixed at the time of the first and second edition. Most reviewers admonish the lack of structure in the book, the repetitions, the author's verbosity and, not in the least, the fact that Gabelentz does not seem to heed all findings of the *Junggrammatiker* in the domain of historical linguistics with the support of a psychological theory of language. Recall that the psychologistic perspective was still dominant in linguistics at the time. The fact that Gabelentz maintains a certain aloofness towards psychology undoubtedly also contributed to the tone of some of the reviews.[12] The

12 Even critical reviewers of Gabelentz's *Sprachwissenschaft* praise its scope and richness, which must have been striking to those who worked within the *junggrammatische* paradigm

next, and historically much more successful, breakthrough of a theory of linguistic enquiry which is not based on psychology was not realized until the publication of Saussure's *Cours de linguistique générale* (1916).[13]

The Husserlian *a priori* is not 'aprioristic' in the sense of metaphysics but in the Kantian sense of the philosophy of science. Otherwise, *die Sache selbst* would remain a theoretical construct, a figment of a scholar's imagination. From a phenomenological perspective, it is important to distinguish regions of scientific enquiry which correspond to experiential domains. These regions call for different methods, ranging from general to specific ones, thus accounting for the diversity of the objective world (HUA III, pp. 23-39; cf. Lembeck, 1994, pp. 68-74). This sense for the theoretical unity in methodological diversity also applies to the study of language as it is conceived by Gabelentz. This is made especially clear by his distinction between three types of linguistic analyses on which the structure of his book is loosely based: *einzelsprachliche Forschung*, the study of an individual language, *genealogisch-historische Sprachforschung*, the genealogical-historical study of languages, and *allgemeine Sprachwissenschaft*, general linguistics. Gabelentz is conscious of the fact that this distinction is a methodological one and that general linguistics plays a role in the other two types of analysis as well (*SPRW*, p. 317). The same holds for another important distinction he introduces, viz. the one between two perspectives in linguistic analysis which he terms the 'analytic' and the 'synthetic' system (*SPRW*, pp. 93-110). I return to this distinction in the next section.

3.2 Gabelentz's theory of meaning

As pointed out in Section 2, the theory of meaning is another thematic domain of general importance to the phenomenology of language. It is intimately linked to the philosophy of science. From a phenomenological perspective, the study of language is by definition the study of meaning.

(e.g. Behaghel, 1892; Streitberg, 1892). Most reviewers commend the fact that Gabelentz provides examples from many lesser-known languages, and his strong plea for the study of still unknown languages and language typology does not go unnoticed. Schmidt-Wartenberg (1892), whose judgement is overall negative (although he scarcely deals with anything else than phonetics), concedes: 'The author's purpose has been, not so much to furnish a handbook as to lead the student through a linguist's workshop' (1892, p. 117); cf. Gabelentz's own reference to 'unsere Werkstatt' (*SPRW*, p. iii).

13 Compare Stawarska's (2015) account of a number of similarities between Saussure's theory ('philosophy') of language and phenomenology. Cf. also the discussion in Coseriu (1967) and Koerner (1978 [1974]).

This is evident from Husserl's focus on meaning in his I. Investigation in the *Logische Untersuchungen* which has the status of a preliminary chapter to the ensuing investigations into the phenomenology of consciousness and knowledge.

In Gabelentz's *Sprachwissenschaft* we find relatively few passages in which the author explicitly addresses his concept of meaning in language. However, the book does contain important insights with respect to meaning and semantics, which show some interesting similarities with passages in Husserl's work. A case in point is Gabelentz's discussion of the relationship between linguistic meaning and reference to objects. As forerunners of his own views, Gabelentz (*Sprw*, p. 344-345) mentions in particular Wilhelm von Humboldt (1767–1835), August Friedrich Pott (1802–1887) and H. Steinthal (Humboldt and Steinthal are quoted extensively). This tribute already provides a general idea of the thrust of Gabelentz's position, given that both Pott and Steinthal explicitly draw on Humboldt's theory of language and meaning despite differences of opinion concerning the relation between linguistics and (Kantian) philosophy, logic and psychology (cf. Knobloch, 1988). From Humboldt, Gabelentz adopts the view that in language one does not find a representation of things (i.e. referents) but concepts of things the mind creates through language (*Sprw*, p. 351; cf. p. 384, 'the creative mind that generates language' [*der sprachbildende Geist*]): 'human language [...] *anthropomorphizes*, i.e. language transfers human being and activity to the extrahuman world' (*die menschliche Sprache* [...] anthropomorphisirt, *d. h. sie überträgt menschliches Sein und Thun auf die aussermenschliche Welt*) (*Sprw*, p. 2, emphasis in original). It is by means of language that thought is 'articulated' (*gegliedert*) in particular ways (*Sprw*, p. 324; also pp. 3, 85, 341, 364). This echoes Humboldt's insistence that language is a 'synthesis' of sound and thought: a synthetic process which creates something that does not exist in the connected parts taken separately ('wo die Synthesis etwas schafft, das in keinem der verbundenen Theile für sich liegt', Humboldt, 1998 [1830–1835], § 12, p. 218). This view is also expressed in Steinthal's work (e.g. succinctly in Steinthal, 1866, pp. 473-474). Thus, because different languages have different form-meaning pairings, linguists should pay heed to the relationship between language-particular categories and crosslinguistic comparative concepts. Gabelentz is most explicit about this with regard to grammar:

> One can boldly assume that no two formal grammatical means in different languages are completely equal in meaning. The German genitive differs greatly from the Latin and Greek genitive, so do the other cases,

> the genders, or the tenses and moods of the verb. The names only approximately designate the essence of an object, they are convenient but very deficient substitutes for very complicated and circuitous definitions. (*SPRW*, p. 428; cf. also p. 121)[14]

Having acknowledged the fact that meanings are language-specific, Gabelentz points out that different meanings can be used to refer to different referents and still convey the same message. For example, someone may say *Der Blaue!* [The blue one!] and refer to a jockey at the races while another person refers to the jockey's horse with *Die Fuchsstute!* [The chestnut mare!], but 'both "mean" the same thing' (*beide meinen dasselbe*) (*SPRW*, p. 386). Similarly, in syntax the difference between the active and the passive construction is evident in terms of two different meanings but they can be 'synonymous' to the extent that 'the same content of thought is conveyed to the hearer' (*dem Hörer derselbe Gedankeninhalt übermittelt wird*) (*SPRW*, p. 383). Conversely, differences in form such as the order of the constituents in a noun phrase can be deployed to convey different *Vorstellungen* even though the words are the same. For example:

> In French, those adjectives which may assume both positions should follow the noun when they could be replaced, e.g. by an equivalent relative clause: *un fidèle ami* conveys the unitary representation of a true friend, *un ami fidèle*, on the other hand, speaks first of a friend, and then says of him that he is a real true friend. Similarly: *un savant professeur – une femme savante*. (*SPRW*, pp. 482-483; cf. also extensively Gabelentz, 1875, pp. 158-165)[15]

These observations are reminiscent of Husserl's well-known discussion of pairs of expressions such as *der Sieger von Jena* (the victor at Jena) and *der Besiegte von Waterloo* (the loser at Waterloo) or *a ist größer als b* (a is larger

14 Original: 'Man darf kühnlich annehmen, dass nicht zwei grammatische Formmittel verschiedener Sprachen einander in ihrer Bedeutung vollständig decken. Der deutsche Genitiv ist vom lateinischen und griechischen sehr verschieden, und ebenso sind es die übrigen Casus, die Genera, Tempora und Modi des Verbums. Die Namen bezeichnen das Wesen der Sache nur annähernd; sie sind bequeme, aber sehr mangelhafte Ersatzmittel für sehr schwierige und weitschweifige Definitionen.'

15 Original: 'Im Französischen haben jene Adjectiven, denen beide Stellungen gestattet sind, dann nachzutreten, wenn sie etwa durch einen entsprechenden Relativsatz ersetzt werden könnten: *un fidèle ami* giebt die einheitliche Vorstellung eines echten Freundes; *un ami fidèle* dagegen redet zunächst von einem Freunde und sagt dann von ihm aus, dass er auch wirklich ein echter treuer sei. Ähnlich verhält es sich mit: *un savant professeur – une femme savante*.'

than b) and *b ist kleiner als a* (b is smaller than a). According to Husserl (HUA XIX/1, p. 53), these pairs of expressions have different meanings but they can be used to refer to the same referent. Gabelentz's account has the additional merit that he provides clear evidence for a third semantic level which has to be distinguished besides meaning and reference, viz. the level of the textual meaning of an utterance, which Gabelentz (*SPRW*, pp. 2, 93, 101-102, 390-391) occasionally calls the 'sense of speech' (*Sinn der Rede*). Whereas Husserl's philosophical point of view is basically restricted to the level of concepts and their combinations according to rational norms, Gabelentz's linguistic point of view extends to the level of discourse (which includes 'texts' in the sense of Coseriu, 1987, p. 152). However, Husserl's approach is open to such an extension, if only because of his concept of a 'horizon' which is intimately linked to the realm of preliminary knowledge and the 'passive genesis' of consciousness and experience in the 'life world'. As noted in Section 2, this part of Husserl's phenomenology accounts for the fact that intentionality, in whatever mode it is realized (i.e. linguistic or non-linguistic), manifests itself in embedded activities of individual actions and social interactions. Interestingly, Gabelentz's observations tie in with such an analysis. He points out that much of what speakers mean in speech is not actually expressed but inferred from context, or as Gabelentz puts it, from the 'the mutual situation in life' (*gegenseitige Lebenslage*) (*SPRW*, p. 386). Gabelentz (*SPRW*, p. 246) also refers to this room for inference with the term *Spielraum* (leeway, margin). Note that the concept of *Spielraum* will later become a central concept in, for instance, Karl Bühler's (1879–1963) theory of meaning and communication (Bühler, 1934, p. 66), which explicitly builds on Husserl's theory of language.[16]

Another noteworthy aspect of Gabelentz's account of meaning is that it differs in some important respects from the standard psychological theory of meaning that can be found in the writings of some of his contemporaries. For authors such as Hermann Paul, Philipp Wegener (1848–1916) and Karl Otto Erdmann (1858–1931), among others, the meaning of a word is essentially a cluster of representations, a *Vorstellungsgruppe* (Wegener, 1885, p. 9; cf. also Wegener, 1885, pp. 47-50; and Paul, 1886 [1880], pp. 66-84). Erdmann (1900, p. 47) maintains 'that words mean only very vague complexes

16 It is unclear whether the use of the (fairly general) term *Spielraum* by Bühler is a coincidence or whether he borrowed it, in the context of his theory of meaning, from Gabelentz. In his *Sprachtheorie* (1934), Bühler only refers to Gabelentz's *Chinesische Grammatik* on p. 240; another reference to Gabelentz on p. 211 is too generic to draw any conclusions. For a discussion of Bühler's view on linguistic meaning with regard to language use, cf. Willems (2018).

of representations' (*daß Worte nur sehr vage Complexe von Vorstellungen bedeuten*). Most of the time Gabelentz, too, conceives of linguistic meanings with reference to *Vorstellungen*. The terms *Vorstellung* and *Bedeutung* are often used interchangeably in the *Sprachwissenschaft*. However, Gabelentz also says that the lexicon of a language is the 'instrument for expressing representations' (*Mittel zum Ausdrucke der Vorstellungen*) (*SPRW*, p. 130), which means that the realm of representations and the linguistic means to designate them are not identical. On another occasion he makes a distinction between a particular 'individual representation' (*Einzelvorstellung*), such as of some word, and an 'overall representation' (*Gesammtvorstellung*) (*SPRW*, pp. 325-326, 342, 383). For example, different utterances can convey the same overall representation, the content of which can be put into words in three different ways:

> { A dog is barking/A dog barks,
> A barking dog,
> The barking of a dog. }
> (*SPRW*, p. 325)[17]

Similarly, active and passive sentences (*SPRW*, p. 388), predicative and attributive structures (*SPRW*, p. 478) etc. differ but give rise to identical *Gesammtvorstellungen*. This means that complete utterances, understood as units of discourse, rather than single lexical items are the bearers of the 'representational function' of language called *Vorstellung*. The *Einzelvorstellungen* associated with individual words often seem to be such only in hindsight, being the result of actual discourse in which words are combined by the speaker to share *Gesammtvorstellungen* with the hearer (Gabelentz, *SPRW*, p. 45). This is also expressed in Gabelentz's 1894 article on linguistic typology. There Gabelentz says that it is the task of general linguistics to account for linguistic diversity in Humboldt's sense (*die Verschiedenheiten des menschlichen Sprachbaues*). This is achieved by reconstructing the relationship between linguistic phenomena and psychology in the following way:

> general linguistics chooses the most striking types of construction – styles of construction –, analyses, characterizes them, explains the meaning of

17 Original: 'Gesammtvorstellung, deren Inhalt wir auf dreierlei Weise in Worte fassen können:
{ Ein Hund bellt,
Ein bellender Hund,
Das Bellen eines Hundes. }

> every perceived feature according to the principle idem per idem, thus translating the linguistic phenomena back into the psychological [...] (Gabelentz, 1894, p. 1)[18]

It is hence the task of general linguists to establish the link that connects linguistic phenomena with the psychological level if the aim is to understand the ways in which the grammars of languages differ. The Latin expression *idem per idem* normally refers to a circular explanation (a definition must not contain the concept that is to be defined). But here Gabelentz uses it in a positive sense, stressing the need to establish language-specific features when the 'psychological' dimensions of linguistic phenomena are determined. This is connected with some important qualifications of psychology from a linguistic point of view. For instance, in Gabelentz (1875, pp. 325-334) there is a lengthy discussion of the many functions of a particle (*ja* in a Malayo-Polynesian language) which Gabelentz explains on the basis of a 'unitary meaning' (*einheitliche Bedeutung*) (1875, p. 333). In the *Sprachwissenschaft*, observations on unitary meanings are limited to the part of the book in which Gabelentz deals with historical linguistics. Whereas such authors as Philipp Wegener – in his *Untersuchungen* (1885), which can be read as a first outline of a *Prinzipienwissenschaft* of linguistic pragmatics (cf. Nerlich & Clarke, 1996, p. 177) – maintain that words are polysemous (cf. in particular Wegener, 1885, p. 50), Gabelentz argues that it is the task of historical linguistics to search for unitary meanings in grammar (*SPRW*, p. 98). He claims that all words and syntactic constructions represent a *Begriff* (or *Grundbedeutung*, i.e. a basic meaning from a diachronic point of view) which he defines as 'quintessence of representations, whose delimitation we call the definition' (*Inbegriff von Vorstellungen, deren Umgrenzung wir die Definition nennen*) (*SPRW*, p. 241; compare also p. 379). He continues:

> all meanings of an expression form a closed unit and emanate radially from a common centre. To discover this centre is sometimes difficult. There are cases where it is obliterated, where the basic meaning is forgotten, and where the rays only glow at isolated points. But in these cases the unity is also lost in one's language awareness. (*SPRW*, p. 241)[19]

18 Original: 'sie [die allgemeine Sprachwissenschaft] wählt die auffälligsten Bautypen – Baustile –, analysiert, charakterisiert sie, erklärt nach dem Satze Idem per Idem, was jede wahrgenommene Eigentümlichkeit bedeute, übersetzt also die Spracherscheinungen zurück ins Psychologische [...]'.

19 Original: 'alle Bedeutungen des Ausdruckes bilden eine geschlossene Einheit und entfliessen strahlenförmig einem gemeinsamen Mittelpunkte. Diesen zu entdecken fällt manchmal schwer.

In the same vein, Gabelentz distinguishes – in a way that anticipates Jakobson's famous distinction between what a language must convey and what it can convey (Jakobson, 1971 [1959], p. 264) – that which is semantically encoded in a language (and which '*must* be expressed' [ausgedrückt werden *muss*], *Sprw*, p. 471, emphasis in original) from the message a speaker wants to communicate.

In Section 2, I indicated that Husserl pays special attention to combinatorial rules ('syntax') in his account of meaning and language. This is to be expected from a phenomenological point of view because syntax, as Husserl understands it, is not simply a matter of language but also of the structure of experience of objects and relationships (HUA XVII, pp. 219-220; cf. Holenstein, 2005 [1975]; De Palma, 2008). This entails that Husserl deliberately uses the term syntax in a systematically ambiguous way. There are several observations in the *Sprachwissenschaft* that point in a similar direction, even though one has to keep in mind that Gabelentz's linguistic explanation and Husserl's philosophical one have different objects of study (as explained in Section 1). First of all, it is striking that Gabelentz attaches particular importance to syntax (*Sprw*, p. 86: 'language is first of all syntax, and then of course also first and foremost syntax' [*Sprachbau ist zunächst Satzbau, und dann natürlich wieder zuhöchst Satzbau*]; pp. 93, 114, 144-145, 333: 'language is speech, is expression of thought, is sentence/proposition' [*Sprache ist Rede, ist Ausdruck des Gedankens, ist Satz*]; pp. 378-379 etc.; cf. also Gabelentz, 1875, pp. 131-132). He claims that 'the grammatical formation of speech' (*die grammatische Formung der Rede*) is mainly determined by internal (*innere*) factors, whereas 'the creation of words' (*die Wortschöpfung*) is more determined by external factors (*äussere Mächte*) (*Sprw*, p. 509). Second, according to Gabelentz (*Sprw*, p. 267), nature itself gives rise to certain categories. Take, for instance, the category dual: 'In this regard, nature itself is the teacher, with things coming in pairs, especially specific parts of the body' (*Schon die Natur mit ihren paarweisen Dingen, zumal den betreffenden Körpertheilen, ist hierin Lehrmeisterin*).[20] The same holds for the major word categories:

Es giebt Fälle, wo er verwischt, die Grundbedeutung vergessen ist, und jene Strahlen nur noch an vereinzelten Punkten leuchten. Dann aber ist auch im Sprachbewusstsein die Einheit gelöst.' The image that the senses of an expression 'radiate' from a unifying centre shows similarities with Darmesteter's (1886, pp. 73-76) organicistic concept of *rayonnement* (and similar views in modern prototype semantics). However, there is no evidence that Gabelentz borrowed the image from Darmesteter. Gabelentz's line of reasoning moreover does not support an organicistic view of language as a natural object.

20 In the second edition of the book, Schulenburg added similar reasoning with regard to the origin of the category of quadral in certain languages (*Sprw*, p. 415).

> In short, man only needed to look at the world to become aware of the difference between object, property and activity. [...] Thus substance determined form, found its expression in form: logical categories pressed themselves to be formed in grammatical classes, to be divided into parts of speech. (*SPRW*, p. 403)[21]

Gabelentz combines this view, which shares similarities with Husserl's account of the 'passive genesis' (*Vorkonstitution*, Husserl, 1939, p. 398) of concepts, with an unambiguous rejection of innatism. To claim that linguistic categories simply are 'innate ideas' (*angeborene Ideen*) is tantamount to saying that they cannot be explained at all (*SPRW*, p. 402) – phenomenologists could not agree more. At the same time, although it is possible that 'the external world itself [can] guide human beings in classifying their representations and concepts' (*die Aussenwelt selbst den Menschen zu einer Classification seiner Vorstellungen und Begriffe anleiten [kann]*) (*SPRW*, p. 406), this by no means entails that languages are 'logical' constructs, which was also the position of H. Steinthal. Thus 'grammar is not logic' (*[d]ie Grammatik ist nicht Logik*) (*SPRW*, p. 473) and it is not uncommon that logic is overruled:

> But grammar and psychology are not bound by the rules of formal logic, – they want to treat the thing as a property and the property as a thing, according to their need and mood. And what they want to do, they can do, and what they label carries the name of the label they have given to it. (*SPRW*, p. 479)[22]

21 Original: 'Kurz, der Mensch brauchte nur die Welt zu betrachten, um des Unterschiedes zwischen Ding, Eigenschaft und Thätigkeit inne zu werden. [...] So bedingte der Stoff die Form, fand in ihr seinen Ausdruck: die logischen Kategorien drängten dazu, sich in grammatische Classen zu gestalten, in Redetheile zu sondern.'

22 Original: 'Aber die Grammatik und die Psychologie binden sich eben nicht an die Regeln der formalen Logik, – sie wollen nach Bedarf und Laune auch das Ding als Eigenschaft, auch die Eigenschaft als Ding behandeln. Und was sie wollen, das können sie auch, und was sie stempeln, das trägt den aufgestempelten Namen.'
Compare also the following passage, which is moreover typical of Gabelentz's at times ornate style: 'The eternal either–or of logic by no means exhausts what the *anima vagula blandula* wants. The latter, a counterpart of the world in which the world is reflected, requires shades in-between, intermediate levels' (*Die Logik mit ihrem ewigen Entweder – Oder erschöpft aber keineswegs das, was die* anima vagula blandula *will. Diese, ein Widerspiel der Welt, die sich in ihr spiegelt, verlangt schillernde Mitteltöne, Zwischenstufen*) (*SPRW*, p. 481). The Latin phrase 'anim[ul]a vagula blandula' is a reference to the Emperor Hadrian's famous farewell poem.

This observation ties in with the view that the two 'grammatical systems' which according to Gabelentz (*SPRW*, pp. 89-91) make up the full grammatical description of a specific language, viz. the 'analytic' and the 'synthetic' system, are complementary but not exactly parallel. The analytic system gives priority to linguistic elements and structures and logically precedes the synthetic system. The analytic point of view takes the grammatical phenomena (*grammatische Erscheinungen*) as a starting point: 'The phenomenon is given, its interpretation is what we are looking for' (*Gegeben ist also die Erscheinung, und gesucht wird ihre Deutung*) (*SPRW*, p. 98). Conversely, the synthetic point of view takes the thought which the speaker wants to express as its starting point and asks which linguistic means are available to give it a grammatically adequate form. Crucially, the meanings corresponding to the grammatical forms (the object of the analytic perspective) and the thoughts expressed by linguistic structures (the object of the synthetic perspective) should not be confused with each other: meanings are 'linguistic' in the narrow sense of the term, whereas thoughts are psychological objects.[23] By and large, this distinction coincides with the modern distinction between a semasiological and an onomasiological approach to grammar (compare, e.g., Coseriu, 1987, pp. 133-143). A full phenomenological account of language not only requires that both perspectives are clearly distinguished but also that both are properly pursued. A Husserlian account, with its focus on universal aspects of grammar and meaning, emphasizes the synthetic point of view, even if it cannot disregard the analytical one. Although Gabelentz stresses the importance of the synthetic point of view, his primarily linguistic approach of language leads him to consider the language-particular 'inner language form' (*innere Sprachform*) (*SPRW*, pp. 158, 344-356, 363, 376, 379), and hence the analytic perspective, as the basis of linguistic investigations. Recall that Gabelentz is to be considered one of the founders of modern language typology (Plank, 1991). His focus on the collection of linguistic facts from a wide range of languages goes hand in hand with careful consideration of the relationship between language-particular categories and crosslinguistic comparison and terminology. Gabelentz (1894, p. 7) summarises his view of language typology succinctly as follows: 'a truly general grammar, wholly philosophical and yet wholly inductive' (*eine wahrhaft allgemeine Grammatik, ganz philosophisch und doch ganz induktiv*).

23 The two systems are already clearly distinguished in Gabelentz's early remarks on comparative syntax (Gabelentz, 1875, p. 130): 'A grammar must therefore answer two questions: 1) What is the meaning of a particular linguistic phenomenon? and 2) What forms of expression does a language possess for the thoughts?' (*Darum muss eine Grammatik zweierlei beantworten: 1) was bedeutet eine jede sprachliche Erscheinung? und 2) welche Ausdrucksformen besitzt die Sprache für die Gedanken?*)

4 Concluding remarks

Both Husserl and Gabelentz aim to provide detailed descriptions of their objects of enquiry, which are consciousness and intentionality (Husserl) and the broad field of language and the study of language (Gabelentz). Because Husserl approaches his research with a keen awareness of the importance of natural language, his phenomenology of language provides an excellent background to discuss phenomenological aspects in works of general linguistics such as Gabelentz's *Sprachwissenschaft*. In this chapter, I have argued that the comprehensive descriptive stance is itself a phenomenological property of Gabelentz's work, which tallies with his observations regarding a non-reductionist approach to linguistic meaning. Gabelentz no longer endorses without qualification the psychologistic theory of meaning typical of the nineteenth century. Similar to Husserl, he conceives of linguistic meaning as a specific, 'original phenomenon' (Benoist, 2008, pp. 222-223) that exists by virtue of an intuitive cultural knowledge, not simply as some content human beings associate in their minds with linguistic activity (cf. his distinction between the analytic and synthetic system). Gabelentz's descriptive approach also ties in with the wide scope of the investigations presented in his *Sprachwissenschaft*. There is in principle no property of language that falls outside the purview of the book or that could not be assigned a proper place in Gabelentz's system of two approaches to linguistic research (*Sprachforschungen*) (i.e. 'research into individual languages' [*die einzelsprachliche Sprachforschung*] and 'genealogical-historical language research' [*die genealogisch-historische Sprachforschung*]) and one overarching 'general linguistics' (*allgemeine Sprachwissenschaft*) (compare in this context Reichling's early appraisal of Gabelentz's work, Reichling, 1966, p. 11).

The objective of a descriptive, presuppositionless scientific approach in which a comprehensive study of the object of enquiry goes hand in hand with a constant reflection on the theoretical and methodological premises of the study itself is an ideal which is difficult to achieve. Just how difficult can also be gleaned from Gabelentz's *Sprachwissenschaft*. This chapter would be incomplete if it did not draw attention to a particularly moot, and perplexing, issue in Gabelentz's work. He himself considers the section on the appraisal or evaluation of languages (*Sprachwürderung. Gesichtspunkte für die Werthsbestimmung der Sprachen*; SPRW, p. 409) in the fourth and last part of the book (pp. 409-502) as a centrepiece of his efforts in general linguistics. It has been claimed that Gabelentz is more cautious in his conclusions than his revered predecessors Humboldt and Steinthal

(Harweg, 2014, p. 221; Elffers, 2012, p. 66). It is true that Gabelentz does not hold back criticism of scholars who draw conclusions about the alleged 'cultural value' (*Culturwerth*, *SPRW*, pp. 409-410) of languages on the basis of particular linguistic features. For example, Gabelentz criticizes some of his predecessors for emphasizing the importance of grammatical features that are present in their preferred languages but downplaying the value of other features that can be found in other, less highly regarded languages (*SPRW*, pp. 413-414). He also rejects Humboldt's 'harsh dualism between Hellenes and barbarians, civilized people and savages' (*schroffen Dualismus zwischen Hellenen und Barbaren, Culturmenschen und Wilden*) (*SPRW*, p. 450). As already pointed out above, it is interesting to note that Gabelentz links these criticisms to the observation that one has to be extremely careful with the use of 'interlinear translations and analyses' (*zwischenzeilige Übersetzungen und Analysen*) (*SPRW*, p. 428) because they gloss over the language-specific differences between the categories in different languages (cf. the current discussion among typologists about 'categorial particularism' and the status of language-specific categories versus merely terminological comparative concepts; Haspelmath, 2010).

However, although Gabelentz calls for caution when establishing the 'interaction between language and the disposition of mind of a people' (*Wechselwirkung zwischen Sprache und Geistesanlage eines Volkes*) (*SPRW*, p. 430), which is at the heart of the *Völkerpsychologie* of his time, he definitely assumes a correspondence between the 'value' of a language and the degree of civilization of a linguistic community:

> But the fact remains: to the extent that the civilized behaviour or coarseness of peoples depends on their disposition of mind, there must be a correspondence between peoples and the value of languages. (*SPRW*, p. 417)[24]

Gabelentz establishes the relationship primarily in terms of language and race (*Rasse*) (*SPRW*, pp. 417, 421, 423, 430 etc.; also Gabelentz, 1894, p. 3). He believes that the 'racial diversity of spiritual abilities' (*rassenweise Verschiedenheit der geistigen Begabung*) can be demonstrated historically (*geschichtlich nachweisbar*) (*SPRW*, p. 417). This position is surprising in view of the fact that 'race' had long been discredited as a scientific category

24 Original: 'Und doch bleibt es dabei: soweit die Gesittung oder Rohheit der Völker von ihrer geistigen Beanlagung abhängig ist, muss jenen der Werth der Sprachen entsprechen'. Gabelentz refers to James Byrne (1820–1897) as an important inspiration (see also Gabelentz, 1894). Cf. Plank (1991) for discussion.

in the language sciences at the time Gabelentz published his *Sprachwissenschaft*. For example, in a remarkable treatise August Friedrich Pott had refuted Arthur de Gobineau's pseudoscientific theory of the inequality of races with regard to languages and linguistic diversity (Pott, 1856; cf. Ricken, 1990, and Morpurgo Davies, 1998, pp. 158-159). Although Pott is one of Gabelentz's favourite linguists in the *Sprachwissenschaft*, he never mentions this particular book by Pott. The sweeping claims about the relationship between language and race which have found their way into the last part of Gabelentz's *Sprachwissenschaft* are predictably unconvincing. For example, according to Gabelentz, black people (e.g. Bantu) are impetuous, superficial, fitful, talkative, irritable and conceited (*SPRW*, p. 445), and this is reflected in their speech, which is sensual and vivid; hence the 'marvellous forceful development of linguistic competence' (*wunderbare Kraftentfaltung des Sprachvermögens*) among the Bantu must in some way be causally linked to their slow-wittedness (*SPRW*, p. 446). Conversely, French exhibits features which correspond to 'the aesthetic sensitiveness of the nation' (*dem ästhetischen Feingefühle der Nation*) (*SPRW*, p. 471). Even specific linguistic phenomena are explained on the basis of racial character (*Charakter der Rasse*), e.g. umlaut in German and Old-Norse. Because the speakers of these languages are more interested in the future than in the past, incessantly pushing things forward (*hastiges Vorwärtsdrängen*) (*SPRW*, pp. 423-424), there is regressive assimilation in their languages instead of progressive assimilation. Languages can also reflect the 'existence in form of phases' (*Schüblingsdasein*) (*SPRW*, p. 440) of communities. For example, while in Indo-European languages, such as German, a sentence has to be construed as a full sentence or a constituent clause from the outset, in Ural-Altaic languages conjunctions and markers of subordination are sentence-final. This is so, according to Gabelentz, because:

> On the one hand, there is an energy on the part of the speaker and listener which must not wane until the entire building is finished. On the other hand, there is a virtually nomadic combination of unsteadiness and sloppiness; everywhere one can choose to stop or move on. (*SPRW*, p. 442)[25]

25 Original: 'Dort auf Seiten des Redenden wie des Hörenden eine Energie, die nicht erlahmen darf, bis das ganze Gebäude fertig dasteht. Hier eine ganz nomadenmässige Vereinigung von Unstätigkeit und Schlendrian; man hat überall die Wahl, ob man Station machen oder weitergehen will.'

These few quotes may suffice.[26] They should not be swept under the carpet in any comprehensive reappraisal of Gabelentz's work which neither apologizes for his views on language and race, nor diminishes his importance to the history of general linguistics. Many of Gabelentz's observations on the 'value' of languages prove that even a scholar of the calibre of Gabelentz, with his keen sense for a thoroughly realistic, open-minded approach to languages and the study of language, has difficulties overcoming preconceived notions that were apparently of particular importance in his education and scholarly views. Gabelentz's claims about the relationship between language and race go far beyond Humboldt's views on the relationship between language and nations (which are not to be confused with linguistic relativism or Whorfian determinism, cf. Di Cesare, 1998, pp. 63-64). Not surprisingly, such claims of Gabelentz's as those I have quoted were met with scepticism at the time of the first and second editions of the book. Sütterlin (1904, p. 320), in his brief review of the second edition of *Die Sprachwissenschaft* (1901), comments that the book is 'a relic of past times' (*ein Ueberbleibsel aus vergangener Zeit*) (cf. Elffers, 2012, p. 65). If this remark is directed to the passages in which Gabelentz deals with language and race, it seems entirely justified (compare also, albeit more circumspectly, Streitberg, 1892, p. 6). Yet, as I hope to have shown in this chapter, Gabelentz's *Sprachwissenschaft* as a whole is not a remnant of the past. On the contrary, in many respects Gabelentz proves that he is actually ahead of his time: the 'air of modernity' which characterizes the arrangement of topics in Gabelentz's book according to Morpurgo Davies (1998, p. 299) also applies to Gabelentz's very approach to language as a phenomenon which is both the object and the instrument of linguistic enquiry.

Acknowledgement

I thank James McElvenny (Edinburgh), Konrad Koerner (Berlin) and Thomas Belligh (Gent) for providing valuable comments on an earlier version of this chapter. Of course, any remaining errors are mine alone.

26 Schulenburg added some further similar claims in the second edition of the book, e.g. the claim that the geography of an area and its climate determine whether a language has soft sounds or hard sounding gutturals, dentals and fricatives (*SPRW*, p. 438); or that people living in regions with steep and high mountains (e.g., Maya, Aztec etc.) also develop 'higher' civilized behaviour (*SPRW*, p. 410).

Works cited

Behaghel, Otto. 1892. Review of Georg von der Gabelentz, *Die Sprachwissenschaft* (1891). *Literaturblatt für germanische und romanische Philologie* 13, pp. 257-258.

Benoist, Jocelyn. 2008. 'Linguistic phenomenology?' In: Mattens (2008), pp. 215-235.

Bernet, Rudolf, Donn Welton & Gina Zavota (eds.) 2005. *The web of meaning: language, noema and subjectivity and intersubjectivity.* London/New York: Routledge.

Brentano, Franz. 1874. *Psychologie vom empirischen Standpunkt.* Leipzig: Duncker & Humblot.

Bühler, Karl. 1934. *Sprachtheorie. Die Darstellungsfunktion der Sprache.* Jena: Fischer.

Coseriu, Eugenio. 1967. 'Georg von der Gabelentz et la linguistique synchronique'. *Word* 23 (= Linguistic Studies Presented to André Martinet, I), pp. 74-100. German translation in: *Beiträge zur Gabelentz-Forschung*, Kennosuke Ezawa, Franz Hundsnurscher & Annemete von Vogel (eds.) (2014), pp. 3-37. Tübingen: Narr.

Coseriu, Eugenio. 1987. *Formen und Funktionen. Studien zur Grammatik.* Tübingen: Niemeyer.

Darmesteter, Arsène. 1887. *La vie des mots étudiée dans leurs significations.* Paris: Ch. Delagrave.

De Palma, Vittorio. 2008. 'Die Syntax der Erfahrung: Zu den sachhaltigen Voraussetzungen des Logischen und des Sprachlichen'. In: Mattens (2008), pp. 127-148.

Di Cesare, Donatella. 1998. 'Einleitung'. In: Wilhelm von Humboldt. *Über die Verschiedenheit des menschlichen Sprachbaues und ihren Einfluß auf die geistige Entwicklung des Menschengeschlechts* (1836), Donatella Di Cesare (ed.), pp. 11-128. Paderborn: Schöningh.

Edie, James M. 1977. 'Husserl's conception of "the grammatical" and contemporary linguistics'. In: Mohanty (1977), pp. 137-161.

Elffers, Els. 2012. 'The rise of general linguistics as an academic discipline. Georg von der Gabelentz as a co-founder'. In: *The making of the humanities. Vol. II: From early modern to modern disciplines*, Rens Bod, Jaap Maat & Thijs Weststeijn (eds.), pp. 55-70. Amsterdam: Amsterdam University Press.

Erdmann, Karl Otto. 1900. *Die Bedeutung des Wortes.* Leipzig: Eduard Avenarius.

Frege, Gottlob. 1892. 'Über Sinn und Bedeutung'. *Zeitschrift für Philosophie und philosophische Kritik, Neue Folge* 100, pp. 25-50.

Gabelentz, Georg von der. 1869. 'Ideen zu einer vergleichenden Syntax. Wort- und Satzstellung'. *Zeitschrift für Völkerpsychologie und Sprachwissenschaft* 6, pp. 376-384.

Gabelentz, Georg von der. 1875. 'Weiteres zur vergleichenden Syntax. Wort- und Satzstellung'. *Zeitschrift für Völkerpsychologie und Sprachwissenschaft* 8, pp. 129-165, 300-338.

Gabelentz, Georg von der. 2016 [[1]1891, [2]1901]. *Die Sprachwissenschaft, ihre Aufgaben, Methoden und bisherigen Ergebnisse*, Manfred Ringmacher & James McElvenny (eds.). Berlin: Language Science Press.

Gabelentz, Georg von der. 1894. 'Hypologie [Typologie] der Sprachen, eine neue Aufgabe der Linguistik'. *Indogermanische Forschungen* 4, pp. 1-7.

Harweg, Roland. 2014. 'Aus der Lektüre von Georg von der Gabelentzens Buch "Die Sprachwissenschaft, ihre Aufgaben, Methoden und bisherigen Ergebnisse" (1901)'. In: *Beiträge zur Gabelentz-Forschung*, Kennosuke Ezawa, Franz Hundsnurscher & Annemete von Vogel (eds.), pp. 215-227. Tübingen: Narr.

Haspelmath, Martin. 2010. 'Comparative concepts and descriptive categories in crosslinguistic studies'. *Language* 86, pp. 663-687.

Hjelmslev, Louis. 1928. *Principes de grammaire générale*. Copenhagen: Bianco Lunos Bogtrykkeri.

Holenstein, Elmar. 2005 [1975]. 'Jakobson and Husserl. A contribution to the genealogy of structuralism'. In: Bernet, Welton & Zavota (2005), pp. 11-48.

Humboldt, Wilhelm von. 1998 [1830–1835]. *Über die Verschiedenheit des menschlichen Sprachbaues und ihren Einfluß auf die geistige Entwicklung des Menschengeschlechts*, Donatella Di Cesare (ed.). Paderborn etc.: Schöningh.

Husserl, Edmund. 1939 [1938]. 'Die Frage nach dem Ursprung der Geometrie als intentional-historisches Problem'. *Revue internationale de philosophie* 2, pp. 203-225. [Also in HUA VI, pp. 365-386.]

Husserl, Edmund. 1939. *Erfahrung und Urteil*, Ludwig Landgrebe (ed.). Prague: Academia Verlagsbuchhandlung.

Husserl, Edmund. HUA I. 1973. *Cartesianische Meditationen und Pariser Vorträge*, Stephan Strasser (ed.). The Hague: Martinus Nijhoff.

Husserl, Edmund. HUA III. 1950. *Ideen zu einer reinen Phänomenlogie und phänomenlogischen Philosophie. Erstes Buch: Allgemeine Einführung in die reine Phänomenologie*, Walter Biemel (ed.). The Hague: Martinus Nijhoff.

Husserl, Edmund. HUA VI. 1976. *Die Krisis der europäischen Wissenschaften und die transzendentale Phänomenologie*, Walter Biemel (ed.). The Hague: Martinus Nijhoff.

Husserl, Edmund. HUA XII. 1970. *Philosophie der Arithmetik*, Lothar Eley (ed.). The Hague: Martinus Nijhoff.

Husserl, Edmund. HUA XIII–XV. 1973. *Zur Phänomenologie der Intersubjektivität*, Iso Kern (ed.). The Hague: Martinus Nijhoff.

Husserl, Edmund. HUA XVII. 1974. *Formale and transzendentale Logik*, Paul Janssen (ed.). The Hague: Martinus Nijhoff.

Husserl, Edmund. HUA XVIII. 1975. *Logische Untersuchungen. Erster Teil. Prolegomena zur reinen Logik*, Elmar Holenstein (ed.). The Hague: Martinus Nijhoff.

Husserl, Edmund. HUA XIX/1-2. 1984. *Logische Untersuchungen. Zweiter Teil. Untersuchungen zur Phänomenologie und Theorie der Erkenntnis* (2 vols.), Ursula Panzer (ed.). The Hague: Martinus Nijhoff.

Husserl, Edmund. HUA XX/1-2. 2002-2005. *Logische Untersuchungen. Ergänzungsband* (2 vols.), Ulrich Melle (ed.). The Hague: Kluwer.

Husserl, Edmund. HUA XXV. 1986. *Aufsätze und Vorträge (1911-1921)*, Thomas Nenon & Hans Rainer Sepp (eds.). The Hague: Martinus Nijhoff.

Husserl, Edmund. HUA XXVI. 1987. *Vorlesungen über Bedeutungslehre. Sommersemester 1908*, Ursula Panzer (ed.). The Hague: Martinus Nijhoff.

Jakobson, Roman. 1971 [1959]. 'On linguistic aspects of translation'. In: Roman Jakobson, *Selected Writings*, Vol. II, pp. 260-266. The Hague/Paris: Mouton.

Knobloch, Clemens. 1984. *Sprachpsychologie: Ein Beitrag zur Problemgeschichte und Theoriebildung*. Tübingen: Niemeyer.

Knobloch, Clemens. 1988. *Geschichte der psychologischen Sprachauffassung in Deutschland von 1850 bis 1920*. Tübingen: Niemeyer.

Koerner, E.F. Konrad. 1978 [1974]. 'Animadversions on some recent claims regarding the relationship between Georg von der Gabelentz and Ferdinand de Saussure'. In: *Toward a Historiography of Linguistics: selected essays*, E.F. Konrad Koerner (ed.), pp. 137-152. Amsterdam: Benjamins. (First published in *Studi saussuriani per Robert Godel*, R. Amacker, T. de Mauro, & L.J. Prieto (eds.), pp. 165-180. Bologna, Italia: Il Mulino.)

Lembeck, Karl-Heinz. 1994. *Einführung in die phänomenologische Philosophie*. Darmstadt: Wissenschaftliche Buchgesellschaft.

Luckmann, Thomas. 1980. 'Aspekte einer Theorie der Sozialkommunikation'. In: *Lexikon der Germanistischen Linguistik*, Hans Peter Althaus, Helmut Henne & Herbert Ernst Wiegand (eds.), pp. 28-41. Tübingen: Niemeyer.

Mattens, Filip. 2008. 'Introductory remarks: new aspects of language in Husserl's thought'. In: Mattens (2008), pp. ix-xix.

Mattens, Filip (ed.). 2008. *Meaning and language: phenomenological perspectives*. Dordrecht: Springer.

Mohanty, Jitendra Nath. 1977. 'Husserl's thesis of the ideality of meanings'. In: Mohanty (1977), pp. 76-82.

Mohanty, Jitendra Nath (ed.). 1977. *Readings on Edmund Husserl's Logical Investigations*. The Hague: Martinus Nijhoff.

Moran, Dermot. 2000. *Introduction to phenomenology*. London/New York: Routledge.

Morpurgo Davies, Anna. 1998. *History of linguistics, vol. IV: Nineteenth-century linguistics*. London/New York: Longman.

Nerlich, Brigitte & David D. Clarke. 1996. *Language, action and context: The early history of pragmatics in Europe and America 1780–1930*. Amsterdam/Philadelphia: John Benjamins.

Paul, Hermann. 1886 [1880]. *Principien der Sprachgeschichte*. Halle: Niemeyer.
Plank, Frans. 1991. 'Hypology, typology: the Gabelentz puzzle'. *Folia Linguistica* 25, pp. 421-458.
Pott, August Friedrich. 1856. *Die Ungleichheit menschlicher Rassen hauptsächlich vom sprachwissenschaftlichen Standpunkte, unter besonderer Berücksichtigung von des Grafen von Gobineau gleichnamigen Werke. Mit einem Überblicke über die Sprachverhältnisse der Völker. Ein ethnologischer Versuch*. Lemgo/Detmold: Meyer'sche Hofbuchhandlung.
Reichling, Anton. 1966. *Verzamelde studies over hedendaagse problemen der taalwetenschap*. Zwolle: Tjeenk Willink.
Ricken, Ulrich. 1990. 'Sprachtheoretische und weltanschauliche Rezeption der Aufklärung bei August Friedrich Pott (1802–1887)'. In: *History and historiography of linguistics*, vol. II, Hans-Josef Niederehe & Konrad Koerner (eds.), pp. 619-631. Amsterdam/Philadelphia: John Benjamins.
Rollinger, Robin D. 1999. *Husserl's position in the School of Brentano*. Dordrecht: Springer.
Saussure, Ferdinand de. 1916. *Cours de linguistique générale*, Charles Bailly, Albert Séchehaye & Albert Riedlinger (eds.). Paris: Payot.
Schmidt-Wartenberg, Hans. 1892. Review of Georg von der. Gabelentz, *Die Sprachwissenschaft* (1891). *Modern Language Notes* 7, pp. 116-118.
Sinha, Debabrata. 1969. *Studies in phenomenology*. The Hague: Martinus Nijhoff.
Sokolowski, Robert. 2000. *Introduction to phenomenology*. Cambridge: Cambridge University Press
Spiegelberg, Herbert. ^{3}1982. *The phenomenological movement. A historical introduction*. The Hague/Boston/London: Martinus Nijhoff.
Stawarska, Beata. 2015. *Saussure's philosophy of language as phenomenology. Undoing the doctrine of the Course in General Linguistics*. Oxford: Oxford University Press.
Steinthal, Heymann. 1866. 'Zur Stylistik'. *Zeitschrift für Völkerpsychologie und Sprachwissenschaft* 4, pp. 465-480.
Streitberg, Wilhelm. 1892. Review of Georg von der. Gabelentz, *Die Sprachwissenschaft* (1891). *Anzeiger für indogermanische Sprach- und Altertumskunde* 2, pp. 1-6.
Sütterlin, Ludwig. 1904. Review of Georg von der Gabelentz, *Die Sprachwissenschaft* (1901). *Literaturblatt für germanische und romanische Philologie* 25, pp. 319-320.
Vandevelde, Pol. 2008. 'An unpleasant but felicitous ambiguity: *Sinn* and *Bedeutung* in Husserl's revisions of the *Logical Investigations*'. In: Mattens (2008), pp. 27-48.
Wegener, Philipp. 1885. *Untersuchungen über die Grundfragen des Sprachlebens*. Halle: Max Niemeyer.

Welton, Donn. 2005 [1973]. 'Intentionality and language in Husserl's phenomenology'. In: Bernet, Welton & Zavota (2005), pp. 81-111.

Whitney, William Dwight. 1875. *The Life and Growth of Language: An Outline of Linguistic Science*. New York: Appleton & Co.

Willems, Klaas. 2002. 'Von der Sprachforschung zur Sprachwissenschaft. Phänomenologische Aspekte in der Sprachtheorie von Georg von der Gabelentz und ihre Relevanz für die moderne Linguistik'. In: *Linguistik jenseits des Strukturalismus*, Kennosuke Ezawa, Wilfried Kürschner, Karl H. Rensch & Manfred Ringmacher (eds.), pp. 55-71. Tübingen: Gunter Narr.

Willems, Klaas. 2016. 'Georg von der Gabelentz and "das lautsymbolische Gefühl". A chapter in the history of iconicity research'. In: *From variation to iconicity. Festschrift for Olga Fischer on the occasion of her 65th birthday*, Anne Bannink & Wim Honselaar (eds.), pp. 439-452. Amsterdam: Pegasus.

Willems, Klaas. 2018. 'Bühler, Reichling, Coseriu und die Vieldeutigkeit von Sprachzeichen'. *Prague Linguistic Circle Papers/Travaux du Cercle Linguistique de Prague. Nouvelle Série* 7, pp. 55-82.

About the author

Klaas Willems is Professor of General Linguistics at Ghent University. His areas of expertise include the epistemology and historiography of the language sciences, philosophy and phenomenology of language, as well as semiotics, semantics and syntax.
Klaas.Willems@UGent.be

6 Content and Form of Speech

Georg von der Gabelentz

Translation, introduction and commentary by
James McElvenny and Manfred Ringmacher

Abstract

This chapter offers an English translation of 'Content and Form of Speech' – one of the core theoretical chapters of Gabelentz's *Die Sprachwissenschaft* – which addresses the notion of 'form' in language and speech, and its opposites, 'content' and 'matter'. 'Form' is a key concept in much nineteenth-century linguistic scholarship, which was employed in various senses to a number of different ends, playing a particularly important role in Humboldtian approaches to typology. In this chapter, Gabelentz provides a wide-ranging survey of existing views on 'form' in language – including extensive quotations from other authors – and develops his own position. This text is therefore useful not only as a statement of Gabelentz's own thinking, but also as an overview of opinions in the field at the time. The translation is accompanied by a short introduction and notes that contextualize the text and help the present-day reader to follow the historical debates.

Keywords: Wilhelm von Humboldt, H. Steinthal, form, typology, speech act theory

McElvenny, James (ed.). *Georg von der Gabelentz and the Science of Language.*
Taylor & Francis Group, 2019
DOI: 10.5117/9789462986244/CH06

1 Form in language: an introduction to Gabelentz's 'Content and Form of Speech'

The essence of language consists in pouring the material of the phenomenal world into the form of thoughts.
– Wilhelm von Humboldt, 'Über das vergleichende Sprachstudium in Beziehung auf die verschiedenen Epochen der Sprachentwicklung'.[1]

'Content and Form of Speech' is one of the core theoretical chapters of Gabelentz's (2016 [1891]) magnum opus, *Die Sprachwissenschaft*. In this chapter, Gabelentz critically reviews various notions of 'form' in the tradition of language study that arose in the wake of Wilhelm von Humboldt (1767–1835) and develops his own conception. Against the strict dichotomies and hierarchies proposed by his predecessors, Gabelentz recognizes diverse and incommensurable types of form across the world's languages, which can be seen to have emerged gradually from other, non-formal features of human language.

The chapter opens in Section I with an examination of content and form in 'speech' (*Rede*); that is, actualized language in use.[2] Through an analysis of the putative structure of propositions and by setting up a taxonomy of sentence types, Gabelentz develops an account of pragmatic aspects of language that is similar in many respects to modern speech act theory (Staffeldt, 2017 [2014]). While the details of this account are unique to Gabelentz, he was not alone in treating language from a pragmatic perspective: in this period there was already a seemingly proto-pragmatic turn perceptible across different approaches to academic language study (Nerlich & Clarke, 1996, pp. 168-169; cf. Elffers, 1993).

In Section II, Gabelentz turns to the division of language into the two components of 'matter' (*Stoff*) and 'form' (*Form*). For the Humboldtians, there is a sense in which all language is form and nothing else, in that language is simply a representation of thoughts.[3] The Humboldtians still, however,

1 Original: 'Das Wesen der Sprache besteht darin, die Materie der Erscheinungswelt in die Form der Gedanken zu giessen' (Humboldt, 1905 [1820], p. 17).

2 Coseriu (1967) saw in Gabelentz's terms *Sprache* (language), *Einzelsprache* (individual language) and *Rede* (speech) the immediate predecessors of Saussure's *langage*, *langue* and *parole*, a proposal that was the source of much controversy. See Koerner (1978 [1974]; 2008) for the main counterarguments and Scheerer (1980, pp. 134-137) for a summary of the debate.

3 An emphasis on the formal nature of language was also a key feature of the structuralist revolution. The *Cours de linguistique générale* of Ferdinand de Saussure (1857–1913) contains the observation: 'la langue est une forme et non une substance' (Saussure 1995 [1916]: 169).

recognized a distinction between material elements of language, usually identified with word roots, which refer to the contents of thoughts, and formal elements, whose prototypical manifestation is inflection, which have no function other than to indicate the relations between items of content (McElvenny, 2016, pp. 31-33). Gabelentz's discussion of matter focuses on the mental processes he holds to underlie the articulation of thoughts in language. Here he draws on the psychological framework of Johann Friedrich Herbart (1776–1841), as elaborated by such figures as H. Steinthal (1823–1899), Hermann Paul (1846–1921) and Wilhelm Wundt (1832–1920) (cf. Knobloch, 1988). Gabelentz concludes his discussion of matter by suggesting there is a fluid boundary between material and formal elements in language.

In turning to 'form' in language, Gabelentz first provides a survey of existing views through a long series of direct quotations from his main Humboldtian predecessors: Humboldt, Steinthal, August Pott (1802–1887), Franz Misteli (1841–1903), and Friedrich Müller (1834–1898).[4] This survey is organized around one of the most characteristic features of Humboldtian form, its subdivision into 'inner linguistic form' (*innere Sprachform*) and 'outer linguistic form' (*äußere Sprachform*). While Humboldt first used these terms, it was Steinthal who expanded them into theoretical concepts (Borsche, 1989), assigning them a range of roles in his philosophy of language and linguistic research (Bumann, 1965, pp. 116-139; Ringmacher, 1996, pp. 99-110).[5] At the end of his survey of inner and outer form, Gabelentz seems to imply that there is little agreement and even confusion in the field about these terms, especially 'inner form'. He advocates a position which eschews the philosophical entanglements of his Humboldtian predecessors and is more immediately directed towards the description of languages. For Gabelentz, inner form describes the categories that achieve grammatical expression in a language, while outer form is the specific grammatical device – such as inflection, word position or particles – through which this expression is realized.

A key task of Humboldtian notions of form was to capture, compare and explain the structural properties of languages, and in this way these notions relate to linguistic typology as that field is understood today.[6] The

4 Friedrich Müller should not be confused with Friedrich Max Müller (1823–1900) of the University of Oxford.

5 In the generative era, Noam Chomsky (b. 1928) saw in 'inner form' and 'outer form' a precursor of his own notions of 'deep structure' and 'surface structure' (Chomsky, 1965, pp. 198-199). Coseriu (1969) strongly opposed Chomsky's suggestion.

6 In fact, Gabelentz (1894) was the first to use 'typology' in more or less its modern linguistic sense (see McElvenny, 2017a; 2017b, § 5).

nineteenth century knew numerous classificatory schemes for languages, which generally concentrated on morphology (Morpurgo Davies, 1975). While Humboldt did not devise any overarching classification himself, his discussion of morphological processes and their assumed relation to cognitive processes served as an inspiration to many later morphologically based schemes (Ringmacher, 2001). In his discussion of outer form, Gabelentz explores and critiques the alleged morphological classes isolating, agglutinative, inflectional and incorporating, which had become standard by his time.

Morphological classifications had already received much criticism before Gabelentz: Steinthal disliked their narrow focus on the outer form of languages and instead looked to the inner form, ultimately setting up his own scheme that distinguished between languages that possess inner form (*Formsprachen*) and those that do not (*formlose Sprachen*). Gabelentz's conclusion to Section II is in essence a response to Steinthal's typology: against Steinthal's absolute dichotomy of *Formsprachen* and *formlose Sprachen*, he proposes a continuum of more formal and less formal elements in languages that recapitulates the conclusion to his discussion of matter. Moreover, linguistic form for Gabelentz is in the first instance the product of an aesthetic 'drive to formation' rather than a reflection of cognitive processes, a view seemingly related to the near-contemporary idealist movement in linguistics represented by such figures as Benedetto Croce (1866–1852) and Karl Vossler (1872–1949) (see McElvenny, 2016).

In Sections III–VII, Gabelentz goes on to offer an overview of what he sees as the most primitive and fundamental means of formation in languages, looking at word order, stress and other modulations of the voice, and the emergence of word classes. In earlier work, Gabelentz (1869; 1875) had already embarked on the cross-linguistic comparison of syntactic categories, as a counterpoint to morphological classifications. These efforts were motivated both by his engagement with Chinese, a language poor in morphology but rich in syntax (McElvenny, 2017b, § 2), and by contemporary debates on the nature of the category 'subject' (Elffers, 1991; Elffers, this volume). These sections expand on this earlier work and place it within a scheme of language evolution which sees morphology and other higher-order grammatical devices as gradually emerging from these more primitive means of linguistic formation.

The last section, Section VIII, draws the threads of the chapter together to reach a final conclusion: grammar is merely solidified style. We want to express ourselves freely, but through habit certain forms become established first as rules and then eventually as inviolable laws. But desire for freedom of expression provokes us to break the laws and thereby set in motion a renewal of the linguistic system (see McElvenny, 2016).

2 A note on the text

The text reproduced and translated in this chapter is based on the recent critical edition of *Die Sprachwissenschaft* (Gabelentz 2016 [1891], pp. 333-407), which brings together and highlights the differences between all previously published versions. Gabelentz initially delivered Section II as a speech to the Königlich-Sächsische Gesellschaft der Wissenschaften (Royal Saxon Society of Sciences), which was subsequently published in the society's proceedings (Gabelentz, 1889). In a note to this first version, Gabelentz (1889, p.185) indicated that it was intended as a chapter of a forthcoming book, 'whose completion [he could] not yet foresee'. A slightly revised version of this text then became part of the chapter 'Content and Form of Speech' in the 1891 first edition of *Die Sprachwissenschaft*. This chapter was in turn revised and expanded at some points in the posthumous second edition of 1901 by Albrecht Graf von der Schulenburg (1865–1902), a nephew and pupil of Gabelentz (see Vogel & McElvenny, this volume).

Each page opening of the following chapter corresponds to one page of the 2016 critical edition. The original German text is given on the right side and the English translation on the left. The page number from the critical edition is provided in the top left corner. To assist in comparison with the sources, page numbers for each of the original publications are also provided throughout the German text: inside curly braces for the 1889 paper (e.g. {185}), inside single vertical bars for the 1891 first edition of *Die Sprachwissenschaft* (e.g. |316|), and inside double vertical bars for the 1901 second edition (e.g. ||324||).

Insertions, deletions and modifications made between the different versions of the text are indicated using different backgrounds and fonts. Passages that are the same in all versions are unmarked. In all sections except Section II, material present in the 1891 first edition of *Die Sprachwissenschaft* but deleted in the 1901 second edition is printed with a grey background, while text added in the second edition is reproduced in a sans-serif font. The mark-up in Section II is slightly more complex, since three sources are in play: the two editions of *Die Sprachwissenschaft* and the 1889 paper. Here underlined text in a serif font is found in the 1889 paper but is different in later versions. A grey background indicates text which was added in the first edition of *Die Sprachwissenschaft* – that is, text not present in the 1889 paper – and which is retained in the second edition, while material present in both the 1889 paper and the first edition but deleted in the second edition is printed in a sans-serif font and with a grey background. In all sections, corrections made by the editors to the second edition are underlined and printed in a sans-serif font. Variants not included in the main text of the critical edition are reproduced in the margins, marked with the year of the publication in which they appear.

Footnotes are used to clarify aspects of the translation and at some points to provide background knowledge that Gabelentz could assume of his readers but which cannot necessarily be assumed of present-day readers of the translation. A comprehensive bibliography of all writings Gabelentz cites is appended at the end. Where English translations are available for sources quoted by Gabelentz, references to the translations with page numbers are provided in square brackets. These translations were consulted in preparing this chapter, but they are rarely followed directly here.

Works cited

Borsche, Tilman. 1989. 'Die innere Form der Sprache. Betrachtungen zu einem Mythos der Humboldt Herme(neu)tik'. In: *Wilhelm von Humboldts Sprachdenken. Symposion zum 150. Todestag*, Hans-Werner Scharf (ed.), pp. 47-65. Essen: Reimer Hobbing.

Bumann, Waltraud. 1965. *Die Sprachtheorie Heymann Steinthals: dargestellt im Zusammenhang mit seiner Theorie der Geisteswissenschaft*. Meisenheim am Glan: Hain.

Chomsky, Noam. 1965. *Aspects of the Theory of Syntax*. Cambridge, Mass.: MIT Press.

Coseriu, Eugenio. 1967. 'Georg von der Gabelentz et la linguistique synchronique'. *Word* 23, pp. 74-110.

Coseriu, Eugenio. 1969. *Semantik, innere Sprachform und Tiefenstruktur*. Tübingen: Romanisches Seminar.

Elffers, Els. 1991. *The historiography of grammatical concepts: 19th and 20th-century changes in the subject-predicate conception and the problem of their historical reconstruction*. Amsterdam: Rodopi.

Elffers, Els. 1993. 'Philipp Wegener as a proto-speech act theorist'. In: *Linguistics in the Netherlands 1993*, Frank Drijkoningen & Kees Hengeveld (ed.), pp. 49-59. Amsterdam: Benjamins.

Gabelentz, Georg von der. 1869. 'Ideen zu einer vergleichenden Syntax. Wort- und Satzstellung'. *Zeitschrift für Völkerpsychologie und Sprachwissenschaft* 6, pp. 376-384.

Gabelentz, Georg von der. 1875. 'Weiteres zur vergleichenden Syntax. Wort- und Satzstellung'. *Zeitschrift für Völkerpsychologie und Sprachwissenschaft* 8, pp. 129-165, 300-338.

Gabelentz, Georg von der. 1889. 'Über Stoff und Form in der Sprache'. *Berichte über die Verhandlungen der königlich-sächsischen Gesellschaft der Wissenschaften zu Leipzig, philologisch-historische Classe* 21, pp. 185-216.

Gabelentz, Georg von der. 1894. 'Hypologie [Typologie], eine neue Aufgabe der Linguistik', *Indogermanische Forschungen* 4, pp. 1-7.

Gabelentz, Georg von der. 2016 [1891]. *Die Sprachwissenschaft, ihre Aufgaben, Methoden und bisherigen Ergebnisse*, Manfred Ringmacher & James McElvenny (eds.). Berlin: Language Science Press.

Humboldt, Wilhelm von. 1905 [1820]. 'Über das vergleichende Sprachstudium in Beziehung auf die verschiedenen Epochen der Sprachentwicklung'. In: *Wilhelm von Humboldts gesammelte Schriften*, vol. IV, Albert Leitzmann (ed.), pp. 1-34. Berlin: Behr.

Knobloch, Clemens. 1988. *Geschichte der psychologischen Sprachauffassung in Deutschland von 1850 bis 1920*. Tübingen: Niemeyer.

Koerner, E.F. Konrad. 1978 [1974]. 'Animadversions on some recent claims regarding the relationship between Georg von der Gabelentz and Ferdinand de Saussure'. In: *Toward a Historiography of Linguistics: selected essays*, E.F. Konrad Koerner (ed.), pp. 137-152. Amsterdam: Benjamins. (First published in *Studi saussuriani per Robert Godel*, R. Amacker, T. de Mauro, & L.J. Prieto (eds.), pp. 165-180. Bologna, Italia: Il Mulino).

Koerner, E.F. Konrad. 2008. 'Hermann Paul and general linguistic theory'. *Language Sciences* 30, pp. 102-132.

McElvenny, James. 2016. 'The fate of form in the Humboldtian tradition: the *Formungstrieb* of Georg von der Gabelentz'. *Language and Communication* 47, pp. 30-42.

McElvenny, James. 2017a. 'Grammar, typology and the Humboldtian tradition in the work of Georg von der Gabelentz'. *Language and History* 60:1, pp. 1-20.

McElvenny, James. 2017b. 'Georg von der Gabelentz'. *Oxford Research Encyclopedia of Linguistics*. http://linguistics.oxfordre.com/view/10.1093/acrefore/9780199384655.001.0001/acrefore-9780199384655-e-379, DOI: 10.1093/acrefore/9780199384655.013.379, accessed 2 August 2018.

Morpurgo Davies, Anna. 1975. 'Language Classification in the Nineteenth Century'. In: *Current Trends in Linguistics*, vol. 13, Thomas A. Sebeok (ed.), pp. 607-716. The Hague: Mouton.

Nerlich, Brigitte & David D. Clarke. 1996. *Language, action and context: The early history of pragmatics in Europe and America 1780–1930*. Amsterdam/Philadelphia: John Benjamins.

Ringmacher, Manfred. 1996. *Organismus der Sprachidee: Steinthals Weg von Humboldt zu Humboldt*. Paderborn: Schöningh.

Ringmacher, Manfred. 2001. 'Die Klassifizierung der Sprachen in der Mitte des 19. Jahrhunderts'. In: *History of the Language Sciences*, vol. 2, Sylvain Auroux, E.F.K. Koerner, Hans-Josef Niederehe & Kees Versteegh (eds.), pp. 1427-1436. Berlin: Walter de Gruyter.

Saussure, Ferdinand de. 1995 [1916]. *Cours de linguistique générale*, ed. by Tullio de Mauro. Paris: Payot & Rivages.

Scheerer, Thomas M. 1980. *Ferdinand de Saussure: Rezeption und Kritik*. Darmstadt: Wissenschaftliche Buchgesellschaft.

Staffeldt, Sven. 2017 [2014]. 'Speech act theory and Georg von der Gabelentz'. *History and Philosophy of the Language Sciences*, https://hiphilangsci.net/2017/09/06/speech-act-theory-and-georg-von-der-gabelentz/, accessed 2 August 2018. (Translation of Sven Staffeldt. 'Die Sprechakttheorie und Georg von der Gabelentz'. In: *Beiträge zur Gabelentz-Forschung*, Kennosuke Ezawa, Franz Hundsnurscher & Annemete von Vogel (eds.), pp. 229-238. Tübingen: Narr Verlag).

About the authors

James McElvenny is an intellectual historian specialising in the history of linguistics. He is currently Newton International Fellow in the Department of Linguistics and English Language at the University of Edinburgh.
james.mcelvenny@mailbox.org

Manfred Ringmacher is a Romanist and specialist in the history of linguistics. Most recently, he was involved in the edition project 'Wilhelm von Humboldt: Schriften zur Sprachwissenschaft' at the Berlin-Brandenburg Academy of Sciences.
Manfred.Ringmacher@gmx.de

333

Chapter III. Content and Form of Speech

I. Speech

Language is speech, is expression of thought, is in sentences.[1] Our first question concerns the content of speech: What type of thought is expressed – that is, of what type is the speech, the proposition?

Here we have to once again broaden the scope of our discussion a little. Thinking amounts to connecting mental representations. For the moment, we will leave the division of mental representations into intuitions and concepts to the philosophers.[2] Speech is the expression of this connection of representations; that is, expression of the representations to be connected as well as their connection, even if this connection is loose, even if its expression is very weak.

If we were to stop here and attempt to classify the different types of speech from this standpoint, we would have no basis for classification other than the types of connection: actual, possible, necessary, certain, desired, etc. In this case, each of the following sentences would represent a different form of speech:

You are coming,
You should come,
You can come,
You must come,
Hopefully you are coming,
You are probably coming.

1 In this translation, 'speech' is the default equivalent for *Rede* in the original German. When *Rede* refers to a specific instance of speech, however, it is translated as 'utterance'. 'Utterance' is also used here as the usual translation for *Äußerung* and *Sprachäußerung*.

2 'Representation' here renders *Vorstellung*, 'intuition' *Anschauung*; 'concept' *Begriff* (cf. Ritter et al., 1971–2007).

333

III. Inhalt und Form der Rede.

I. Die Rede.
Sprache ist Rede, ist Ausdruck des Gedankens, ist Satz. Wir fragen zunächst nach dem Inhalte der Rede: Welcher Art Gedanke wird ausgedrückt, welcher Art ist also die Rede, der Satz?

Wir müssen hier wieder etwas weiter ausholen. Denken heisst Vorstellungen verknüpfen. Die Eintheilung der Vorstellungen in Anschauungen und Begriffe überlassen wir vorläufig der Philosophie. Die Rede ist Ausdruck jener Verknüpfung von Vorstellungen, also Ausdruck sowohl der zu verknüpfenden Vorstellungen, als auch ihrer Verknüpfung, mag diese Verknüpfung noch so lose, ihr Ausdruck noch so schwächlich sein.

Wollten wir hierbei stehen bleiben, von diesem Standpunkte aus die Arten der Rede zu classificiren versuchen, so wüsste ich keinen anderen Eintheilungs||318||grund, als jene verschiedenen Verknüpfungsweisen: ob thatsächlich, möglich, nothwendig, gewiss, gewollt u. s. w. Dann würden also die Sätze:

Du kommst,
Du sollst kommen,
Du kannst kommen,
Du musst kommen,
Du kommst hoffentlich,
Du kommst wahrscheinlich,
ebensoviele verschiedene Formen der Rede darstellen.

334
But obviously each of these does not represent a different type of speech; the standpoint from which we attempted our classification was premature. What was our mistake? Our mistake was that we thought only of the speaking 'I' and not of the 'you' that is spoken to. Such a 'you' is either present or not. The only case in which it is not present is that of the genuine exclamation. In all other cases it is present, at least in the mind of the speaker.[3] Speech, in the broader sense, is sharing thoughts, whether I include the one spoken to in a 'we' or whether I talk to myself as if I were looking in the mirror. Let us begin by examining communicative speech in this broader sense. Since the following investigations will take us into confusing terrain, it is good to start where things seem clearest.

Now every act of speech contains in the first instance a demand on the 'you': 'Listen to me!' In some regions and languages this demand finds particular expression and emphasis: 'Listen – *écoutez*'.

If I ask for nothing more than attention, my speech is *declamation* in the broader sense of the word. But in this case I imagine someone other than my actual audience as the 'you', and I make different demands on this 'you'. If, however, I direct my speech to my audience then the declamation becomes an address and we are back where we started: I ask for something more from those spoken to than mere attention. We will now consider this 'something more'.

1. I express my thought and request that my interlocutor think in the same way: 'Believe me!' The corresponding form of speech is *communicative speech in the narrow sense*: I share my thought with you so that it also becomes your thought.

2. I cannot share anything with you that I do not have or that I pretend to have. You cannot believe anything I say that I do not myself believe or know, or at least pretend to believe or know. However, if I am uncertain myself, you can perhaps tell me what I am missing. Now I want this from you: 'Tell me…!' The corresponding form of speech is the *interrogative*. The answering form, by contrast, is by its nature a subtype of the communicative.

3 To maintain terminological parallels from the source text in the translation, the following equivalents are used in most instances: 'mind' for *Geist*, 'mental' for *geistig*, 'soul' for *Seele*, and 'in the soul' for *seelisch*. Note that Gabelentz uses *Geist* and *Seele* – and their corresponding adjectives *geistig* and *seelisch* – more or less interchangeably. His usage of *Seele* and *seelisch* accords with general conventions in German psychology of the time and should not be interpreted in a spiritual or religious sense.

334
Offenbar thun sie dies nicht; der Standpunkt, von dem aus wir die Eintheilung versuchten, war verfrüht. Worin lag nun der Fehler? Darin, dass wir nur an das redende Ich dachten, nicht auch an das angeredete Du. Ein solches ist entweder vorhanden oder nicht vorhanden. Nicht vorhanden ist es nur im Falle des eigentlichen Ausrufes. Sonst ist es überall, mindestens im Geiste des Redenden, gegenwärtig, die Rede ist im weiteren Sinne Gedankenmittheilung, mag ich nun den Angeredeten in ein Wir mit einschliessen, mag ich mir im Selbstgespräche mich selbst |309| wie im Spiegelbilde gegenüberstellen. Wir wollen zunächst die mittheilende Rede in diesem weiteren Verstande in's Auge fassen. Denn die folgenden Untersuchungen werden uns in ein wirres Gebiet führen, und es ist gut, da anzufangen, wo noch die Dinge am Klarsten zu liegen scheinen. thuen 1891

Nun enthält jede Rede unmittelbar eine Aufforderung an das Du: „Höre mich!" Bekanntlich giebt man in manchen Gegenden und Sprachen dieser Aufforderung noch besonderen Aus- und Nachdruck. „Höre, – *écoutez*."

Verlange ich nichts weiter als Aufmerksamkeit, so ist meine Rede Declamation im weiteren Sinne des Wortes. Dann stelle ich mir aber unter dem Du jemand anderes vor, als meine Zuhörerschaft, und an dies Du stelle ich noch andere Anforderungen. Wende ich mich aber an die Zuhörerschaft, wird die Declamation zur Ansprache, so stehen wir wieder auf dem alten Punkte: ich verlange von den Angeredeten noch etwas mehr als blosse Aufmerksamkeit. Um dieses Mehrere wird es sich hier handeln.

1. Ich spreche meinen Gedanken aus und verlange, dass der Angeredete nun ebenso denke: „Glaube mir!" Die entsprechende Form der Rede ist die mittheilende im engeren Sinne: ich theile Dir meinen Gedanken mit, damit er hinfort auch Dein Gedanke werde.

2. Ich kann Dir nichts mittheilen, was ich nicht habe oder zu haben vorgebe. Du kannst mir nichts glauben, was ich nicht – mindestens vorgeblich – glaube oder weiss. Bin ich aber selbst im Ungewissen, so kannst Du vielleicht mir mittheilen, was mir fehlt. Nun begehre ich dies von Dir: „Sage ||319|| mir...!" Die dem entsprechende Redeform ist die fragende. Die antwortende dagegen ist ihrer Natur nach nur eine Unterart der mittheilenden.

3. I want neither to tell you something nor to learn something from you, but rather I want you to do or not do something, regardless of what it is. This form of speech is called the *imperative* (forbidding, requesting, rejecting).

4. Human language is by its nature a means of communication; as a rule, a 'you' is present at the very least in the mind of the speaker. But it is just as much the rule that I not only want to say something to you but also express myself; I want to hear myself what I think and how I feel.

3. Ich will weder, dass Du etwas durch mich erfahrest, noch dass ich etwas durch Dich erfahre, sondern dass Du etwas, gleichviel was, thuest oder unterlassest Die Form dieser Rede nennen wir die gebietende (verbietende, bittende, verbittende).

4. Die menschliche Sprache ist ihrem Wesen nach Verkehrsmittel; dass dem Redenden ein Du mindestens im Geiste gegenwärtig sei, bildet die Regel. Allein ebenso ist es die Regel, dass ich nicht nur Dir etwas sagen, sondern auch mich aussprechen will; ich will selbst hören, was ich denke, und wie ich empfinde.

335 In such moods an individual finds himself under the influence of strong excitation that pushes to be discharged. What he does in the end is in a broad sense pathological, whether he cries, laughs, claps, stamps his foot, calls out, clicks his tongue or produces a piece of human speech. For through continual practice language has become so natural to us that it can break out from within us unconsciously and unintentionally. Such acts of speech we call *exclamatives*. This kind of speech deserves our attention in particular for two reasons: firstly, because it is in fact based on a different foundation in the soul from the kinds of speech that serve communication; and, secondly, because it stands to reason that where language has escaped the shackles of communication it has created new forms.

At this point we have to test our taxonomy of speech types for its effectiveness and completeness, and here appear immediately some serious concerns that we must resolve.

A. We have more or less stayed with the conventional names and contrasted communicative speech with interrogative and imperative, but in reality interrogative and imperative speech also communicate a speaker's thought.

B. We have placed interrogative speech in a category alongside the imperative. The imperative demands from the interlocutor a positive or negative response through action: *Do this, stop that!* But the verbal answer that is demanded of someone who is asked a question is also a kind of response through action. It might therefore seem as if a question were just a subtype of command.

C. Finally, language often clothes its thoughts in borrowed garments. The *communicative* form of speech might contain a question: 'I would like to know whether ...' – or it might include a command, a request or a prohibition: 'You must ..., You must not ..., You would be doing me a favour if ...' etc. The interrogative form might conceal a judgement already reached. This is the form of the rhetorical question, which says: 'Don't give me the answer; give it to yourself, ask yourself the question and you will come to the same conclusion as me!' Or it might be a demand expressed in interrogative form: 'Will you hurry up and come now?! Would you be so kind as to ... ?' It seems natural that the exclamation also tends to take on the interrogative form: 'How great is that!' This is no doubt because the kind of emotional excitation in this case is related to that of the incomplete judgement. The *commanding* and *requesting* form is immediately suited to asking for information: 'Name your accomplices! Tell me the exact time!' Finally, the exclamation can serve to convey a genuine message: 'A lovely picture!' Or a question: 'If I only knew ... !' Or a demand: 'If you would only help me!'

335 In solchen Stimmungen befindet sich |310| der Mensch unter dem Einflusse mächtiger Erregungen, die nach Entladung drängen. Was er zu dem Ende thut, ist im weiteren Sinne pathologisch, es sei Weinen, Lachen, Händeklatschen, Aufstampfen mit dem Fusse, ein Aufschrei, ein Schnalzen mit der Zunge oder ein Stück menschlicher Rede. Denn auch die Sprache ist uns durch Übung so zur Natur geworden, dass sie unbewusst und absichtslos aus unserm Innern hervorbrechen kann. Solche Reden nun nennen wir *ausrufende*. Wir mussten ihrer besonders gedenken, zunächst weil sie in der That auf anderer seelischer Grundlage beruhen, als jene, welche dem Verkehre dienen; dann aber auch, weil es nahe liegt, dass sich die Sprache, wo sie die Fesseln des Verkehres abgestreift, neue Formen geschaffen habe.

Jetzt gilt es, die gewonnene Eintheilung auf ihre Durchschlägigkeit und Vollständigkeit zu prüfen, und da erstehen auf den ersten Blick ernstliche Bedenken, die beseitigt werden müssen.

A. Wir haben uns an die herkömmlichen Benennungen gehalten, der mittheilenden Rede die fragende und die befehlende entgegengesetzt, die doch in Wahrheit auch einen Gedanken des Redenden mittheilen sollen.

B. Wir haben die fragende Rede der befehlenden nebengeordnet. Der Befehl verlangt vom Angeredeten eine positive oder negative Thatäusserung: Thue das, unterlasse jenes! Eine solche Thatäusserung ist aber doch auch die Antwort, zu der der Gefragte aufgefordert wird. Also könnte es scheinen, als wäre die Frage nur eine Unterart des Befehles.

C. Endlich kleidet die Sprache oft ihre Gedanken in geborgte Gewänder. Die *mittheilende* Redeform mag jetzt eine Frage enthalten: „Ich wüsste gern ob …“, – jetzt mag sie einen Befehl, eine Bitte, ein Verbot in sich schliessen: „Du musst …, Du darfst nicht …, Du würdest mir einen Gefallen thun, thuen, 1891
wenn Du …“ u. s. w. Die fragende Form mag ein fertiges Urtheil verhüllen. Es ist dies der Fall der rhetorischen Frage, die besagen will: „Gieb die Antwort nicht mir, sondern Dir, stelle Dir die Frage, so wirst Du urtheilen wie ich!“ Oder es mag eine Aufforderung in fragender Form ausgesprochen werden. „||320|| Du gleich kommen?! Wärest Du wohl so freundlich …?“ Es scheint naturgemäss, dass auch der Ausruf gern die fragende Form annehme: „Wie schön ist das!“ Denn der Zustand der Gemüthserregung ist jenem des unfertigen Urtheils verwandt. Die *befehlende* und *bittende* Form eignet sich ohne Weiteres da, wo Auskunft erfordert wird: |311| „Nenne mir Deine Gehilfen! Sagen Sie mir doch die genaue Zeit!“ Endlich kann der Ausruf jetzt Gehülfen! 1891
eine thatsächliche Mittheilung bezwecken: „Ein schönes Bild!“, – jetzt eine Frage: „Wüsste ich doch …!“ – jetzt wohl auch eine Aufforderung: „Wenn Du mir doch hülfest!“

336

We will orient ourselves using the forms above and now compare the exclamative form with the other three, which are communicative in the broader sense.

The thought to be shared can be a judgement, a wish or both at the same time.

If it is a judgement, it can be complete or incomplete. If it is complete then the form of speech is *communicative in the narrower sense of the word*.

If the judgement is incomplete then I share it with you as something that is incomplete, so that you can fill it out. Both of these aspects, the sharing of a judgement and the indication of the wish for it to be filled out, come together in the *question*.

Finally, if the thought is complete but represents a wish that I want you to fulfil rather than a judgement then the conditions are met for *imperative* speech and its relatives: requesting, forbidding, advising, etc.

The following diagram provides an overview of the three communicative speech forms:

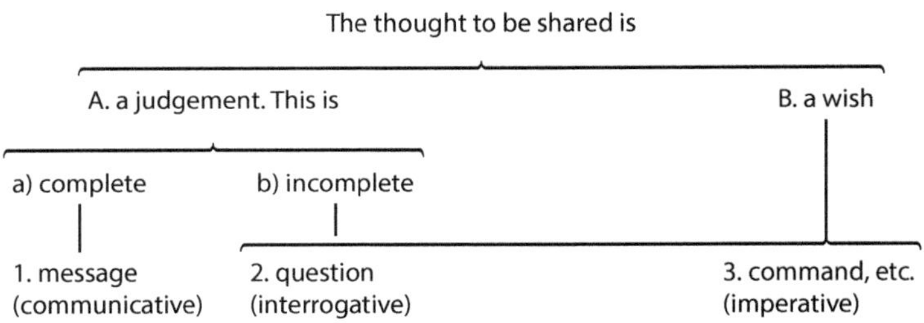

I believe we can claim completeness for this diagram and that we will be able to apply it analogously to the subtypes of *exclamative* speech.

Let us distinguish here once more between content and form.

Depending on its form, an exclamation can be:

A. A complete sentence or a single word serving as a sentence, such as an optative: 'If he would but come!' Or an interrogative: 'How great is that!' An imperative: 'Halt!' We may wish to remind ourselves in passing of the special modal verb forms or particles that serve exclamative speech in some languages.

336
Wir halten uns an die Formen und stellen nun der ausrufenden die drei übrigen als mittheilende im weiteren Sinne gegenüber.

Der mitzutheilende Gedanke kann sein ein Urtheil, ein Wunsch, oder Beides zugleich. beides 1891

Er sei ein Urtheil, so kann dieses vollständig oder unvollständig sein. Ist es vollständig, so ist die Rede mittheilend im engeren Sinne des Wortes.

Ist das Urtheil unvollständig, so theile ich es dir als unvollständiges mit, um es von dir ergänzen zu lassen. Beides, jene Mittheilung eines Urtheils und jener ausgesprochene Wunsch nach dessen Vervollständigung, vereinigt sich in der Frage.

Endlich: der Gedanke sei vollständig, aber er sei nicht ein Urtheil, sondern ein Wunsch, dessen Erfüllung ich von dir begehre: so sind die Voraussetzungen gegeben zur befehlenden Rede und ihren Verwandten: der bittenden, verbietenden, rathenden u. s. w. verbittenden, 1891

Zur Übersicht stellen wir für die drei mittheilenden Redeformen folgendes Schema auf:

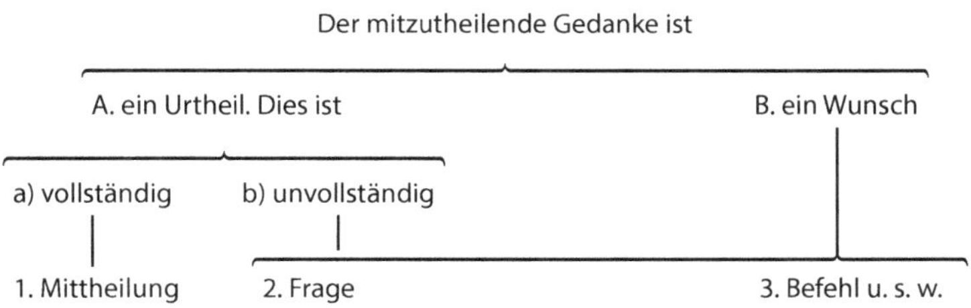

Ich glaube, diese Tabelle trägt die Gewähr der Vollständigkeit in sich, und sie wird noch ein zweites Mal analog anwendbar sein, wenn es gilt, die ausrufende Rede in Unterarten zu theilen.

Unterscheiden wir auch hier wieder zwischen Inhalt und Form.

Der Form nach kann der Ausruf sein:

A. ein vollständiger Satz bezw. ein Satzwort, z. B. ein optativer: „Käme er doch!" oder ein fragender. „Wie schön ist das!" Ein Imperativ: „Halt!" An die besonderen modalen Verbalformen oder Hilfswörter, die in manchen Sprachen der ausrufenden Rede dienen, mag hier nur im Vorübergehen erinnert werden. bez. 1891 Hülfswörter, 1891 dienen; 1901

|312| ||321||

337
B. Elliptical, replacing a sentence, as in 'Good boy!' 'Rejected!' 'And yet another one!' Sometimes such ellipses are supplemented and made into complete sentences, in which case the process in the soul may be manifested in a unique ordering of the parts of the sentence: 'And yet another one – is coming!' 'A great person – is so-and-so.' In Chinese, we find: *šén tsāi* (善哉 *shàn zāi*) 'Good, verily!' But we also find this sentence with an explicit subject appended to the end: *šén tsāi wén* (善哉問 *shàn zāi wèn*) 'Good, verily, [was your] question!' In these cases we speak of inversions, but we have to remember that unlike proper sentences these are not already formed in the soul, but only become wholes through a later extension, so to speak. The previous two forms are exclamative and would be perceived as such by the listeners. Indeed, such exclamations only escape our lips when we know that someone is listening to us and we want them to hear us. So it may seem as if these sentences were communicative in the broader sense. But it strikes me how often such exclamations are on the tips of our tongues and stay unexpressed, are suppressed. Why is this? Because it would not do any good to let them out. If we utter them in the presence of a third party, the situation is as follows: we intend to communicate something, but we do it in such a form as if we were speaking simply with the intention of opening up a valve to release the pressure in our soul. We would say: 'It had to be said, I had to let it out.' We then feel a double sense of relief, since now others are also affected by it. Imperatives like 'Hat off!' 'Stand still!' are calls, that is instances of communication in the broader sense of the word, not exclamations.

C. Vocatives in the broader sense of the word – I mean those in the second and third person – are absolute substantives not connected to the rest of the sentence, which are however not directed at a present or imagined second person. To this group belong in many cases invocations of supernatural forces or phenomena, such as 'Dear God!' 'The Devil!' 'Blazes!' These are emotional outbursts through which absolutely no prayer or incantation is intended: there is no thought of a second person. With these exclamations there is generally no thought of any possible listener, either. This is what distinguishes them from those vocative and other kinds of calls that are addressed to a 'you': *Police! Help! Fire!* These are in the broader sense communicative – whether the interjectional accusative in Latin, as in *Me miserum!* 'Poor me!', belongs here or among the ellipses I will not attempt to say.

337
B. Elliptisch, einen Satz ersetzend, z. B. „Brav gemacht!" „Abgeblitzt!" „Schon wieder Einer!" Manchmal werden wohl solche Ellipsen durch einen nachträglichen Zusatz ergänzt zu vollständigen Sätzen, und dann mag sich der seelische Hergang in einer eigenartigen Anordnung der Satzglieder ausprägen: „Schon wieder Einer – kommt da an!" „Ein wunderlicher Mensch, – ist NN." So finden wir im Chinesischen: *šén tsāi!* Gut, traun! Aber auch, mit nachträglicher Nennung des Subjectes: *šén tsāi wén*, gut, traun, (war deine) Frage! Wir reden dann von Inversionen, müssen aber bedenken, dass diese nicht, wie rechtschaffene Sätze, in der Seele fertig vorher geformt waren, sondern erst hinterdrein, so zu sagen durch einen späteren Anbau, zu einem Ganzen geworden sind. – Die beiden bisherigen Formen sind wohl ausrufend und werden auch von den Zuhörern so aufgefasst. In der That pflegen aber solche Ausrufe nur dann über unsere Lippen zu kommen, wenn wir wissen, dass man uns hört, und wollen, dass man uns höre. So könnte es scheinen, als wären sie doch im weiteren Sinne mittheilend. Ich denke indessen daran, wie oft uns derlei Ausrufe auf den Lippen schweben und ungeäussert bleiben, unterdrückt werden. Warum das? Weil es doch zu nichts frommen würde, sie verlauten zu lassen. Sprechen wir sie Dritten gegenüber aus, so ist der Thatbestand dieser: Wir bezwecken eine Mittheilung, machen diese aber in solcher Form, als sprächen wir lediglich mit der Absicht, unserer übervollen Seele ein Ventil zu öffnen. Man sagt wohl: „Es musste heraus, ich musste mich aussprechen", und dann empfindet man es als eine doppelte Erleichterung, wenn nun Andere in Mitleidenschaft gezogen werden. – Imperative wie: „Hut ab!" „Still gestanden!" sind Zurufe, also Mittheilungen im weiteren Sinne des Wortes, nicht Ausrufe.

C. Vocative im weiteren Verstande des Wortes, ich möchte sagen in zweiter und dritter Person, sind absolute, der Satzverknüpfung entbehrende substantivische Redetheile, die sich aber nicht an eine vorhandene oder vorgestellte zweite Person richten. Dahin gehören in vielen Fällen die Anrufungen übersinnlicher Mächte oder fürchterlicher Ereignisse, z.B. „Herr Gott!" „Den Teufel!" „Donnerwetter!" „Schwere Noth". Es sind das Stimmungsäusserungen, mit denen keinerlei Gebet oder Beschwörung beabsichtigt wird: man denkt an gar keine zweite Person. Auch pflegt man bei solchen Ausrufen nicht an etwaige Zuhörer zu denken. Das unterscheidet sie von jenen Vocativen und sonstigen |313|Zurufen, die an ein Du gerichtet sind: Polizei! Hülfe! Feuer! Diese sind im weiteren Sinne mittheilend. – Ob der interjectionelle Accusativ im Lateinischen, z. B. Me miserum! hierher oder unter die Ellipsen gehört, wage ich nicht zu unterscheiden.

Noth." 1891

entscheiden. 1891

338
D. Pure interjections, that is words that do not belong to any other part of speech. These can be divided into

a) more objective, imitative interjections, such as *Whoops! Wham!*

b) more subjective interjections, that is those that express only an individual sensation, pain, joy, amazement, and so on: *Ouch! Hey! Ah! Hm!* These unformed words can also occupy a phonologically exceptional position in relation to the formed language. In these interjections High German still has an initial *p*, and German dialects that would otherwise turn *ö* and *ü* into *e* and *i* preserve the sounds in their original form. Similar to interjections in terms of both their syntactic and phonological properties are certain vocal appeals and calls: the *Hö* and *Hüist* of wagon drivers and ploughboys in Saxony, the *Shhh!* that calls for silence and also the interrogative *Hm?* should be counted here, and no doubt also originally the deictic sounds that are the basis of demonstrative pronouns. Such 'natural sounds', as they are called, most likely harbour older organic structures that have been mangled or misunderstood over time. The Saxon *Jemine!* is probably a replacement for the name Jesus, perhaps also with similarity in sound to *domine* (Latin 'O Lord'). In Spain they innocently call out *Caramba!* instead of the vulgar *carajo* 'dick', just as in many parts of Germany they say *Gottstrambach* or *'strambach* instead of *Gott straf' mich* (lit. 'God punish me!'). Conversely, these sounds can undergo grammatical formation as roots or stems. We say: *the cat miaows, die Katze miaut (maunzt), le chat miaule*. In German, from *ach!* 'argh!' and *pfui!* 'ugh!' we form *ächzen* and in some dialects *pfuzen* (both meaning approximately 'to groan'), just as we form *duzen* and *siezen* from *Du* and *Sie*. It is surely the case universally that animals are named for their calls. The cuckoo has looked after itself and made sure that the Germanic sound shift could not do any harm to its name. But we do not need to mention again here the onomatopoeic forms that appeared in the prehistory of language.

As far as I can see, what we have said above exhausts the possible forms of exclamative speech, indeed the possible forms of human speech altogether. Human speech is either grammatically formed or unformed. When formed, it is either a complete sentence or not a complete sentence. In the latter case it is either a fragment of a sentence that calls for completion – i.e. elliptic – or it formally refuses syntactic connection, i.e. it is absolute. The fourth category, that of the unformed linguistic expression, we divided up in such a way that the taxonomy simultaneously addresses the origin of the expression and its content. If we were to apply this taxonomy directly to communicative utterances, we would run into difficulties that would be better to avoid for the moment. But let us summarise what we have established so far in a diagram.

338
D. Reine Interjectionen, das heisst Wörter, die keinem anderen Redetheile angehören. Sie dürften einzutheilen sein

a) in mehr objective, nachahmende, z. B. Pardauz! hopsa! puff!

b) in mehr subjective, d. h. solche, die nur ein individuelles Empfinden, ||322|| Schmerz, Freude, Erstaunen u. dgl. ausdrücken: Au! Ei! Ach! Hm! Es sind dies ungeformte Wörter, die der geformten Sprache gegenüber auch in lautlicher Hinsicht eine Ausnahmestellung einnehmen können. In ihnen hat das Hochdeutsche noch anlautende *p*, haben Dialekte, die sonst *ö* und *ü* in *e* und *i* verwandeln, diese Laute rein bewahrt. In beiden Hinsichten stehen ihnen gewisse Auf- und Zurufwörter gleich, z. B. das Hö und Hüist der Fuhrleute und Ackerknechte in Sachsen, das Ruhe gebietende St! Auch das fragende Hm? ist hierher zu rechnen, und gewiss ursprünglich auch die Deutelaute, die den Demonstrativpronominibus zu Grunde liegen mögen. Solche „Naturlaute", wie man sie wohl genannt hat, mögen zuweilen verstümmelte oder unverstandene ältere organische Gebilde in sich enthalten. Unser Jemine! ist wohl ein Ersatz für den Namen Jesus, vielleicht mit Anklang an *domine*. In Spanien ruft man arglos: *Caramba!* statt des unfläthigen *carajo*, wie man in manchen Gegenden Deutschlands Gottstrambach oder 'strambach statt Gott straf' mich sagt. Umgekehrt können jene Laute als Wurzeln oder Stämme grammatische Formung erfahren. Man sagt: die Katze miaut (maunzt), le chat miaule. Von ach! und pfui! bilden wir ächzen ächsen 1901
und (mundartlich) pfuzen, ganz wie duzen und siezen von Du und Sie. Dass Thiere nach ihren Rufen benannt werden, ist wohl allgemein verbreitet. Der Kukuk hat selbst dafür gesorgt, dass die germanische Lautverschiebung seinem Namen nichts anhaben konnte. Von jenen Onomatopöien aber, die in die Urgeschichte der Sprache fallen, brauchen wir hier nicht wieder zu reden.

Mit Obigem sind meines Wissens die möglichen Formen der ausrufenden Rede erschöpft. In der That die möglichen Formen der menschlichen Rede überhaupt. Diese ist entweder grammatisch geformt oder ungeformt. Die geformte ist entweder ein vollständiger Satz oder kein vollständiger Satz. Letzteren Falles ist sie entweder ein zur Ergänzung aufforderndes Bruchstück eines Satzes – elliptisch, – oder sie lehnt |314| die syntaktische Verknüpfung formell ab, ist absolut. Die vierte Kategorie, jene der ungeformten Sprachäusserungen, theilten wir so, dass die Eintheilung zugleich den Ursprung und den Inhalt betraf. Soll sie nun auch auf die mittheilenden Sprachäusserungen bezogen werden, so gerathen wir in Schwierigkeiten, die wir besser noch vermeiden. Wir fassen aber das bisher Gewonnene in ein Schema zusammen.

339

The linguistic expression is

A. grammatically formed

a) complete sentence, b) ellipse, c) absolute

B. unformed

a) imitative,
b) expressing sensations,
c) indicating,
d) asking ...

But now, since we want to consider communicative speech in the broader sense, we have to extend the concept of ellipsis to every kind of sentence fragment, even to the case found in dialogue where the completion of the sentence comes from the speech of the interlocutor, as in *Who was there? – I* [*was there*]. – *When* [*were you there*]*? – Yesterday. – Well? And ... ?* [*what happened there?*]

Let us now turn back to exclamative speech and examine its possible content.

An exclamation is an expression of lively excitement; what is expressed is either just the type of excitement or also the reason for it. The reason can be either a wish or an actual accomplished fact. If the latter, the excitement may have to do with the known part of the fact, or it may be because there is a part of the fact that we do not know. The relevant diagram, which will once again give the impression of completeness, is structured as follows:

Expression of excitement,

A. according only to its type, B. according also to its reason

a) wish, b) fact

a) known, b) unknown

339

Der sprachliche Ausdruck ist

A. grammatisch geformt

a) voller Satz, b) Ellipse, c) absolut

B. ungeformt

a) nachahmend,
b) Empfindungen äussernd,
c) deutend,
d) fragend …

||323||

Nun aber, da wir die im weiteren Sinne mittheilende Rede mit in Betracht ziehen, müssen wir auch den Begriff der Ellipse ausdehnen auf jederlei eigentliches Satzfragment, auch auf den Fall, wo im Zwiegespräche die Ergänzung zum Satze aus der Rede des Anderen zu erholen ist, z. B. Wer war dort? – Ich (war dort). – Wann (warst Du dort)? – Gestern. – Nun? Und … ? (was geschah da?) z.B., 1901

Jetzt kehren wir zur ausrufenden Rede zurück, um ihren möglichen Inhalt zu untersuchen.

Im Ausrufe äussert sich eine lebhafte Erregung, entweder nur die Art dieser Erregung, oder auch ihr Grund. Der Grund kann sein ein Wunsch oder eine vollendete Thatsache. Ist er eine solche, so kann die Erregung entweder in dem bekannten Theile der Thatsache, oder darin beruhen, dass wir einen Theil der Thatsache nicht kennen. Das Schema, das wieder den Eindruck der Vollständigkeit machen würde, gestaltet sich demnach folgendermassen :

Äusserung der Erregung, und zwar

A. nur ihrer Art nach,

B. auch ihrem Grunde nach

a) Wunsch,

b) Thatsache

a) Bekanntes,

b) Unbekanntes

340
If we align fact and judgement, then an analogy with our first diagram becomes plausible. We see the following parallels:

I. Wish – Command etc.

II. Fact – Judgement.

a) Known – Message.

b) Unknown – Question.

If I were to examine this matter further in either an empirical or an *a priori* way, it would not be possible for me to devise a new diagram nor for me to make any additions to the existing diagrams; I will find neither an additional category for my taxonomy nor a gap in the existing taxonomy. If I am not mistaken, the logical task has been completed and it would seem that we have found a means to easily classify every kind of speech.

But things are not that simple. We do not find ourselves on the cleanly swept ground of logic, but in the middle of the abundant tangle of psychological possibilities. Here, as in a jungle, the roots and branches are ensnarled in one another, and the vines are entwined from trunk to trunk. But it is precisely in botanic gardens, where the plants are neatly arranged in beds and rows according to their species and subspecies, that one cannot see the forest for the trees. And the human soul is under even less restraint in its variations than nature in its creations; it does not shy away from hybrid forms. In our case, this means we have to be prepared to sometimes take an apparently communicative mode of speech as having an exclamative sense, or to see an exclamation being used in place of a message, question or command. It is conceivable that in the life of a language the exclamative forms of speech might oust the communicative ones, might replace them. And so in many cases the art of classification may fail, since the state of affairs in the soul cannot be firmly established, perhaps because it was an uncertain, mixed one. We can plot stages, stations, we can at a pinch calculate the possible combinations, but any effort to count all the points of a line would be in vain.

Our investigation was however not fruitless, even if the results were not as we expected. It led to the development of a range of concepts that linguists will have to continue to use.

340

Setzen wir hier Thatsache und Urtheil auf eine Stufe, so ist die Analogie mit dem ersten Schema einleuchtend: Es laufen parallel:

I. Wunsch – Befehl u. s. w.

II. Thatsache – Urtheil.

|315|

a) Bekanntes – Mittheilung.

b) Unbekanntes – Frage.

Mag ich nun meine Erfahrungen mustern, mag ich versuchen, der Sache auf apriorischem Wege beizukommen, so finde ich weder die Möglichkeit zu einem neuen Schema noch zu einer Ergänzung in einem der vorhandenen Schemata, ich finde weder einen weiteren Eintheilungsgrund, noch eine Lücke in den gemachten Eintheilungen. Täusche ich mich hierin nicht, so wäre also die logische Aufgabe gelöst, und nun könnte es scheinen, als wäre ein Mittel gefunden, um jederlei Rede mit Leichtigkeit zu classificiren.

So einfach liegen indessen die Dinge nicht. Wir befinden uns nicht auf dem glattgefegten Boden der Logik, sondern mitten drinnen in dem üppigen Gewirre psychologischer Möglichkeiten. Da sind wie in einem Urwalde die Wurzeln und Gezweige der verschiedensten Pflanzen ineinander verfitzt, und Schlinggewächse ranken von Stamme zu Stamme. Jene dendrologischen Gärten aber, wo die Pflanzen säuberlich nach Arten und Unterarten in Beete und Reihen geordnet sind, sind recht eigentlich die Stätten, wo man den Wald vor lauter ||324|| Bäumen nicht sieht. Und die Menschenseele schaltet schrankenloser als die schaffende Natur; vor keiner Zwitterform scheut sie zurück. Wir in unserm Falle müssen darauf gefasst sein, jetzt (unserem 1891) die eine oder andere mittheilende Redeweise in ausrufendem Sinne, jetzt diese oder jene Art des Ausrufes statt der Mittheilung, der Frage oder des Befehles angewandt zu sehen. Es ist denkbar, dass im Leben einer Sprache die ausrufenden Redeformen die mittheilenden geradezu verdrängen, ersetzen. (ersetzen, und 1891) Und so mag in vielen Fällen die Kunst der Classification überhaupt versagen, weil der seelische Thatbestand nicht festzustellen ist, vielleicht weil er an sich ein unsicherer, gemischter war. Stufen, Stationen konnten wir zeichnen, die möglichen Combinationen können wir zur Noth ausrechnen, aber alle Punkte einer Linie zu zählen, wäre vergebliches Bemühen.

Fruchtlos aber war die Untersuchung, wenn anders sie gelungen, darum doch nicht. Sie hat zur Entwickelung einer Reihe von Begriffen geführt, mit denen die Sprachwissenschaft fort und fort hantiren muss.

341

Herr v. d. Gabelentz spoke about matter and form in language.

Preliminary remark.

The following essay is intended to appear as a chapter of a larger work whose completion I cannot yet foresee. The topic that it addresses is among the most controversial, and indeed the issue in dispute is a fundamental one. But my views are quite different from those of my outstanding predecessors and contemporaries, and I felt that I should develop my views independently rather than turn them into a polemic.

II. Division of speech into matter and form

§ 1. A. Matter

We will now concentrate on the specifically human language – that is, the language in which thought receives articulated expression – and ask ourselves: What does its matter consist in? The answer is obvious: in everything that arouses human thinking. We do not need to discuss here all the things this could possibly be. For the moment it is irrelevant to our purposes whether the thought to be expressed contains a complete, imagined, expected or desired fact, an inner or outer fact, a conclusion or whatever else it might contain. The key issue for us is how the thought articulates its matter.

In order to articulate its matter, the thought must first decompose it and then combine it again. Matter can only be decomposed into matter, and it can only be formed through combination; formation is exclusively the product of combination, and combination is exclusively directed to the purpose of formation. By 'combination' we mean both simple concatenation and mutual penetration, since both are means of formation. But as we have already observed, formation is not always directed to the purpose of decomposing and combining; it can also modify individual pieces of matter. In this connection, we might think of our German diminutives and, as an example of the interpenetration of matter and form, we might think of the dialect expressions *Kietze* = *Kätzchen* 'kitty', *Zicke* = *kleine Ziege* 'little goat, kid'.[4] But we will not concern ourselves with this sort of formation just yet.

4 Standard German has a highly productive diminutive suffix -*chen*, which umlauts the vowel in the root to which it attaches. *Kätzchen* is thus *Katze* 'cat' + *chen* dimin. (English has a cognate suffix in such forms as 'lambikin', G. *Lämmchen* = 'little lamb'; 'maiden', G. *Mädchen* = 'girl', but the English suffix is not productive.) The dialect forms *Kietze* and *Zicke* are diminutives of *Katze* and *Ziege* respectively with no clearly separable diminutive suffixes.

341
|316| {185}

Herr v. d. Gabelentz sprach über Stoff und Form in der Sprache.

Vorbemerkung.

Die folgende Abhandlung ist bestimmt, als Capitel eines grösseren Werkes zu erscheinen, dessen Abschluss ich noch nicht absehen kann. Der Gegenstand den sie bespricht, gehört zu den vielumstrittenen, und die Streitfrage ist recht eigentlich eine grundsätzliche. Meine Anschauungen aber weichen von denen hervorragender Vorgänger und Zeitgenossen erheblich ab, und ich habe geglaubt, sie lieber selbständig entwickeln, als zu einer Polemik ausspinnen zu sollen.

Eintheilung der Rede in Stoff und Form.
§ .1.
A. Der Stoff.
Wir halten uns nun an die specifisch menschliche Sprache, das heisst an diejenige, in welcher der Gedanke gegliederten Ausdruck findet, und fragen: worin besteht deren Stoff? Die Antwort giebt sich von selbst: in Allem was des Menschen Denken erregt. Was Alles dies sein kann, braucht hier nicht weiter erörtert zu werden. Uns ist es für jetzt gleichgültig, ob der auszudrückende Gedanke eine fertige, eingebildete, erwartete oder gewollte Thatsache, eine innere oder äussere, eine Schlussfolgerung oder was immer enthält. Darauf kommt es uns an, wie er seinen Stoff gliedert. Worin 1889 and 1901 alles 1889 Und 1901

Um ihn zu gliedern, muss er ihn zunächst zerlegen, dann wieder verbinden. Stoff lässt sich nur in Stoff zerlegen, nur in der Verbindung formen; die Formung ist ausschliessliches Erzeugniss der Verbindung, und die Verbindung dient ausschliesslich dem Zwecke der Formung. Unter Verbindung aber haben wir sowohl die blosse Aneinanderfügung als auch die gegenseitige Durchdringung zu verstehen, denn Beide sind Mittel der {186} Formung. Allein, dies sei schon jetzt bemerkt, – nicht immer dient die Formung dem Zwecke der Zergliederung und Verknüpfung, sie kann auch den Einzelstoff für sich bearbeiten, – man denke an unsere Diminutiven und, als Beispiele der Durchdringung von Stoff und Form, an die dialektischen Ausdrücke Kietze = Kätzchen, Zicke = kleine Ziege. Mit Formungen dieser Art haben wir es jetzt noch nicht zu thun. beide 1889 and 1891

342
Now we have to remember that, as a rule, a thought appears before us all at once as a complete image. I say 'as a rule' because there are exceptions where the parts of the thought come to us piece by piece. In any case, the thought appears before our soul complete and whole before it is expressed in speech. And even when I, for instance, hesitate and say, 'Six times seventeen is ... a hundred and two', there was still from the beginning the idea of a result that was yet to be calculated, and the thought was therefore formally complete. It is as this original whole entity that we want to call a thought a (complete) representation. The decomposition of this thought into parts and its subsequent reconstitution through combination is the task of speech-producing thinking. Here we are concerned solely with these parts: these we will call individual representations, in contrast to the complete representations which thinking has to decompose. We understand the difference between the two only in this sense. Based on its content, a complete representation can be very simple – for example, the complete representation of lightning – and an individual representation can be multifaceted; for example, the individual representation of a war. The greater or lesser definiteness of the content is not a reason for the difference, either: Caesar's death and the sentence 'the whole is the sum of its parts' can both be complete representations, but Caesar just as much as death, sum, whole and part are individual representations.

In its language a people undertakes the task of processing the totality of its individual representations; it is its world that a people decomposes and reconstitutes in language. The people decomposes its world into portions of matter and reconstitutes it in forms. Both matter and form are determined by the character of the people, which is in turn determined by the people's inner balance of talents and external fate. We speak justifiably of a mental point of view, mental perspective and mental horizon. As in optics, the first of these provides the conditions for the other two: only so much or so little falls within my field of view and forms my world; this is closest to me, that has slipped away into the foggy distance. But also my eye – my mental eye as well as my physical eye – plays a role: I can be short-sighted or long-sighted, or perhaps colour-blind. My vision might be better suited to getting an overview of a large area than to examining fine details; through practice it may be sharpened for doing one, and through neglect it may be blunted for the other.

But the mental eye sees more than the physical eye. With greater or lesser sharpness it recognizes and distinguishes the relations between material individual representations, such as the 'and' or 'next to' that exists between

342
||325||
Nun müssen wir daran denken, dass allerdings in der Regel der Gedanke mit Einem Schlage wie ein fertiges Bild vor uns steht. Ich sage: in der Regel; denn es giebt Ausnahmen, wo uns die Bestandtheile des Gedankens Stück für Stück kommen. Jedenfalls steht der Gedanke fertig und ganz vor unserer Seele, ehe er in der Rede zum Ausdrucke kommt; und wenn ich etwa, zögernd innehaltend, sage: „Sechs mal siebzehn ist … hundertundzwei", so hat mir doch von Anfang an die Idee eines noch zu bestimmenden Productes vorgeschwebt, und der Gedanke war mithin formell vollständig. In dieser ursprünglichen Ganz|317|heit wollen wir ihn eine Vorstellung (Gesammtvorstellung) nennen. Ihn in seine Bestandtheile zu zerlegen und diese Theile zum Wiederaufbaue zu verbinden ist Sache des redebildenden Denkens. Nur mit diesen Bestandtheilen haben wir es hier zu thun: wir wollen sie Einzelvorstellungen nennen im Gegensatze zu jener Gesammtvorstellung, die das Denken zerlegend zu bearbeiten hatte. Wir begreifen den Unterschied Beider nur in diesem Sinne. Ihrem Inhalte nach kann eine Gesammtvorstellung ganz einfach sein, z. B. die eines Blitzes, – und eine Einzelvorstellung kann sehr vieltheilig sein, z. B. die eines Krieges. So begründet auch die grössere oder geringere Bestimmtheit des Inhaltes keinen Unterschied: Caesar's Tod und der Satz, dass das Ganze gleich ist der Gesammtheit seiner Theile: Beide können Gesammtvorstellungen sein; Caesar aber sowohl als Tod, Gesammtheit, Ganzes und Theil sind Einzelvorstellungen.

Produktes 1889

beider 1889 and 1891

sein: 1889

Nun ist es die Gesammtheit seiner Einzelvorstellungen, die ein Volk in seiner Sprache darstellend zu verarbeiten hat, es ist seine Welt, die es in der Sprache zerlegt und wieder aufbaut. Es zerlegt sie in Stoffe, baut sie auf in Formen. Beides ist bestimmt durch die Eigenart des Volkes, die ihrerseits bestimmt wird durch seine innere Beanlagung und seine äusseren Schicksale. Man redet mit Recht von geistigem Standpunkte, geistiger {187} Perspective und geistigem Horizonte. Wie in der Optik bedingt der erste die beiden anderen: so viel oder so wenig fällt innerhalb meines Gesichtskreises, bildet meine Welt; dies steht mir am Nächsten, jenes ist mir in nebelhafte Ferne gerückt. Aber auch mein Auge, das geistige wie das leibliche, ist mit entscheidend: ich kann kurzsichtig sein, oder fernsichtig, vielleicht farbenblind; mein Blick mag sich besser zum Überschauen grosser Bildflächen als zur Prüfung enger Einzelheiten eignen, er mag durch Übung für das Eine geschärft, durch Vernachlässigung für Anderes abgestumpft sein.

soviel 1889

nächsten, 1889 and 1891

farbenblind, 1889

Ueberschauen 1889

Uebung 1889

Nun sieht aber das geistige Auge mehr als das leibliche. Mit mehr oder minderer Schärfe erkennt und unterscheidet es auch die Beziehungen der materiellen Einzelvorstellungen untereinander, z. B. zwischen dem Baume

a tree and a house, or the relation of belonging between a horse and its mane. And, depending on the degree and direction in which this occurs, such categories jostle for linguistic representation.

und dem Hause das „und“ oder „neben“, zwischen dem Pferde und seiner Mähne die Zugehörigkeit. Und je nach dem Masse und der Richtung, in der dies geschieht, drängen auch solche Kategorien zur sprachlichen Darstellung. Maasse 1889

343 At this point they are, as a rule, not yet forms for the language-creating mind, but matter to be formed. I consider as original forms only those where the word (the stem) itself is phonetically modified, through reduplication or inner modification, as for instance in the Semitic languages, or the verb in Tibetan or in Grebo, assuming of course that such changes are not the mechanical aftereffect of lost external formative elements. So long as languages were created this way, they treated at least for the moment the corresponding categories as matter. Matter is also a binding agent, also cement and putty, but when it is applied properly it is no longer felt to be matter, but instead simply a binding, forming force. Certainly it is so in the logical sense. But we cannot expect the naive mind that creates languages to feel more than a quantitative difference between pure and empirical concepts. For this naive mind, generality and indistinctness merge into one; and when in places this mind manages to give especially ephemeral expression to a particular category, we may ask: Was the reason for this largely mechanical, phonetic wearing down of the unemphasized part of the word? And – inasmuch as it was in the soul – was the reason not the same as that which leads us to describe the mind and soul as mist, breath or shadow? For it is not the same representations that strive for formal expression in all languages. Even representations very clearly related to the senses can be among those that do, such as representations of size, intensity, number, time, closeness, distance or direction. On the other hand, pure concepts in the logical sense, such as being and becoming, may be treated in the linguistic expression in the same way as those most clearly related to the senses: being as standing, as in Spanish *estar*, or inhabiting, as in German *war, gewesen*, Sanskrit √*vas*; becoming as turning, Sanskrit √*vṛt*, Latin *vertere*, English *to turn pale*; the completed action as possession: 'he *has* slept'.

As incontestable as it is important seems to me the following: as soon as a material word becomes a relational expression through generalization of its sense, it effects also a transformation in the soul. The more general meaning is thereafter the predominant meaning. This change can occur rather rapidly, but still of course unnoticed. For example, while their parents still spoke metaphorically of the 'face' of the house, thinking of it as being like the face of a person, the children might understand these words as referring to front of the person and the front of the house. From here the way to becoming a truly formal preposition or postposition is not far.[5] Such

5 In the 1889 and 1891 versions of this passage, the example of 'face' *Antlitz* and 'front' *Vordere* is *Bauch* 'belly' und *Innere* 'inside'.

343 Insoweit sind auch sie in der ||326|| Regel für den sprachschaffenden Geist zunächst nicht Formen, sondern zu formender |318| Stoff. Von ursprünglichen Formen möchte ich nur da reden, wo das Wort (der Stamm) selbst in seinem lautlichen Bestande verändert wird, durch Doppelung oder durch inneren Wandel, wie etwa in den semitischen Sprachen, im Verbum des Tibetischen und des Grebo, vorausgesetzt noch immer, dass Solches nicht mechanische Nachwirkung verschwundener äusserer Formativelemente ist. Soweit die Sprachen dergleichen geschaffen, haben sie mindestens für den Augenblick die entsprechenden Kategorien als Stoff behandelt. Stoffe sind ja auch die Bindemittel, auch Leim und Kitt; sie aber werden, wenn sie richtig angewandt sind, nicht mehr als Stoff, sondern nur als bindende, formende Kräfte empfunden. Gewiss sind sie dies im logischen Sinne. Nur muthe man dem naiven Geiste, der die Sprachen schafft, nicht zu, dass er zwischen reinen Begriffen und den empirischen einen mehr als quantitativen Unterschied verspüre. Für ihn laufen Allgemeinheit und Unbestimmtheit auf Eins hinaus; und wenn er stellenweise dazu gelangt ist, jenen Kategorien besonders flüchtigen Ausdruck zu verleihen, so wird man fragen dürfen: war der Grund nicht guten Theils mechanisch, lautliche Abnutzung des Unbetonten, und, – soweit er seelisch, – war er nicht etwa derselbe, der es auch veranlasst, dass {188} man Geist und Seele so gern als Dunst, Odem, Schatten bezeichnet? So sind es denn auch nicht in allen Sprachen dieselben Vorstellungen, die zur Formenbildung drängen, und auch sehr sinnliche können darunter sein, wie die der Grösse, der Intensität, der Zahl, der Zeit, der Nähe, Ferne oder Richtung, während reine Begriffe im logischen Sinne, wie die des Seins, Werdens, im sprachlichen Ausdrucke den sinnlichsten völlig gleich behandelt sein mögen: sein als stehen, – spanisch *estar*, oder wohnen, – deutsch w a r, g e w e s e n, sanskrit √*vas*; – werden als drehen, – sanskrit √*vr̥t*, lateinisch *vertere*, englisch *to turn pale*, – die vollendete Handlung als Besitz: „er h a t geschlafen".

Ebenso unbestreitbar wie wichtig deucht mir aber dies: Ist einmal das Stoffwort durch Verallgemeinerung seiner Bedeutung als Beziehungsausdruck in Dienst genommen worden, so hat sich auch in der Seele ein Umschlag vollzogen: die allgemeinere Bedeutung ist hinfort die vorwiegende. Dieser Wechsel mag ziemlich schnell und doch natürlich unvermerkt geschehen. Und wenn z. B. die Eltern noch bildlich vom Antlitze des Hauses sprachen, Bauche 1901
wie vom Antlitze des Menschen, so mögen die Kinder sich unter denselben Bauche 1901
Worten das Vordere des Menschen und das Vordere des Hauses denken, und Innere 1901
von da an ist der Weg zur wahrhaft formalen Prä- oder Postposition nicht

generalizations occurred for example in French: *face* ‘face’: front; *côté* (from Latin *costatum*): side;

mehr weit. Solche Verallgemeinerung erfuhren z. B. im Französischen: *face*, Antlitz: Vorderseite, *côté* (*costatum*): Seite;

344 *près*, actually = 'crowded' has become a preposition meaning 'close, near'. Low German *daal* 'valley' has become an adverb meaning 'downwards'. In Siamese *k'ōni* 'thing' and in Mafoor *ro* = *roi* 'thing' became signs of the genitive case; the spirit of the language combined them with the logical concept of belonging. If we did not know about its etymological relation with the Low German and Dutch *achter* 'aft, behind', the corresponding High German word, *After* 'anus', could seriously mislead us. We would group the German preposition *mit* 'with' alongside *Mittel* 'means' and *vermittelst* 'by means of', if our investigations into historical linguistics had not taught us that in this case the *t* is not to be derived from *dh*, but from *t*.

§ 2 B. Form

After what was said above, I cannot consider any language completely formless. Rather, I have to attribute both *outer form* and *inner form* to every language. The only questions that arise are: What is formed in a language, and by what means does this occur? The former question asks about the inner form, the latter about the outer form.

1. Inner linguistic form

A dictum of my father ('Über das Passivum' [On the passive]. *Abhandl. d. K. Sächs. Ges. d. Wiss.* VIII, S. 452-453) may serve as a good introduction to the following discussion: Language is 'not an expression of what is to be represented, but of what does the representing. In the shape that it shows itself to us, language is not to be grasped objectively, but subjectively. If we want to treat it objectively, simply in terms of its content, as is done in some so-called general grammars, then we get lost in the field of logic. It is, however, not the physical objects or the concepts in themselves that are expressed in language, but the impressions they make on the human mind, the representations the mind makes of them, and the way in which and the point of view from which the mind looks at them.'

So said my father, who – surely with good reason – preferred to describe the issue rather than name it. The concept of *inner linguistic forms* is indeed simultaneously one of the most difficult and most fertile in our science. Alongside Wilhelm von Humboldt, whom we have to thank for its introduction,

344 *près* eigentlich = gedrängt, ist Präposition geworden; *daal* = Thal, im Niederdeutschen Adverb: hinunter. Indem im Siamesischen *k'ōni* Sache, im Mafoor *ro* = *roi*, Sache, zu Zeichen des Genitivs wurden, verband der Sprachgeist mit ||327|| ihnen den logischen Begriff der Zugehörigkeit. Wüssten wir nicht, welche etymologische Bewandtniss es mit dem niederdeutschen und holländischen „achter" = hinter hat, so könnte uns das Hochdeutsche recht garstig missleiten. Unsere Präposition „mit" würden wir im guten Glauben zu „Mittel, vermittelst" ziehen, wenn uns nicht die Sprachgeschichte belehrte, dass diesmal das *t* nicht aus *dh*, sondern nur aus *t* herzuleiten ist.

||319||
§ .2
B. Die Form.
Nach dem Gesagten kann ich keine Sprache für gänzlich formlos halten. Vielmehr muss ich einer jeden Beides zusprechen, die äussere Form und die innere. Es fragt sich nur: was wird in einer Sprache geformt, und durch welche Mittel geschieht die Formung? Jenes ist die Frage nach der inneren, dieses die Frage nach der äusseren Form.

§ .3.
1. Die innere Sprachform.
Ein Ausspruch meines Vaters (Über das Passivum. Abhandl. d. K. Sächs. Ges. d. Wiss. VIII, S. 452–453) möge zur Einleitung in das Folgende dienen: Die Sprache ist „nicht Ausdruck des Darzustellenden, sondern des Darstellenden, sie ist in der Gestalt, in welcher sie sich uns zeigt, nicht objektiv, sondern subjectiv zu fassen. Wollen wir sie objektiv, ihrem blossen Inhalt nach, betrachten, wie dies in manchen sogenannten allgemeinen Grammatiken geschehen ist, so verlieren wir uns auf das Gebiet der Logik; aber nicht die Gegenstände oder Begriffe an sich, sondern die Eindrücke, welche sie auf den menschlichen Geist machen, die Vorstellungen, welche sich derselbe von ihnen macht, die Art und Weise, wie, und die Gesichtspunkte, unter denen er sie betrachtet, kommen in der Sprache zum Ausdruck".

Soweit mein Vater, der, wohl mit gutem Grunde, die Sache lieber beschreibt als benennt. In der That ist der Begriff der inneren Sprachformen in unserer Wissenschaft zugleich einer der schwierigsten und der fruchtbarsten. Nächst Wilhelm von Humboldt, dem wir seine Einführung verdanken,

Der Begriff der inneren Sprachform ist
1889 and 1891

345 two other men have served to further develop and exploit the concept: August Friedrich Pott and H. Steinthal.

Humboldt's view of language was directed first and foremost, and almost exclusively, to the structure of languages, hence to their grammatical side. Here his initial questions were: What is the relation between the phonetic expression and the content of thoughts? What are the relative values of the various means of expression? His essay 'Ueber die Entstehung der grammatischen Formen und ihren Einfluss auf die Ideenentwicklung' (On the origin of grammatical forms and their influence on the development of ideas) (1997 [1822]) is dedicated to these questions. Here we find views like the following expressed:

(p. 404 [26]) 'It does not follow simply because all grammatical relationships can be indicated in almost every language that every language possesses *grammatical forms* in the sense that they are known in highly developed languages.' He makes similar comments on p. 421 [42-43].

(p. 407 [29]): 'For the point at which this development (of ideas) can successfully commence is to be found when man is concerned not only with the material purpose of his speech, but also with the formal quality of what he says; and this point cannot be reached without the influence and reaction of the language.'

(p. 408 [29-30]): 'The whole aim of a grammar of a language is the depiction of an intellectual process by means of sounds. The grammatical signs cannot also be words which denote objects, for in that case they are again isolated and require new links. If, in terms of the genuine denotation of grammatical relationships, one excludes both a group of words where the relationship has to be understood and also the mere denotation of objects, one is left with the *modification of the words which denote objects,* and this alone is the true concept of a grammatical form.' Earlier, on pp. 405ff [27-28], he provided examples from American languages, such as Carib *a-veiri-daco* = 'on the day of your being' = 'if you were'; Lule *a-le-ti pan* = 'earth from they make' = 'made from earth'; Tupi *caru* = 'to eat' and 'food'; *che caru ai-pota* = 'my food I want', or *ai-caru-pota* = 'I eat want' = 'I want to eat'; Nahuatl: *ni-nequia* = 'I wanted', *tlaçotlaz* = 'I shall love': *ni-tlaçotlaz nequia* = 'I, I shall love, wanted', or *ni-c-nequia tlaçotlaz* = 'I *that* wanted, (namely) I shall love.' He defines:

(p. 411 [33]): 'A grammatical form is that part of a language which is a characteristic denotation of a grammatical relationship, such that it always occurs in the same case.' And so he lists, pp. 417-418 [39], indicators of languages whose grammatical forms are of a less formal nature than the inflecting languages:

a) The formal elements are separable or can change their position, phonetically invariable;

345 haben sich um seine Entwickelung und Ausbeutung zumal zwei Männer verdient gemacht: AUGUST FRIEDRICH POTT und H. STEINTHAL.

[Heinrich 1889 and 1891]

HUMBOLDT's Sprachbetrachtung richtete sich in erster Reihe und fast ausschliesslich auf den Bau der Sprachen, somit auf ihre grammatische Seite, und hier fragte er wieder zunächst: {189} Wie verhält sich der lautliche Ausdruck zum gedanklichen Inhalte, wie verhalten sich die Ausdrucksmittel untereinander in ||328|| Rücksicht auf ihren Werth? Seine Abhandlung „Über die Entstehung der grammatischen Formen und ihren Einfluss auf die Ideenentwickelung" (1822) ist diesen Fragen gewidmet. Hier finden wir Aussprüche wie diese:

[wie 1889] [1889 and 1901]

(S. 404): „Darum, dass sich mit den Bezeichnungen fast jeder Sprache alle grammatischen Verhältnisse andeuten lassen, besitzt noch nicht auch jede g r a m m a t i s c h e F o r m e n in demjenigen Sinne, in dem sie die hochgebildeten Sprachen kennen." Ähnlich S. 421.

(S. 407): „Der Punkt, auf dem diese (die Ideenentwickelung) besser zu gelingen beginnt, ist der, wo dem Menschen, ausser dem materiellen Endzwecke der Rede, ihre formale Beschaffenheit nicht länger gleichgültig bleibt, und dieser Punkt kann nicht ohne die Ein- und Rückwirkung der Sprache erreicht werden."

[(S. 407:) 1889] [werden". 1889 werden. 1901]

(S. 408): „In der Darstellung der Verstandeshandlung durch den Laut liegt das ganze g r a m m a t i s c h e S t r e b e n der Sprache. Die grammatischen Zeichen können aber nicht auch Sachen bezeichnende Wörter sein; denn sonst stehen wieder diese isolirt da und fordern neue Verknüpfungen. Werden nun von der echten Bezeichnung grammatischer Verhältnisse die beiden Mittel: Wortstellung mit hinzugedachtem Verhältniss, und Sachbezeichnung ausgeschlossen, so bleibt zu derselben nichts als M o d i f i c a t i o n d e r S a c h e n b e z e i c h n e n d e n W ö r t e r, und dies allein ist der wahre Begriff einer grammatischen Form." – Zuvor, S. 405 flg., hatte er Beispiele aus amerikanischen Sprachen angeführt, wie Caraibisch: *a-veiri-daco* = am Tage Deines Seins |320| = wenn Du wärest; Lule: *a-le-ti pan* = Erde-aus sie machen = aus Erde gemacht; Tupi: *caru* = essen und Speise; *che caru ai-pota* = mein Essen ich-will, oder *ai-caru-pota* = ich essen will = ich will essen; Nahuatl: *ni-nequia* = ich wollte, *tlaçotlaz* = ich werde lieben: *ni-tlaçotlaz nequia* = ich, ich werde lieben, wollte, oder *ni-c-nequia tlaçotlaz* = ich das wollte (nämlich:) ich werde lieben. – Er definirt

[(S. 408:) 1889] [grammatische Streben 1889 and 1891] [Verknüpfungen." 1889 and 1891] [und = Speise; 1889 and 1891]

(S. 411): „Was in einer Sprache ein grammatisches Verhältniss charakteristisch (so, dass es im gleichen Falle immer wiederkehrt) bezeichnet, ist für sie grammatische Form." – Und so stellt er, S. 417 bis 418, Kennzeichen der Sprachen auf, deren grammatische Formen nicht so formaler Natur sind, wie die flectirenden:

a) Die Formenelemente sind trennbar oder verschiebbar, lautlich unveränderlich;

[(S. 411:) 1889] [Form". 1889] [417–418, 1889] [Form 1901] [die 1889]

346

b) They appear as independent units elsewhere in the language or serve multiple grammatical purposes;

c) The uninflected words have no sign of part of speech;

d) The same grammatical relationships can be implied through phonetic forms, or through mere juxtaposition, with the connection being understood.

At another point in the same essay, pp. 418-419 [40], he says: 'As long as the indications of the grammatical relationships are seen to consist of individual and more or less separable elements, it can be said that the speaker constructs the forms himself in everything he says, rather than using forms which are already available. As a result, a much greater variety of such forms will arise, for it is the natural propensity of the human mind to strive for completeness, and any relationship, however rarely it occurs, becomes a grammatical form as does every other relationship. When, on the other hand, a form is seen in a stricter sense and has been developed through usage, and when, furthermore, normal speech does not involve the need for new formations, there will only be forms for what frequently has to be indicated. What occurs less frequently must be paraphrased and indicated by separate words. Two other circumstances play a part in this process: firstly, the uncultured man will present any particular thing in all its particularities, not simply in those relevant to his aim on a given occasion; secondly, certain nations have the custom of contracting whole sentences into what appear to be forms, incorporating within the verb, for example, the object governed by the verb, especially if it is a pronoun. As a result, it is precisely those languages essentially lacking the *genuine concept of form* which possess an astonishing plethora of what appear to be strictly analogous forms which together produce completeness.'

(p. 422 [43]): 'What is decisive in an investigation of the development and influence of grammatical formality is the ability to distinguish correctly between the denotation of objects and relationships, of things and forms. Speech, which is material and the consequence of an actual need, is directly concerned with the denotation of things; thought, which is ideal, is concerned with form. A predominant capacity for thought therefore gives a language formality, and a predominant formality in a language heightens the capacity for thought.' He now proposes four stages of development. *Grammatical denotation* is effected namely:

a) through fixed expressions, phrases, sentences;

b) through fixed word groups and words that vacillate between denotative and formal meanings;

346

b) Sie sind auch selbständig vorhanden oder dienen zweierlei grammatischen Zwecken; sie 1889

c) Die noch unflectirten Wörter tragen nicht Zeichen des Redetheils; die 1889

d) Dieselben grammatischen Verhältnisse werden bald durch lautliche Formen, bald durch blosses Nebeneinanderstellen, mit Hinzudenken der Verknüpfung, angedeutet. dieselben 1889 angedeutet". 1889

{190}

An einer anderen Stelle derselben Abhandlung, S. 418–419, sagt er: „Solange ||329|| die Bezeichnungen der grammatischen Verhältnisse, als aus einzelnen mehr oder weniger trennbaren Elementen bestehend angesehen werden, kann man sagen, dass der Redende mehr die Formen in jedem Augenblick selbst bildet, als sich der vorhandenen bedient. Daraus pflegt eine bei Weitem grössere Vielfachheit dieser Formen zu entstehen. Wo dagegen die Form in einem engeren Sinne genommen und durch den Gebrauch gebildet wird, nun aber fernerhin das gewöhnliche Reden nicht in neuem Bilden besteht, da giebt es Formen nur für das häufig zu Bezeichnende, und das seltener Vorkommende wird umschrieben, und durch selbständige Wörter bezeichnet. Zu diesem Verfahren gesellen sich noch die beiden andern Umstände, dass der noch uncultivirte Mensch gern jedes Besondere in allen seinen Besonderheiten, nicht bloss in den, zu dem jedesmaligen Zweck nothwendigen darstellt und dass gewisse Nationen die Sitte haben, ganze Sätze in angebliche Formen zusammenzuziehen, z. B. den vom Verbum regierten Gegenstand, vorzüglich wenn er ein Pronomen, mitten in den Schooss des Verbums aufzunehmen. Hieraus entsteht, dass gerade die Sprachen, denen es an dem wahren Begriff der Form wesentlich gebricht, doch eine bewunderungswürdige Menge in strenger Analogie, zusammen Vollständigkeit bildender, angeblicher Formen besitzt.

(S. 422): „Dasjenige, worauf Alles bei der Untersuchung des Entstehens und des Einflusses grammatischer Formalität hinausläuft, ist richtiges Unterscheiden zwischen der Bezeichnung der Gegenstände und Verhältnisse, der Sachen und Formen. Das Sprechen, als materiell, und Folge realen Bedürfnisses, geht unmittelbar nur auf Bezeichnen von Sachen; das Denken, als ideell, immer auf Form. Überwiegendes Denkvermögen verleiht daher einer Sprache Formalität, und überwiegende Formalität in ihr erhöhet das Denkvermögen". Nun stellt er vier Entwickelungsstufen auf. Es geschieht nämlich die grammatische Bezeichnung: Bezeichung. 1901

a) durch Redensarten, Phrasen, Sätze;

b) durch feste Wortstellungen und zwischen Sach- und Formbedeutung schwankende Wörter;

347

c) agglutinatively, through apparent forms, affixes;

d) (p. 423 [44]): 'Formality finally prevails. The word becomes one, modified in its grammatical relationships only by the alternation of the inflectional element; every word belongs to a particular part of speech and has not only lexical, but also grammatical individuality; the words denoting form no longer have any distracting additional meaning but serve purely to express relationships. At the ultimate stage, therefore, grammatical denotation is effected by genuine forms, by *inflection, and by purely grammatical words*.' (See also p. 425 [46-47])

'The *essence of form* lies in its unity and in the primacy of the word to which it belongs over the subsidiary element which is added. This is facilitated by the loss of meaning of the elements and by the erosion of the sounds through long use. However, the development of a language can never fully be explained by the mechanical effect of lifeless forces, and one must never lose sight of the influence of the strength and individuality of the capacity for thought.'

(p. 424 [45]): ... 'and it remains irrefutably certain that, whatever happens to a language, it can never acquire an exemplary grammatical structure if it does not have the good fortune to be spoken at least once by a nation characterized by intelligence or profound thought. Nothing else can save it from the *inadequacies of lazily constructed forms* which make an unclear impression on the mind.'

(pp. 426-427 [47-48]): 'In terms of the effect of a language on the mind, genuine grammatical form creates the impression of form and encourages formal development even when the attention is not deliberately focused upon it. As it contains the pure expression of a relationship without material elements which might distract the understanding, the understanding itself recognizes that the original concept of the word has been altered and must therefore make use of the form. In the case of a form which is *not genuine*, this cannot happen, as the understanding does not see the concept of the relationship clearly enough and is distracted by subsidiary concepts. In both cases, this is what happens in the most normal of speech and throughout all classes of the nation. Where the effect of the language is positive, a general clarity and precision is developed in the concepts as well as a general propensity for an easier understanding of the purely formal. The nature of the mind is also such that this propensity, once present, always develops further, for, if a language presents the intellect with impure and defective grammatical forms, the longer such an influence lasts, the more difficult it will be to escape from this obscuring of the purely formal aspect.'

347
c) agglutinirend, durch Analoga von Formen, Affixe;

d) (S. 423): „Die Formalität dringt endlich durch, das Wort ist Eins, nur durch umgeänderten Beugungslaut in seinen grammatischen Beziehungen modificirt; jedes gehört zu einem bestimmten Redetheil, und hat nicht bloss lexikalische, sondern auch grammatische Individualität; die formbezeichnenden Wörter haben keine störende Nebenbedeutung mehr, sondern sind reine Ausdrücke von Verhältnissen. So geschieht auf der höchsten Stufe die grammatische Bezeichnung durch wahre Formen, durch Beugung, und rein grammatische Wörter." (Ähnlich S.425.)

„Das Wesen der Form besteht in der Einheit und der vorwaltenden Herrschaft des Worts, dem sie angehört, über die ihm beigegebenen Nebenlaute. beigegeben 1901
||330|| Dies wird wohl erleichtert durch verloren gehende Bedeutung der Elemente, und Abschleifung der Laute in langem Gebrauch. Allein das Entstehen der Sprache ist nie ganz durch so mechanische Wirkung todter Kräfte erklärbar, und man muss niemals darin die Einwirkung, der Stärke und Individualität der Denkkraft aus den Augen setzen."

(S. 424): [...] „und so bleibt es unumstösslich gewiss, dass, welche Schicksale auch eine Sprache haben möge, sie nie zu einem vorzüglichen grammatischen Bau gelangt, wenn sie nicht das Glück erfährt, wenigstens einmal von einer geistreichen, oder tiefdenkenden Nation gesprochen zu werden. Nichts kann sie sonst aus der Halbheit träge zusammengefügter, die Denkkraft nirgends mit Schärfe ansprechender Formen retten."

(S. 426–427): „In der Rückwirkung der Sprache auf den Geist macht die echt grammatische Form, auch wo die Aufmerksamkeit nicht absichtlich auf sie gerichtet ist, den Eindruck einer Form, und bringt formale Bildung hervor. Denn da sie den Ausdruck des Verhältnisses rein, und sonst nichts Stoffartiges enthält, worauf der Verstand abschweifen könnte, dieser aber den ursprünglichen Wortbegriff darin verändert erblickt, so muss er die Form selbst ergreifen. Bei der unechten Form kann er dies nicht, da er den Verhältnissbegriff nicht bestimmt genug in ihr erblickt, und noch durch Nebenbegriffe zerstreut wird. Dies geschieht in beiden Fällen bei dem gewöhnlichen Sprechen, durch alle Classen der Nation, und wo die Einwirkung der Sprache günstig ist, geht allgemeine Deutlichkeit und Bestimmtheit der Begriffe, und allgemeine Anlage, auch das rein Formale leichter zu begreifen, hervor. Es liegt auch in der Natur des Geistes, dass diese Anlage, einmal vorhanden, sich immer ausbildet, da, wenn eine Sprache dem Verstande die grammatischen Formen unrein und mangelhaft darbietet, je länger diese Einwirkung dauert, je schwerer aus dieser Verdunkelung der rein formalen Ansicht herauszukommen ist."

348
Further in this vein, we read in a manuscript (in Steinthal [1883], *Die sprachphilos. Werke W. v. H.* [The works on philosophy of language of Wilhelm von Humboldt], p. 616): '*Grammatical form* must exist completely for and through language. Understanding must be led by it and at its hand alone; the view presented in the sequence of speech must not be created only from the combination of the thoughts expressed. Nothing foreign, derived from external reality, nothing that is not exclusively calculated for grammatical purposes, should be allowed to penetrate into grammatical form.'

Further (ibid., p. 348): 'Language, as sensualization of thought outside the human mind, a world of individual words, concepts stamped by sounds, stands opposed to things. In the same way, language creates a certain implicit indication of the connection of thoughts that comes only from it and belongs only to it, and this indication, conceived in the unity of infinite diversity, is the *form of grammar.*'

Everywhere in these passages we see that the outer form, morphology, is in the foreground, but that the inner need for formation is sought behind it. Despite this, Humboldt still speaks more of the expression of this need than of the thought content of expressions. He also speaks, however, in an unpublished essay, of the realization that 'we have to distinguish the representation of the thing itself from those representations *that the word gives to the thing because of the way the word is constructed and comes into being*' (cf. Steinthal [1883], *Die sprachphilos. Werke W. v. H.* p. 341).

From his essay *Ueber den Dualis* (On the Dual Form) (1997 [1827]), I take the following two quotations:

(p. 20 [129]): 'Language is not, however, merely a means of communication but bears the mark of the intellect and world view of those who speak; sociability is the most indispensable aid to its development but is far from being the only goal it is seeking to achieve; rather, its ultimate goal is to be found in the individual, insofar as the individual can be separated from mankind as a whole. Whatever aspects of the external world and the inner mind may find their way into the grammatical structure of languages can be adapted, used and developed according to the measure of vitality and *purity of the sense of language* and the *particular nature of its attitudes*.'

'Here, however, a remarkable variety is revealed. Certain elements of language suggest that its formation is derived largely from a *sensual world view*, or from *inside thoughts*, where the world view was already influenced by the working of the mind.'

(p. 26 [134]): 'All languages indicating only natural *sexes* and not recognizing the metaphorical indication of gender prove

348
Hierzu aus einer Handschrift (bei STEINTHAL S. 616): „Die grammatische Form muss ganz für und durch die Sprache bestehen, das Verständniss muss bloss durch sie und an ihrer Hand geleitet, die Einsicht in die Redefügung nicht erst aus dem Zusammenhang der Gedanken geschöpft werden, es muss sich überhaupt nichts Fremdes aus der Wirklichkeit Entnommenes, nicht ausschliesslich auf den grammatischen Zweck Berechnetes in sie eindrängen."

Ferner (ebenda S. 348): „Wie die Sprache als Versinnlichung des Gedankens, ausserhalb des menschlichen Geistes, eine Welt einzelner Wörter, durch Laute gestempelter Begriffe, den Gegenständen gegenüberstellt, ebenso schafft sie eine nur aus ihr entspringende und nur ihr angehörende Andeutung der Gedankenverknüpfungen, und diese Andeutung, in der Einheit der unendlichen Mannigfaltigkeit aufgefasst, ist die Form der Grammatik."

Man sieht, wie hier überall die äussere Form, die Morphologie im Vorder-||331||grunde steht, dahinter aber doch das innere Bedürfniss der Formung gesucht wird. Noch aber redet er mehr von den Äusserungen dieses Bedürf- Aeusserungen 1889
nisses, als von dem gedanklichen Inhalte der Äusserungen. Bald jedoch Aeusserungen. 1889
spricht er in einer ungedruckten Abhandlung die Erkenntniss aus, dass „man die Vorstellung des Gegenstandes selbst von derjenigen unterscheiden muss, welche das Wort seiner Bildung und Entstehung nach von ihm giebt" (vergl. STEINTHAL, Die sprachphilos. Werke W. v. H. S. 341). giebt 1889

Seiner Abhandlung: Über den Dualis (1827) entnehme ich zwei Stellen: die 1889

(S. 20): „Die Sprache ist aber durchaus kein blosses Verständigungsmittel, sondern der Abdruck des Geistes und der Weltansicht des Redenden; die Geselligkeit ist das unentbehrliche Hülfsmittel zu ihrer Entfaltung, aber bei Weitem nicht der einzige Zweck, auf den sie hin arbeitet, der vielmehr seinen Endpunkt doch in dem Einzelnen findet, insofern der Einzelne von der Menschheit getrennt werden kann. Was also aus der Aussenwelt und dem Innern des Geistes in den grammatischen Bau der Sprachen übergehen mag, kann darin aufgenommen, angewendet und ausgebildet werden, und wird es wirklich, nach Massgabe der Lebendigkeit und Feinheit des Sprachsinnes und der Eigenthümlichkeit seiner Ansicht."

„Hier aber zeigt sich sogleich eine auffallende Verschiedenheit. Die Sprache trägt Spuren an sich, dass bei ihrer Bildung vorzugsweise aus der sinnlichen Weltanschauung geschöpft worden ist, oder aus dem Innern der Gedanken, wo jene Weltanschauung schon durch die Arbeit des Geistes gegangen war."

(S. 26): „Alle Sprachen, die nur die natürlichen Geschlechter bezeichnen, und kein metaphorisch bezeichnetes Genus anerkennen, beweisen,

349 that they were not infused with *pure linguistic form*, that they did not comprehend the subtle and delicate meaning language lends to the objects of reality, because either originally or at the time when they no longer took account of this distinction or became confused about it they merged masculine and neuter together.'

But (*Kawi-Sprache* [Humboldt, 1838], vol. II, p. 221): 'It is a vain endeavour to try and find, even in a language that is considered to be so "original", truly *unformed* material. The concept of language stands or falls with that of form, since language is entirely form and nothing but form. Grammar does not advance from the root sound, but with the root sound. Every root sound, because it is a sound of language, is mixed with something subjective, something ruled by change. This is the case even for true *root sounds*. What should we say about these sounds that we call root sounds when we only know words of languages, which have rolled on the tongues of the most different peoples over millennia? In our actual understanding they are only artificial entities that, on the way to abstraction and identification, perhaps lose precisely their essential identifying features.'

(Ibid., p. 286, of the Malay languages): 'In this joining [of prefixes, suffixes and infixes] we see a successful effort to bind the word and its appendages into a whole. From here there emerge true *grammatical forms* in this language family, since the appendages are bound with variations in sound and movement of the accent; that is, with visible signs of the striving for *unity in the word*.' – But then:

(pp. 286-287, on the Malay languages): '*Formation* is actually not required in order to capture the constantly moving aspect of speech, the always changing relations of words to one another in terms of subject and object, and the integration of both in the unity of the sentence [...] Since they devote so little care to their speech constructions, they are unable to appreciate the true nature of the *verb* as the soul of the sentence. They treat the verb only materially according to its meaning, go around it as much as they can in expression and leave [...] it very often ambiguous as to what category it should be taken to be, verb or noun? With a higher appreciation of language, we see that this is the main affliction of this family. It is precisely the key element of the speech construction that is left the least definite; precisely at the point where the unity of thought should be symbolically marked in the language by the most inner melting together of sounds, the language does without form, and it is in form alone that symbolic indication can lie.' Cf. p. 325.

349 dass sie entweder ursprünglich, oder in der Epoche, wo sie diesen Unterschied der Wörter nicht mehr beachteten, oder über ihn in Verwirrung gerathend, Masculinum und Neutrum zusammenwarfen, nicht von der *reinen Sprachform* energisch durchdrungen waren, nicht die feine und zarte Deutung verstanden, welche die Sprache den Gegenständen der Wirklichkeit leiht."

Aber (Kawi-Sprache II, 221): „Es ist ein vergebliches Bemühen, auch in einer für noch so ursprünglich gehaltenen Sprache noch wirklich *Ungeformtes* antreffen zu wollen. Der Begriff der Sprache steht und verfliegt mit dem der Form, denn sie ist ganz Form und nichts als Form. Die Grammatik hebt nicht von, sondern mit dem Wurzellaut an, und jedem Wurzellaut ist, weil er Sprachlaut ist, schon Subjectives, mithin der Veränderung Unterworfenes beigemischt. Dies ist selbst bei dem wahren *Wurzellaute* der Fall. Was soll man aber gar von demjenigen sagen, was wir, die wir bloss Wörter der Sprachen kennen, welche schon Jahrtausende hindurch auf der Zunge der verschiedensten Völker gerollt haben, Wurzellaute nennen? Sie sind im eigentlichen Verstande nur ||332|| künstliche Gebilde, die auf dem Wege der Abstraction und Bezeichnung vielleicht gerade das wesentlich Bezeichnende ihrer Individualität verlieren."

(Das. S. 286, von den malaischen Sprachen): „In diesen Zusammenfügungen [Prä-, Sub- und Infigirungen] zeigt sich nun bestimmt ein gelungenes Streben, das Wort und seine Anfügungen zu einem Ganzen zu verbinden. Es entstehen von dieser Seite in dem Sprachstamm wahre *grammatische Formen*. Denn die Anfügungen sind mit Lautveränderungen und Accent-Umstellungen, also mit sichtbaren Zeichen des Strebens nach *Worteinheit* verbunden." – Dann aber (S. 286–287, über die malaischen Sprachen): „Um das der Rede beständig Bewegliche, die immer wechselnden Beziehungen der Wörter aufeinander in Rücksicht auf Subject und Object, und das Zusammenfassen Beider in der Einheit des Satzes zu bezeichnen, wird die *Formung* gar nicht gebraucht ... Da sie der Redefügung so geringe Sorgfalt widmen, so konnten sie nicht dahin gelangen, das *Verbum* in seiner wahren Natur, als die Seele des Satzes zu denken. Sie nehmen dasselbe nur materiell nach seiner Bedeutung, umgehen es, soviel sie können, im Ausdruck und lassen ... es sehr oft zweideutig, in welcher Kategorie, ob als Nomen oder als Verbum? es genommen worden soll. Dies ist bei einer höheren Sprachansicht das hauptsächlichste Gebrechen dieses Stammes. Gerade die Hauptsache in der Redefügung wird am Wenigsten bestimmt, gerade in dem Punkte, wo sich die Gedankeneinheit durch die innigste Lautverschmelzung symbolisch in der Sprache ausprägen sollte, entbehrt sie der Form, in welcher allein symbolische Bezeichnung liegen kann." Vgl. hierzu S. 325.

In 1901

350

(*Über die Verschied. des menschl. Sprachbaues* [On the diversity of human language construction] [Humboldt, 1999 [1836]], p. 43 [50]): 'The *characteristic form of languages* depends on every single one of their smallest elements; however inexplicable it may be in detail, each is in some way determined by that form.'

(Ibid., p. 45 [51-52]): 'In an absolute sense there can be no *unformed matter* within language, since everything in it is directed to a specific goal, the expression of thought, and this work already begins with its first element, the articulated sound, which of course becomes articulate precisely through being formed. The real *matter of language* is, on the one hand, the sound as such, and on the other the totality of sense impressions and spontaneous mental activities which precede the creation of the concept with the aid of language.'

(p. 121 [101]): 'There are languages which regularly add the species concept to the names of living creatures, and some among them where the designation of this species concept has become a true suffix, recognizable only by analysis. These cases, indeed, are still always connected with our earlier statement, in that a two-fold principle, an objective one of designation and a subjective one of logical classification, is also visible in them. But on the other hand they are utterly at variance with it, in that here it is no longer *forms of thought and speech* which enter into the designation but merely different *classes of actual objects*. Words so formed are now becoming quite similar to those in which two elements constitute a compound concept. What corresponds, on the other hand, in the inner shaping, to the concept of *inflection*, differs precisely in that the duality we started from in defining this concept is made up, not of two elements at all, but only one, transposed into a specific category. The characteristic feature here is precisely that this duality, when analysed, is not of the same kind, but a different one, and pertains to different spheres. Only so can *purely organized languages* achieve the deep and firm combination of spontaneity and receptiveness, from which there subsequently proceeds in them an infinity of thought-couplings, all bearing the mark of a *true* form, which satisfies, purely and fully, the demands of language as such.'

(p. 130 [106]): 'Between the *want of all indication of the categories of words*, as we find in Chinese, and *true inflection*, there can be no third possibility compatible with pure organization of languages. The only conceivable intermediate is compounding used as inflection; that is, intended but not perfected inflection –

350
(Über die Verschied. des menschl. Sprachbaues S. 43): „Die charakteristische Form der Sprachen hängt an jedem einzelnen ihrer kleinsten Elemente; jedes wird durch sie, wie unmerklich es im Einzelnen sei, auf irgend eine Weise bestimmt."

(Das. S. 45): „Absolut betrachtet, kann es innerhalb der Sprache keinen ungeformten Stoff geben, da Alles in ihr auf einen bestimmten Zweck, den Gedankenausdruck, gerichtet ist, und diese Arbeit schon bei ihrem ersten Element, dem articulirten Laute, beginnt, der ja eben durch Formung zum articulirten wird. Der wirkliche Stoff der Sprache ist auf der einen Seite der Laut überhaupt, auf der andern die Gesammtheit der sinnlichen Eindrücke und selbstthätigen Geistesbewegungen, welche der Bildung des Begriffes mit Hülfe der Sprache vorangehen."

(S. 121): „Es giebt Sprachen, welche den Benennungen der lebendigen Geschöpfe regelmässig den Gattungsbegriff hinzufügen, und unter diesen solche, wo die Bezeichnung dieses Gattungsbegriffes zum wirklichen, nur durch Zergliederung erkennbaren Suffixe geworden ist. Diese Fälle insofern auch in ihnen ein doppeltes Prinzip, ein objectives der Bezeichnung, und ein subjectives logischer Eintheilung, sichtbar wird ... auf der anderen Seite dadurch ... dass hier ||333|| nicht mehr Formen des Denkens und der Rede, sondern nur verschiedene Classen wirklicher Gegenstände in die Bezeichnung eingehen. So gebildete Wörter werden nun denjenigen ganz ähnlich, in welchen zwei Elemente einen zusammengesetzten Begriff bilden. Was dagegen in der innerlichen Gestaltung dem Begriffe der Flexion entspricht, unterscheidet sich gerade dadurch, dass gar nicht zwei Elemente, sondern nur Eines, in eine bestimmte Kategorie versetztes, das Doppelte ausmacht, von dem wir bei der Bestimmung dieses Begriffes ausgingen. Dass dies Doppelte, wenn man es auseinanderlegt, nicht gleicher, sondern verschiedener Natur ist und verschiedenen Sphären angehört, bildet gerade hier das charakteristische Merkmal. Nur dadurch können rein organisirte Sprachen die tiefe und feste Verbindung der Selbständigkeit und Empfänglichkeit erreichen, aus welcher hernach in ihnen eine Unendlichkeit von Gedankenverbindungen hervorgeht, welche alle das Gepräge echter, die Forderungen der Sprache überhaupt rein und vollbefriedigender Form an sich tragen."

Bezeichung, 1901

(S. 130): „Zwischen dem Mangel aller Andeutung der Kategorien der Wörter, wie er sich im Chinesischen zeigt, und der wahren Flexion kann es kein mit reiner Organisation der Sprachen verträgliches Drittes geben. Das einzige dazwischen Denkbare ist als Beugung gebrauchte Zusammensetzung, also beabsichtigte, aber nicht zur Vollkommenheit gediehene Flexion,

351 a more or less mechanical *appending*, not a truly organic *accretion*.'

(p. 180 [140]): '*Grammatical formation* arises from the *laws of thinking* in language, and rests on the *congruence of sound forms* with these laws. Such a congruence must in some way be present in every language; the difference lies only in degree, and the blame for defective development may attach to an insufficiently plain emergence of these laws in the soul, or to an inadequate malleability of the sound system. But the deficiency on the one point always reacts back at once upon the other. The perfecting of language demands that *every word be stamped as a specific part of speech*, and carry within it those properties that a philosophical analysis of language perceives in it. It thus itself presupposes inflection.' Cf. pp. 187-188 [145-146].

(p. 187 [144]): '[D]istinciton [...] between languages which have evolved forcefully and consistently from pure principle in regulated form, and those which cannot boast of this advantage [...]. The latter have a *deviant form*, in which two things come together: a weakness of the linguistic sense that at first is always surely inherent in man, and a one-sided deformation arising from the fact that to a sound form not emanating necessarily from the language, others, attracted by that sound form, become annexed.'

(p. 255 [187]): 'If the line between those languages (the inflecting), which proceed from the *true concept* of grammatical forms, and those (the agglutinating) that are imperfectly struggling towards it, is drawn on the double principle of either creating from the form a sign quite unintelligible when taken on its own, or merely fastening two significant concepts tightly together [...]'

(p. 260 [191]): 'For I have already often remarked in these pages that where the *linguistic form* is *clearly and vividly* present in the mind, it enters into the outer process of development which otherwise guides external language formation, makes its influence felt and, in the mere elaboration of lines once started, does not permit makeshifts, as it were, to be produced instead of pure forms.'

Finally, § 11 of the same essay bears the title 'inner linguistic form', and here he says (p. 96 [84]): 'For language never represents the objects, but always the concepts that the mind has spontaneously formed from them in producing language; and this is the forming under discussion here, insofar as it must be seen as quite internal, preceding, as it were, the sense of articulation.'

351 mehr oder minder mechanische Anfügung, nicht rein organische Anbildung."

(S. 180): „Die grammatische Formung entspringt aus den Gesetzen des Denkens durch Sprache, und beruht auf der Congruenz der Lautformen mit denselben. Eine solche Congruenz muss auf irgend eine Weise in jeder Sprache vorhanden sein; der Unterschied liegt nur in den Graden, und die Schuld mangelnder Vollendung kann das nicht gehörig deutliche Hervorspringen jener Gesetze in der Seele oder die nicht ausreichende Geschmeidigkeit des Lautsystems treffen. Der Mangel an einem Punkte wirkt aber immer zugleich auf den andern zurück. Die Vollendung der Sprachen fordert, dass jedes Wort als ein bestimmter Redetheil gestempelt sei und diejenigen Beschaffenheiten an sich trage, welche die philosophische Zergliederung der Sprache an ihm erkennt. Sie setzt dadurch selbst die Flexion voraus." Vergl. hierzu S. 187–188.

(S. 187): „Unterschied ... zwischen Sprachen, die sich aus reinem Prinzipe in gesetzmässiger Freiheit kräftig entwickelt haben, und zwischen solchen, die sich dieses Vorzuges nicht rühmen können ... Die Letzteren haben eine abweichende Form, in welcher zwei Dinge zusammentreffen, Mangel an Stärke des ursprünglich immer im Menschen rein liegenden Sprachsinnes, und eine einseitige, aus dem Umstande entspringende Verbildung, dass an eine nicht aus der Sprache nothwendig herfliessende Lautform andere, durch sie an sich gerissene, angeschlossen werden."

||334||

(S. 255): Wenn man als die Scheidewand der von dem wahren Begriff der grammatischen Formen ausgehenden (flectirenden) und der unvollkommen zu ihnen hinstrebenden (agglutinirenden) Sprachen den zwiefachen Grundsatz aufstellt: aus der Form ein einzeln ganz unverständliches Zeichen zu bilden, oder zwei bedeutsame Begriffe nur eng aneinander zu heften, ..."

(S. 260): „Denn ich habe schon oft in diesen Blättern bemerkt, dass, wo die Sprachform klar und lebendig im Geiste dasteht, sie in die, sonst die äussere Sprachbildung leitende äussere Entwickelung eingreift, sich selbst geltend macht, und nicht zugiebt, dass im blossen Fortspinnen angefangener Fäden, statt der reinen Formen, gleichsam Surrogate derselben gebildet werden."

Endlich ist § .11 seiner Abhandlung „Über die Verschiedenheit des menschlichen Sprachbaues" überschrieben: „Innere Sprachform", und hier sagt er: Denn die Sprache stellt niemals die Gegenstände, sondern immer die durch den Geist in der Spracherzeugung selbstthätig von ihnen gebildeten Begriffe dar; und von dieser Bildung, insofern sie als ganz innerlich, gleichsam dem Articulationssinne vorausgehend angesehen werden muss, ist hier die Rede."

Ueber 1889
Sprachbaues 1889
„denn 1889
derselben [Titel der Abhandlung ist hier nicht wiederholt] 1901

352 – That is what Humboldt had to say, whose views have up to the present day been adopted by almost everyone who takes an interest in such questions. We always find his thoughts repeated, his standard applied, when form or formlessness is attributed to languages, or when their forms receive praise or criticism. Ultimately it is due to him when particularly strict judges deny the existence of verbs in most languages, or even the possession of words.

If etymology is to decide the inner linguistic form, as Humboldt would like, then the inner form will show itself at least as clearly in the creation of words as in the creation of forms, in vocabulary no less clearly than in structure. Here, from the lexical side, Pott approached the problem. What are things called? This question can and must be answered from two perspectives, in that we go either from the name to the thing named or the other way round. In the former case, words are ordered etymologically, and the roots and stems pursued through their derivations and applications. In the latter, the perspective is that of a thesaurus, and the question is: According to what characteristics are things named? In this spirit Humboldt had already considered the name of the elephant in Sanskrit: *hastin*, 'the handed one', *dvipa*, 'the twice drinking one', etc. Pott pursued both approaches, well beyond the boundaries of the Indo-European languages. At once he showed how the same representation can be applied to different objects, and then how the same object is named by different representations – characteristics – connected with it.

Perhaps no one has written more extensively and perspicuously on matter and form in language, and on formlessness of languages, than H. Steinthal. The following quotations from his writings should serve to illustrate his position.

(*Charakteristik der hauptsächlichsten Typen des Sprachbaues* [Characterization of the main types of language construction] [Steinthal, 1860], p. 78): 'Whatever is won through sensation is *matter* in the consciousness [...] In the realm of perception, and that is essentially pre-linguistic consciousness, there is only matter and no form. Form is not perceived, but is rather a pure product of the reflexive activity of the soul, and indeed speaking is the first forming activity, and the first matter on which it operates are perceptions.'

(Ibid., p. 79): 'The *essence of the forming activity* is most generally and still quite indefinitely designated as the intuition of the intuition.'

352 – Soviel von HUMBOLDT, dessen Anschauungen noch bis auf den heutigen Tag fast von Allen angenommen worden sind, die solche Fragen eines Interesses würdigen. Immer findet man seine Gedanken wiederholt, seinen Massstab angelegt, wenn den Sprachen Form oder Formlosigkeit, ihren Formen Lob oder Tadel zugesprochen wird. Im letzten Grunde ist es auch auf ihn zurückzuführen, wenn besonders strenge Richter den meisten Sprachen das Verbum, ja den Besitz von Wörtern aberkennen.

Wenn die Etymologie über die innere Sprachform entscheidet, wie |321| HUMBOLDT will, so wird sich die innere Sprachform in der Wortschöpfung mindestens ebenso klar zeigen, wie in der Formenschöpfung, im Sprachschatze nicht weniger klar, als im Sprachbaue. Hier, von der lexikalischen Seite, griff POTT die Sache an. Wie sind die Dinge benannt? Diese Frage kann und will von zwei Gesichtspunkten aus beantwortet sein, indem der Weg entweder von der Benennung zum Benannten genommen wird, oder umgekehrt. Dort werden die Wörter etymologisch geordnet, die Wurzeln und Stämme durch ihre Ableitungen und Anwendungen hindurch verfolgt. Hier ist der Gesichtspunkt der der Synonymik, und die Frage lautet: Nach welchen Merkmalen werden die Dinge benannt? In diesem Sinne führte schon HUMBOLDT die Namen des Elefanten im Sanskrit an: *hastin*, der Behandete, *dvipa*, der zweimal Trinkende u. s. w. POTT hat beide Wege verfolgt, weit über die Grenzen des indogermanischen Sprachstammes hinaus. Jetzt zeigte er, wie dieselbe Vorstellung auf verschiedene Gegenstände angewandt wird, jetzt wieder, wie derselbe Gegenstand nach verschiedenen mit ihm verbundenen Vorstellungen, – Merkmalen – bezeichnet wird.

Sprachbau. 1889 lexicalischen Seite 1889 and 1891

Anleitungen 1901

Mehr und schärfer hat vielleicht Keiner über Stoff und Form und Form||335||losigkeit der Sprachen geschrieben, als H. {191} STEINTHAL. Zur Kennzeichnung seines Standpunktes mögen folgende Stellen aus seinen Schriften dienen.

Heinrich 1889 and 1891

(Charakteristik der hauptsächlichsten Typen des Sprachbaues S. 78): „Was nur immer durch Wahrnehmung gewonnen wird, das ist im Bewusstsein als Stoff ... Es giebt im Reiche der Wahrnehmung, und das heisst soviel wie im vorsprachlichen Bewusstsein, nur Stoff und keine Form. Form wird nicht wahrgenommen, sondern ist reines Erzeugniss der Selbstthätigkeit der Seele, und zwar ist Sprechen die erste formende Thätigkeit, und der erste Stoff, an dem sich diese versucht, sind die Wahrnehmungen."

78:) 1889

.... 1889 and 1891

Wahrnehmungen". 1889

(Das. S. 79): „Das Wesen der formenden Thätigkeit wird am Allgemeinsten und noch ganz unbestimmt bezeichnet als die Anschauung der Anschauung."

79:) 1889 and 1891

Anschauung". 1889

353
(Ibid., p. 81): 'The products of the intuition of intuitions may be called *representations*.'

(Ibid., p. 87): '[*Inner* word form is] that characteristic, that relation, through which the subjectivity of peoples wanted to make the intuition clear to itself and reproduce it.'

(*Sprachphilos. Werke W. v. Humboldt's* [Steinthal, 1883], p. 341): 'In this way the representation or *inner form of the word* is subjective; the conception of the object that lies in it is determined by sensuality, imagination, enduring or momentary inner excitation of the mind.'

(*Charakt.*, p. 89): [*Grammatical formation* is] 'the linguistic foundation and designation of a definite relation between individual representations and words.'

(Ibid.) '[...] just as relational forms beyond individual representations emerge in consciousness, spreading over the individual representations, wrapping around them, then, to the same extent and in a corresponding way, words also sprout forth *phonetically formed* unconsciously and unwilled.'

(Ibid., p. 84): 'Just as we perceive external entities through the senses, so in general is the *inner language form* an intuition or apperception of every possible content that the mind possesses, a means to make this content present to itself, to fix it and reproduce it, indeed even a means to acquire or create new content.'

(Ibid., p. 92): He describes the *inner language form* as 'an inner intuition of the inner content, an apperception of intuitions and concepts.'

(*Mande-Negersprachen* [Steinthal, 1867], p. VII): '[...] Distinction of the *inner language form* – that is the grammatical categories – from the logical forms of intuitions and concepts.'

(Ibid., p. VIII): 'When researching the inner language form, we must start with *comparative etymology* and must never assume the existence of an inner form where there is no corresponding phonetic form, and we must see no other categories than those that the etymology indicates; since Phonesis is the only solid ground, the secure hold of the language researcher, which he may not give up unpunished.' Similarly

353
(Das. S. 81): „Die Producte der Anschauung von Anschauungen mögen *Vorstellungen* heissen.“

(Das. S. 87): „[*Innere Wortform* ist] dasjenige Merkmal, diejenige Beziehung, wodurch die Subjectivität der Völker die Anschauung sich vergegenwärtigen und reproduciren wollte.“

|322|
(Sprachw. Werke W. v. HUMBOLDT's S. 341): „So ist die Vorstellung oder *innere Form des Wortes*, subjectiv; die Auffassung des Objectes, welche in ihr liegt, wird bestimmt von Sinnlichkeit, Phantasie, dauernder oder augenblicklicher Gemüthserregung.“

(Charakt. S. 89): [*Grammatische Formung* ist] „die sprachliche Gründung und Bezeichnung einer bestimmten Beziehung zwischen den einzelnen Vorstellungen und Wörtern.“

(Das.) „… dass, wenn und insoweit und wie im Bewusstsein Beziehungsformen ausser den einzelnen Vorstellungen und sich über sie verbreitend, sie umschlingend, auftauchen, dann auch ebensoweit und in entsprechender Weise, unbewusst und ungewollt, die Wörter auch *lautlich geformt* hervorbrechen werden.“

(Das. S. 84): „Wie wir also durch die Sinne die äusseren Gegenstände wahrnehmen, percipiren: so ist im Allgemeinen die *innere Sprachform* eine Anschauung oder Apperception jedes möglichen Inhaltes, den der Geist besitzt, ein Mittel, diesen Inhalt sich zu vergegenwärtigen, festzuhalten und zu reproduciren, ja sogar ein Mittel, neuen Inhalt zu erwerben oder geradezu zu schaffen“.

{192}
Das. S. 92 bezeichnet er die *innere Sprachform* als „eine innere Anschauung des inneren Inhaltes, eine Apperception von Anschauungen und Begriffen.“

(Mande-Negersprachen S. VII): „… Unterscheidung der *inneren Sprachform*, d. h. der grammatischen Kategorien, von den logischen Formen der Anschauungen und Begriffe.“

(Das. S. VIII): Wir müssen bei der Erforschung der inneren Sprachform ||336|| überall von der *vergleichenden Etymologie* ausgehen und dürfen nie innere Sprachform da annehmen, wo ihr keine phonetische Form entspricht, und dürfen auch keine andere Kategorie sehen, als worauf die Etymologie hinweist; denn die Phonesis ist der einzige feste Boden, der sichere Haltepunkt des Sprachforschers, den er ungestraft nicht aufgeben darf.“ – Ähnlich

81:) 1889 and 1891
Produkte 1889
heissen“. 1889
87:) 1889 and 1891
[Innere 1889
„dasjenige 1889
Beziehung 1889
Subjektivität 1889
(Sprachwiss. Werke W. v. Humboldts 1889
341:) 1889 and 1891
subjektiv; 1889
Objektes, 1889
Gemüthserregung“. 1889
89:) 1889 and 1891
„. . . . 1889
werden“. 1889
84:) 1889 and 1891
schaffen.“ 1889
VII:) 1889 and 1891
Begriffe“. 1889
VIII:) 1889 and 1891
ausgehen, 1889
darf“ 1889

354

(*Abriss der Sprachw.* I [Outline of linguistics] [Steinthal, 1881], p. 428): '[...] the etymon, the characterizing inner linguistic form,' and

(Ibid., p. 429): '[...] the inner language form, the etymon.' But opposed to this

(Ibid., p. 432): '[...] the inner language form, which however does not lie in the etymology.' And similarly

(*Charakteristik* [Steinthal, 1860], p. 234): 'But just because the root and suffix in Egyptian are not tightly bound to one another does not mean that it is not a *form language*, since form is based not on the phonetic connections but on the inner sense. The Egyptian, however, thought formally and for this reason his language is formal.' But contrary to this, once again about the Finnish languages:

(Ibid., p. 203): 'People did not have the need to allow only formed elements into their speech, to allow only elements with a certain property; and for this reason a suffix does not count as form, and a stem with a suffix is not a stem that has become a word-form, it has not ceased to be a stem, that is to be *formless*.'

(Ibid., pp. 318ff, he addresses this issue even more comprehensively): 'Quite apart from the many cases in which it is uncertain, *etymology* is a doubtful witness for or against a formal nature, since an originally material word can be used purely formally, as we see in our languages. The essential point in which a material or formal kind of representation shows itself lies in the treatment of words, in their *construction* [...] But the formal nature of language always lies in the construction, that is in pure activity, synthesis, in itself, in the expression of the predication, attribution, objectification, as the mental function of linguistic representation. Only at this point, where the mind is externalized in the subtlest way, where it is revealed as an activity, freed from the object of the activity, pure and not materialized, grasping and penetrating the object: only here can we assess the formal principle of language structure, and here most securely. Since if here the revelation of the mind is subtle, it is also powerful and effective. If we stick with an individual form of a language, then it is not possible to decide in any case whether we have a real form before us or an agglomeration. The decisive aspect in each individual case lies in the general principle of language. If a language is formless in principle, then it possesses not even a single true form. If there were *one* true form represented in the mind of a people that spoke a formless language, then it would not have passed like lightning in a dark night and have left only thick darkness behind; it would rather have ignited and produced a glow,

354

(Abriss der Sprachw. I, S. 428): „… das *Etymon*, die characterisirende innere Sprachform," und

(Das. S. 429): „… die innere Sprachform, das Etymon." Dagegen

(Das. S. 432): „… die innere Sprachform, welche aber doch nicht in der Etymologie liegt." Und ähnlich

(Charakteristik S. 234): „Darum aber, dass im Ägyptischen Wurzel |323| und Suffix nicht fest miteinander verbunden sind, ist dasselbe nicht etwa keine *Formsprache*; denn nicht auf der Lautverbindung, sondern auf dem inneren Sinn beruht die Form. Der Ägypter aber hat formal gedacht, und darum ist seine Sprache formal." Dagegen wieder von den finnischen Sprachen:

(Das. S. 203): „Man hatte nicht das Bedürfniss, jedes Element nur geformt, nur mit einem bestimmten Characteristicum, in der Rede zuzulassen; und darum gilt auch ein Suffix nicht als Form, und ein Stamm mit einem Suffix ist nicht ein zur Wortform gewordener Stamm, hat nicht aufgehört Stamm, d. h. *formlos* zu sein."

(Das. S. 318flg. spricht er sich noch ausführlicher aus): „Die *Etymologie* ist, abgesehen von der in so vielen Fällen ihr anhaftenden Unsicherheit, auch sonst ein zweifelhafter Zeuge für oder gegen formale Auffassung: denn es kann, wie wir in unseren Sprachen zuweilen sehen, ein ursprüngliches Stoffwort rein formal verwendet werden. Das Wesentliche also, worin sich die materielle oder formelle Vorstellungsweise kundgiebt, liegt in der Behandlung der Wörter, in der *Construction* … Überhaupt aber liegt das formale Wesen der Sprache eben immer in der Construction, d. h. in der reinen Thätigkeit, Synthesis, an sich, im Ausdruck der Prädicirung, der Attribuirung, der Objectivirung, als der geistigen Function sprachlicher Vorstellung. Nur an diesem Punkte, wo der Geist in feinster Weise äusserlich wird, wo er als Thätigkeit, abgelöst von dem Gegenstande der Thätigkeit, rein und nicht materialisirt, den Gegenstand ergreifend und durchdringend, offenbar wird: nur hier ist das formale Prinzip des Sprachbaues zu prüfen, und hier am Sichersten. Denn wenn hier auch die Offenbarung des Geistes fein ist, so ist sie doch mächtig und wirksam. Bleibt man bei einer einzelnen Form einer Sprache stehen, so lässt sich in keinem Falle entscheiden, ob man eine wirkliche Form vor sich hat, oder eine Agglomeration. Das Entscheidende in jedem einzelnen Falle liegt im allgemeinen Princip der Sprache. Ist eine Sprache dem Prinzip nach formlos, so besitzt sie auch keine einzige wahre Form. Wäre nur *eine* wahre Form in dem Geiste eines Volkes, ||337|| welches eine formlose Sprache spricht, vorgestellt worden, sie würde nicht wie ein Blitz in finsterer Nacht vorübergegangen sein und dichte Finsterniss zurückgelassen haben; sie würde vielmehr gezündet und eine Gluth erzeugt haben,

Sprachwiss. 1889
428: 1889
428:) 1891
… „das 1889
charakterisirende 1889
Sprachform", 1889 and 1891
(das. 1889
429:) 1889 and 1891
… „die 1889
das. 1889
432:) 1889 and 1891
… „die 1889
234:) 1889 and 1891
Aegyptischen 1889
Aegypter 1889
formal". 1889
(das. 1889
203:) 1889 and 1891
Charakteristikum, 1889
sein". 1889
aus:) 1891
Princip 1891
Principe 1891

355 which would have remoulded the entire way of thinking of the people.'

(Ibid., p. 223, on *composition* and *appending*): 'The former [composition] combines matter with matter, that is the representation of something material, of a property, activity, substance, with other representations of material things, in order to depict through both of them the concept of something material. The latter [appending] attaches a formal element to a material element, in order to designate a concept through a simple representation and a formal relation.'

(Ibid., p. 186): 'And here I see *materialism* once again, which seeks only to depict the present state of facts in such a way that the listener will reconstruct it as it is; while the spirit of the people, which creates *true forms*, finds joy in formal action in itself and leaves nothing that appears in speech unformed.'

(Ibid., p. 234, on *inflection*): 'In our languages it is the complete word that lives in the spirit of the language, with no distinction between the root and the affix, since the living spirit conceives of the content in the form as one with the form, and only the scientific, analytic spirit separates, through abstraction, the form from the content. If the word is a unit, then it will gradually shrink together over time, without losing comprehensibility, and its end is usually abandoned to weathering.'

(Ibid., p. 251, on *agglutination*): 'Since this sticking of a suffix on the end of another, as we find in Turkish and even more so in American, is precisely *formlessness*.'

(*Mande-Negerspr.* [Steinthal, 1867], p. 234, § 592): 'This is in fact the true character of *formlessness*: that the joining of words, composition and word formation all fall together.'

That is what Steinthal had to say. Where Humboldt speaks of languages 'whose grammatical forms are of not such a formal nature as the inflecting languages' – that is, still possessing grammatical forms and of a formal nature – Steinthal sees an abrupt dualism, divides languages into formless languages and form languages, and takes the grammatical forms of languages to court to test whether they really are formal or not. What for his great predecessor was an indication of a less formal nature is for him a sign of formlessness, and only within these two sharply delineated categories does he allow different levels of lower or higher development.

Friedrich Müller pays homage to similar views. I take the following two passages from his *Grundriss der Sprachwissenschaft* [Outline of Linguistics] [Müller 1876–1888]:

355 welche die ganze Denkweise des Volkes umgeschmolzen hätte."

|324|

(Das. S. 223 über Zusammensetzung und Anbildung): „Erstere verbindet Stoff mit Stoff, d. h. die Vorstellung eines Materialen, einer Eigenschaft, Thätigkeit, Substanz, mit einer anderen Vorstellung eines Materiellen, um durch beide wiederum den Begriff eines Materiellen darzustellen; letztere fügt ein Formelement an ein Stoffelement, um einen Begriff durch eine einfache Vorstellung und eine formale Beziehung zu bezeichnen."

(Das. S. 186): „Und hierin sehe ich wieder den Materialis{193}mus, der nur den wirklich vorliegenden Thatbestand so darstellen will, dass der Hörende ihn so nachbildet, wie er ist; während der Volksgeist, der wahre Formen schafft, am formalen Thun selbst seine Freude findet und nichts ungeformt lässt, was in der Rede auftritt."

(Das. S. 234, Flexion): „Bei uns ist es allemal das ganze Wort, das im Sprachgeiste lebt, ohne Unterscheidung von Wurzel und Affix: denn der lebendige Geist erfasst den Inhalt in der Form als Eins mit der Form, und nur der wissenschaftliche, analytische Geist scheidet durch Abstraction die Form vom Inhalt. Ist nun das Wort eine Einheit, so schrumpft es mit der Zeit allmählich zusammen, ohne an Verständlichkeit zu verlieren, und gerade sein Ende ist am Meisten der Verwitterung ausgesetzt."

(Das. S. 251, Agglutination): „Denn jenes Aneinanderleimen eines Suffixes an das andere, wie wir es im Türkischen und noch mehr im Amerikanischen finden, ist eben Formlosigkeit."

(Mande-Negerspr. S. 234, § .592): „Das ist nun der eigentliche Charakter der Formlosigkeit, dass Wortfügung, Zusammensetzung und Wortbildung zusammenfallen."

Soweit STEINTHAL. Wo HUMBOLDT von Sprachen redet, „deren grammatische Formen nicht so formaler Natur sind, wie die der flectirenden", – also doch immerhin auch grammatische Formen und formaler Natur sind, – da sieht STEINTHAL einen schroffen Dualismus, theilt die Sprachen in formlose und in Formsprachen und geht mit den grammatischen Formen der Sprachen scharf in's Gericht, ob sie auch wirklich formal seien oder nicht. Was seinem grossen Vorgänger als Merkmale minder formaler Natur galt, das ist ihm ein Zeichen der Formlosigkeit, und nur innerhalb dieser beiden scharf geschiedenen Kategorien lässt er verschiedene Stufen niederer oder höherer Entwickelung zu.

Ähnlichen Anschauungen huldigt FRIEDRICH MÜLLER, dessen Grundriss der Sprachwissenschaft ich zwei Stellen entnehme:

Anbildung:) 1889 and 1891
Materialen, 1889 Beide
Materialen 1889
186:) 1889 and 1891
„Und 1889
Und 1891
auftritt". 1889
234 Flexion:) 1889 and 1891
Affix; 1889
ausgesetzt". 1889
Agglutination:) 1889 and 1891
„denn 1889
Denn 1891
Formlosigkeit".
Formlosigkeit. 1891
592:) 1889 and 1891
Merkmal 1889
beiden, 1889 and 1891

356
(Vol. I, p. 104): 'The linguistic process that develops on the basis of roots is of such a kind that the thinking subject, processing (that is, now decomposing and then linking together) representations that become expressible through roots, forms out of them definite words (representatives of more closely determined intuitions). Languages differ widely from one another, however, in the way the primitive intuitions are more closely determined – in the way finished words – parts of a sentence – are made out of raw roots. While those languages that stick to the principled distinction between *matter* (that given from outside) and *form* (that which comes from within to the matter) from the beginning create *two different kinds of sound complexes* for both (so alongside the material roots develop also formal roots), those languages where the principled distinction between matter and form has not penetrated into the consciousness remain stuck with material roots alone and see only *matter* where we are accustomed to see form opposed to matter. We do not want to claim here that those languages that have no understanding for form have not developed other roots that are analogous to our formal roots alongside the material roots (e.g. pronominal and adverbial stems), but these roots are not felt by the language to be different from the material roots – a state of affairs which namely arises from the circumstance that the language does not use pure formal roots where it should (in inflection, conjugation), but rather uses pure material roots here. But we can still say: *All* languages have *material roots*, while only those languages that have an understanding of the difference between material and form have developed a second category of roots, namely *formal roots*, alongside the material roots.'

(p. 131): 'We notice here straight away that languages that really have a sense for form produce *true forms* – that is, constructions that present themselves in their entirety as closed – while languages that show no sense for form cannot produce any true forms. Since the opposition between matter and form does not exist there, the difference between stem and form is not present. What appears to us initially as real form is on closer inspection actually not, since it can still be treated as a stem.'

Hardly any less extensive than Steinthal but, as far as I can see, milder is the editor of the new edition of his *Charakteristik*, Franz Misteli [1893]. I take the following passages from the book:

356

|325|

(I, S. 104): „Der auf Grundlage der Wurzeln vor sich gehende Sprachprozess ||338|| ist derart, dass das denkende Subject die durch die Wurzeln zum Ausdrucke gelangenden Vorstellungen bearbeitend (d. h. bald zerlegend, bald miteinander verknüpfend), daraus bestimmte Worte (Repräsentanten näher be{194}stimmter, determinirter Anschauungen) formt. In der Art und Weise aber, wie die primitiven Anschauungen näher bestimmt – wie aus den rohen Wurzeln die fertigen Worte – Theile eines Satzes – herausgebildet werden, gehen die verschiedenen Sprachen weit auseinander. Während jene Sprachen, welche den principiellen Unterschied zwischen dem *Stoffe* (dem Aussen Gegebenen) und der Form (dem von Innen aus zum Stoffe Hinzutretenden) festhalten, auch von Anfang an *zweierlei verschiedene Lautcomplexe* für Beide ausbilden, also neben den Stoff-Wurzeln auch Form-Wurzeln entwickeln, bleiben jene Sprachen, welchen der principielle Unterschied zwischen Stoff und Form nicht in's Bewusstsein gedrungen ist, bei den Stoffwurzeln stehen und sehen dort, wo wir Form im Gegensatze zu Stoff zu sehen gewohnt sind, nur *Stoff*. Wir wollen damit nicht behaupten, dass jene Sprachen, welche kein Verständniss für die Form besitzen, neben den Stoffwurzeln nicht auch andere Wurzeln entwickelt hätten, welche unseren Formwurzeln analog sind (z. B. Pronominal- und Adverbialstämme), aber diese Wurzeln werden von der Sprache nicht anders denn als Stoffwurzeln gefühlt – was sich namentlich aus dem Umstande ergiebt, dass die Sprache dort, wo reine Formwurzeln zur Verwendung kommen sollten (bei der Flexion, Conjugation), diese nicht zur Anwendung bringt, sondern vielmehr reine Stoffwurzeln dazu verwendet. Wir können also trotzdem behaupten: *Alle* Sprachen kennen *Stoffwurzeln*, während nur jene Sprachen, welche ein Verständniss für den Unterschied zwischen Stoff und Form besitzen, neben diesen noch eine zweite Kategorie von Wurzeln, nämlich *Form-Wurzeln* entwickelt haben."

104:) 1889 and 1891
Subjekt 1889
mit einander 1889
aussen 1889
innen 1889
beide 1889 and 1891
Conjugation) 1889

(S. 131): „Wir bemerken gleich hier, dass nur Sprachen, welche wirklich Sinn für Form haben, *wahre Formen* erzeugen, d. h. Bildungen, die in ihrer Ganzheit abgeschlossen sich darstellen, während Sprachen, die keinen Sinn für Form zeigen, auch keine wahren Formen hervorbringen können. Da der Gegensatz zwischen Stoff und Form dort nicht existirt, so ist auch ein Unterschied zwischen Stamm und Form nicht vorhanden. Das was uns auf den ersten Anblick als wirkliche |326| Form erscheint, ist es bei näherer Betrachtung dennoch nicht, da es weiterhin als Stamm behandelt werden kann."

131:) 1889 and 1891
kann". 1889

Kaum weniger ausgiebig als STEINTHAL, aber, soviel ich sehe, milder ist der Neubearbeiter seiner Charakteristik, FRANZ MISTELI. Ich entnehme dem Buche folgende Stellen:

357
(p. 2): 'Under *form* of speech we understand all the means that either constitute the grammatical frame, or at least contribute to the logical clarity or to the subjective colouring of speech.'

(p. 3): 'Already here we encounter a *formal treatment of the given differences between things*, and we will attribute an advantage to that language which, with a stricter maintenance of the difference between nouns and verbs, can convert as many nouns as possible easily and unforcedly into verbs.' (For example, *Kopf* 'head' – *köpfen* 'to behead'; *Mund* 'mouth', *munden* 'to be tasty'; English *to head* etc.)

(p.4): '*Number words, relation words and pronouns* [...] serve properties that depend on the subject and object equally, and logical relations either not at all, as in the case of the number words and pronouns, or, as in the case of the relation words, in a metaphorical way and not from the beginning, and so are similarly to be seen at first as matter of speech.'

(p. 5): 'In the different shape and conception of pieces of root-like sound material (*you*: *your*, *four*: *fourth*) there lies already *grammatical form*, whether substantive, adjective, verb, just as there does in the exchange of those three word classes among one other.'

(p. 15): French: *concernant, touchant* = concerning, *suivant* = according to, *à partir de* = 'from ... onwards', Greek ἔχων, λαβών, φέρων = 'with' etc. – 'All these figures of speech follow from the construction of finite verb forms and fall outside the area of *prepositions*. Since with verbal prepositions, as in the case of nominal prepositions, we must maintain the principle that form or construction distinguishes them from the relevant noun or verb, if they are supposed to constitute their own class.'

(p. 19): 'Although it is still entirely within the material part of speech, we will distinguish *two* kinds of *lower form*: the possibility of swapping representations relating to things, properties and verbs, especially dissolving a noun in a verb, and the objectification of properties, circumstances and activities in a noun; and the different degrees of clarity with which the relational and circumstantial words set themselves apart from one another and from the substantives, adjectives and verbs, which can serve as material replacements. A *third* kind will show itself in the following' [namely:]

(pp. 19-20): [...] 'Contrary to this, *particles*, *conjunctions* and *negations* correspond to nothing objective, neither things nor determinations or relations of things; they are of a purely formal kind, since they describe partly subjective moods, partly logical relations between representations,

357

(S. 2): „Unter Form der Rede versteht man alle diejenigen Mittel, welche entweder den grammatischen Rahmen bilden, oder wenigstens zur logischen Deutlichkeit oder zur subjectiven Färbung der Rede beitragen.“

(S. 3): Schon hier stossen wir auf formelle Behandlung der gegebenen ||339|| Sachunterschiede, und derjenigen Sprache werden wir einen Vorzug zuschreiben, welche bei strenger Wahrung nominalen und verbalen Unterschiedes möglichst viele Nomina leicht und ungezwungen in Verba auflösen kann.“ (Beispiele: Kopf – köpfen, Mund – munden, englisch to head u. s. w.)

(S. 4): „Zahlwörter, Verhältnisswörter und Pronomina ... dienen Bestimmungen, die vom Subjecte und Objecte gleichmässig abhängen, und logischen Verhältnissen entweder gar nicht, wie die Zahlwörter und Pronomina, oder wie die Verhältnisswörter, in übertragener Weise und nicht von Anfang an, und sind daher gleichfalls zunächst als Stoff der Rede zu betrachten.“

(S. 5): „In der verschiedenen Gestalt und Auffassung des einen wurzelhaften Lautstoffes (du: dein, vier. vierter) liegt schon grammatische Form, ob Substantiv, Adjectiv, Verbum, gerade wie in dem Austausch der letzten drei Wortarten untereinander.“

(S. 15): Französisch: *concernant, touchant* = betreffend, *suivant* = gemäss, *à partir de* = von ... an, griechisch ἔχων, λαβών, φέρων = mit u. s. w. – „Alle diese u. s. w.) 1901 Wendungen folgen durchaus der Construction der finiten Verbalformen und fallen aus dem Bereiche der Präpositionen heraus. Denn daran muss man, wie bei der nominalen, auch bei der verbalen Präposition festhalten, dass Form oder Construction sie vom betreffenden Nomen und Verbum unterscheide, wenn sie eine eigene Classe bilden sollen.“

(S. 19): „Obschon noch ganz im Stoffe der Rede befindlich, unterscheiden wir doch zwei Arten einer niederen Form: Die Möglichkeit des Austausches gegenständlicher, eigenschaftlicher und verbaler Vorstellung, besonders die der Auflösung des Nomens in ein Verbum, und die der Vergegenständlichung von Eigenschaften, Zuständen und Thätigkeiten in ein Nomen; dann die verschiedenen Grade der Deutlichkeit, mit der Verhältniss- und Umstandswörter sich voneinander und von den Substantiven, Adjectiven und Verben abheben, welche als stofflicher Ersatz dienen können. Eine dritte Art wird sich im Folgenden zeigen“ [nämlich:]

(S. 19–20): ...“Dagegen den Partikeln, Conjunctionen und Negationswörtchen entspricht nichts Objectives, weder Dinge noch Bestimmungen oder Verhältnisse von Dingen; sie sind rein formaler Art, weil sie theils subjective Gemüthsstimmungen, theils logische Verhältnisse der Vorstellungen,

358 and partly the non-application or rejection of such; they are acts of feeling and thought which no human speech can do without.'

(p. 28): 'As for pronouns and prepositions, we must consider the most suitable form of *particles* to be those which do not remind us in any way of a noun or verb, and which can be comfortably manipulated phonetically [...]. It deserves to be noticed [...] that it has always been such slightly high-spirited linguistic shapes, discharged from an energetic sense, to which the feeling of language has turned, and in which the most primitive formal elements have been contained. That is precisely the *formal sense*, which works also in the finer embodiment of the concept...' (Cf. p. 38).

(p. 30): 'So it is impossible to determine the content of each form in advance, just as it is impossible to criticize a language for lacking a category that another has developed [...]. But that the categories are consistent, not too narrow, and not phonetically cumbersome, these are the *requirements of a true form*.'

(pp. 35-36, *substantives, adjectives, verbs*): 'Form consists in this case precisely in the fact that this tripartition, which both appearance and logic suggest to us in equal measure, receives the freest treatment through the exchange of its members, since every form is mental activity.'

(p. 39): [...] 'there are no parts of speech in Old Chinese that are equally mentally and phonetically formed, or which are more mentally than phonetically formed. For this reason, we must describe this language as formless, since it does not even seem to carry the minimum of form necessary for clearly distinguishing matter, if it does not completely compensate for the lack of *lower* form through the *higher* form of sentence and thought expression.' Additionally:

(p.43): 'The more a language seeks to secure the comprehensibility of a sentence in a formal way, through word groups, inflection, abstract particles, and not through mere content and context, the more claim it has to possessing *higher form*.'

(p. 46:) '[Chinese] possesses no real verb and no conjugation, but also no false verb and no pseudo-form. We could only admit that the first of these two opposites dissolves if the subject finds a light, general *indication* on the verb, and is united with it *internally*, and more specially as a noun in nominative shape or according to a fixed position, or at least comes from outside as an absolutive. Here the phonetic means through which the indication takes place makes no difference, neither does it make a difference if the indication is achieved only negatively [...] since we represent each form not for itself,

358 theils Nichtsetzen und Ablehnen solcher bezeichnen, lauter Gefühls- und Denkacte, deren keine menschliche Rede entbehrt."

(S. 28): „Auch für die *Partikeln* muss, wie für die Pronomina und Präpositionen, als die geeignetste Form diejenige gelten, welche an ein Nomen oder Verb gar nicht erinnert und auch nach den Lauten sich bequem handhaben lässt ... Interesse verdient es, ... dass es immer solche leichtbeschwingte, eines energischen Sinnes entleerte Sprachgestalten waren, an die das Sprachgefühl sich ||340|| wandte, und in denen wohl die primitivsten Formelemente sich bargen. Das ist eben *Formsinn*, der nun auch in der feineren Ausgestaltung des Begriffes wirkt ..." (Vergl. dazu S. 38).

(S. 30): „So kann man den Inhalt der jeweiligen Form unmöglich von vornherein bestimmen, noch einer Sprache einen Vorwurf daraus machen, dass diese oder jene Kategorien ihr fehlen, welche eine andere entwickelt ... Aber dass die Kategorien consequent, nicht zu eng, noch lautlich schwerfällig seien, das sind die *Erfordernisse einer richtigen Form*."

(S. 35–36, *Substantiva, Adjectiva, Verba*): „Darin, dass diese Dreitheilung, so sehr sie durch den Augenschein und die Logik gleichmässig empfohlen wird, die freieste Behandlung durch Vertauschung ihrer Glieder erfährt, besteht in diesem Falle eben die Form; denn jede Form ist geistige Bethätigung."

(S. 39): ... „ebenso sehr oder noch mehr geistig als lautlich geformte Redetheile giebt es im Alt-Chinesischen nicht, und deswegen müssten wir diese Sprache als formlos bezeichnen, weil sie nicht einmal so viel Form an sich zu tragen scheint, als zur deutlichen Sonderung des Stoffes nöthig ist, wenn sie nicht den Mangel der *niederen* Form durch die *höhere* des Satz- und Gedanken-Ausdruckes vollständig ausgliche." Hierzu:

(S. 43): Je mehr eine Sprache die Verständlichkeit des Satzes auf formelle Weise, durch Wortstellung, Flexion, abstracte Wörtchen nicht durch blossen Inhalt und Zusammenhang, zu sichern sucht, desto mehr darf sie diese *höhere Form* zu besitzen Anspruch erheben."

(S. 46: Das Chinesische) „besitzt kein ächtes Verbum und keine Conjugation, aber auch kein unächtes Verbum und keine Scheinform. Denn dass nur dann, wenn das Subject am Verbum, und *innig* mit ihm vereint, eine leichte allgemeine *Andeutung* findet und specieller als Nomen in Nominativgestalt oder nach fester Stellung, oder wenigstens als Absolutiv von Aussen noch einmal hinzutritt, d. h. in der Vereinigung von Nomen und conjugirtem Verb, der erstere der beiden Gegensätze sich auflöst, dürfte man willig zugeben. Hierbei verschlägt es nichts, durch welches lautliche Mittel die Andeutung stattfindet, oder ob sie gar nur negativ vollzogen wird ... weil man jede Form nicht für sich,

359 but in relation to the others, and in this way the general indication of a subject is transferred to that form where it is not explicitly indicated.'

(p. 51, *Semitic*): 'The closed nature of the word form or the *word unit* comes forth here once again as a not indispensable but nonetheless important condition of *higher formation*.'

(p. 52): [...] 'To restrict the *true verb* to Indo-European and Semitic.'

(p. 60): 'But wherever this mix (of subject and possessive suffixes) appears from the very beginning, there rules also from the very beginning a lax conception of the verbal and nominal category.' [Similarly, Humboldt [1999 [1836]], e.g. *Versch. des menschl. Sprachbaues*, pp. 260-261.]

(p. 66): [The participle ending with *-l*, *-la*, *-lo* of the present-day Slavic languages] 'today no longer has any participial sense, [...] but serves as a marker of narrative tense and may, despite the distinction of genders, serve as a *finite verb*; its mere form cannot rob it of its claim to be counted among the finite verbs; only when it wants to appear as participle and then as finite verb can we properly deny it the latter.'

(p. 68): [...] 'that some finite verbal forms, since they show neither possessive nor abstract nor participle form, must for this reason be counted as nouns, because they append the plural sign to the tense stem.'

(p. 76): 'The *object conjugation* is based on an inability to neatly separate the activity from the subject and object – reality always presents all three as bound together – and as a result the object is always seen as being equally as important as the subject.'

(p. 80): 'That in the predicative, objective and attributive relations the two parts do not mix together to become a whole, but rather show their value separated through position, particles and inflection, this lies in the proper relation of sentence and word. But then *objective conjugation* and *possessive pronominal suffixes* go beyond the word boundary through incorporation of the object and of the possessor; but whoever constructs sentence-words does not need to recoil from possessive suffixes or prefixes, as the case may be.'

(p. 81): 'From the perspective of linguistic form we cannot permit *the object to melt into the verb*, not even in the milder type of incorporation of just the pronoun, because in this case the boundary of word and sentence shifts to the side of the word, and the energy of the subject is weakened, if a whole encompasses subject and object.' [He denies the existence of the objective conjugation in Semitic languages.]

359 sondern im Verhältnisse zu den anderen vorstellt, und somit die allgemeine Hinweisung auf ein Subject von den letzteren auch auf diejenige Form übergeht, wo sie ausdrücklich nicht vorhanden ist."

(S. 51, *Semitisch*): „Die Geschlossenheit der Wortform oder die *Worteinheit* tritt hier wieder als nicht unerlässliche aber wichtige Bedingung *höherer Formung* hervor."

(S. 52): … „Das *ächte Verbum* auf das Indogermanische und Semitische zu beschränken."

(S. 60): „Wo aber diese Mischung (von Subjects- und Possessivsuffixen) von ||341|| Anfang an auftritt, da herrschte eben auch von Anfang an eine laxe Auffassung der verbalen und nominalen Kategorie." [Ähnlich HUMBOLDT, z. B. Versch. des menschl. Sprachbaues S. 260–261.]

(S. 66, Das Particip auf –l, –la, –lo der heutigen slavischen Sprachen) „hat heute keine participiale Geltung mehr, … sondern versieht den Dienst eines Erzählungstempus und darf wohl, trotz der Unterscheidung der Geschlechter als *Verbum finitum* gelten; die blosse Form kann dasselbe kaum des Anspruchs berauben, dem Verbum finitum beigezählt zu werden; nur wenn es bald als Participium bald als Verbum finitum auftreten wollte, könnten wir das Letztere von Rechtswegen ihm nicht zugestehen".

(S. 68): … „dass einige finite Verbalformen, ohne weder possessive noch abstracte noch participiale Form zu zeigen, deswegen als Nomina gelten müssen, weil sie an den Tempusstamm das nominale Mehrheitszeichen hängen."

(S. 76): „Die *Objectsconjugation* beruht auf dem Mangel, die Thätigkeit reinlich von Subject und Object abzusondern – die Wirklichkeit giebt alle drei immer verbunden –, in Folge dessen das Object an Wichtigkeit dem Subjecte mindestens gleichsteht."

(S. 80): „Dass in der prädicativen, objectiven, attributiven Beziehung die beiden Glieder sich nicht in einem Ganzen vermischen, sondern abgetrennt durch Stellung, Partikeln und Flexion ihre Geltung zeigen sollen, das liegt im richtigen Verhältniss von Satz und Wort. – Nun überschreiten eben *objective Conjugation* und *possessive Prononimalsuffixe* durch Einverleibung des Objectes und des Besitzes die Wortgrenze; doch wer Satzworte bildet, braucht vor Possessiv- Sub- resp. Präfixen nicht zurückzuschrecken." Possessiv-, 1901

(S. 81): „Von Seiten der sprachlichen Form kann man es nicht gestatten, *das Object mit dem Verbum zu verschmelzen*, auch nicht in der milderen Art der Einverleibung bloss des Pronomens, weil damit die Grenze von Wort und Satz nach Seiten des Wortes verrückt, und die Energie des Subjects geschwächt wird, wenn ein Ganzes Subject und Object umschliesst." [Dem Semitischen spricht er die Objectivconjugation ab.]

360

(p. 82): 'It is also a mistake when *possession and possessor are united in one word*, since the possessive concept modifies neither content nor grammatical relationships; it rather describes a relationship between things.'

(p. 87): [...] 'that the endings of the nominative and accusative can be dropped according to the needs of comprehensibility [...]. The lack of inflection as such is not blameworthy, because the position still keeps both cases separate, but *true inflection* never abandons the category for which it was created, and strict *word unity* is not compatible with random loss of the endings.'

(pp. 98-99): 'In *objective* and *attributive* relationships we will determine the form in such a way that: 1. both members of each relationship stay clearly separated, so that objective conjugation and possessive constructions or compositional forms do not take their place and damage the unity of the word to the advantage of the unity of the sentence; 2. the categories of object and attribute offer abstractly general sense and are consequently applied; 3. their reproduction proceeds either through position or through a light, handy exponent (inflection, particle).'

(pp. 99-100): 'It has already emerged [...] that, of all languages, only two show the properties of subjectivity and energy in the verb or in the predicative relationship. That is, only two have a verb that consists of a material root and an internally melded affix that indicates the subject in general, and which does not confine itself to the bare proposition but at the same time does not contain any improper side-concepts – these languages are *Semitic* and *Indo-European*. The sentence comes into being when a noun that is recognizable as subject through its position, an accompanying particle or its ending (nominative or absolutive) specifies the general reference of the person affix. But this conception of the verb contains in itself the concept of *inflection* and of the *word*; the word is a representation that is determined by all applicable categories of the language and as such phonetically characterized, and which as a closed whole inserts itself into the sentence; the verb is the word *par excellence*. We will call the two aforementioned languages *inflecting* or *true-word* languages, and view them as *form languages*, because of the proper relationship between word and sentence they exhibit, in that subject and predicate, predicate and object, attribute and head appear separately. Other languages strive to create verbs and form words, and in word forms length is not lacking, but what is lacking is the strength to contract the phonetic fullness and the mental processes; the verb is thought of nominally and the words are not complete. The apparently inflectional elements that terminate words may be lacking, usually when they are not needed for comprehensibility,

360

(S. 82): „Fehlerhaft ist auch die Vereinigung von Besitz und Besitzer in einem Wort, denn der Possessivbegriff verändert weder den Inhalt noch die grammatischen Verhältnisse; er bezeichnet ein Sachverhältniss."

(S. 87): ... „dass die Endungen des Nominativs und Accusativs nach Bedürfnissen der Verständlichkeit wegfallen können ... Nicht die Flexionslosigkeit als solche ist tadelnswerth, weil die Stellung immer noch die beiden Casus auseinanderhalten würde, wohl aber lässt wahre Flexion nie die Kategorie im Stiche, wofür sie geschaffen wurde, und strenge Worteinheit ist mit beliebigem Abfallen der Endungen unvereinbar."

(S. 98–99): „Beim objectiven und attributiven Verhältnisse werden ||342|| wir die Form so bestimmen, dass 1. beide Glieder eines jeden Verhältnisses gesondert bleiben, damit nicht objective Conjugation und Possessivbildungen resp. Composita an ihre Stelle treten und die Worteinheit zu Gunsten der Satzeinheit schädigen; 2. die Kategorien des Objectes und des Attributes abstract allgemeinen Sinn bieten und consequent angewendet werden; 3. ihre Wiedergabe entweder durch Stellung oder durch einen leichten, handlichen Exponenten (Flexion, Partikel) erfolge."

(S. 99–100): „Nun ergab sich schon ... dass von sämmtlichen bloss zwei beim Verbum oder im prädicativen Verhältnisse die Eigenschaften der Subjectivität und Energie aufweisen, d. h. ein Verbum besitzen, welches aus einer Stoffwurzel und innig angeschmolzenen, das Subject allgemein andeutenden Affixen besteht und sich nicht auf die nackte Aussage einschränkt und doch, auch keine ungehörigen Nebenbegriffe enthält – der semitische und indogermanische. Dadurch, dass ein durch Stellung, Partikel oder Endung als Subject (Nominativ oder Absolutiv) kennbares Nomen die allgemeine Andeutung des Personalaffixes specialisirt, kommt der Satz zu Stande. Diese Auffassung des Verbums schliesst aber den Begriff der Flexion und des Wortes in sich; Wort ist eine nach allen anwendbaren Kategorien der Sprache bestimmte und als solche lautlich charakterisirte Vorstellung, welche als geschlossenes Ganze in den Satz sich einfügt; das Verbum ist das Wort par excellence. Wir werden die beiden genannten Sprachen die flektirenden oder ächtwortigen nennen, und sie wegen des richtigen Verhältnisses von Wort und Satz, insofern nämlich Subject und Prädicat, Prädicat und Object, Attribut und Regens gesondert auseinandertreten, als Formsprachen betrachten. Andere Sprachen bestreben sich, Verba zu schaffen und Worte zu bilden, und die Länge fehlt nicht, aber die Kraft fehlt, die lautliche Fülle und die geistigen Processe abzukürzen; das Verbum ist nominal gedacht, und die Wörter sind nicht geschlossen; die uns flexivisch erscheinenden wortschliessenden Elemente können fehlen, meist nach Rücksichten der Verständlichkeit,

361 or they can appear where the word already seems closed; these are the *agglutinative* or *pseudo-word* languages: Finnish, Magyar, Yakut, Kannada. Because of the unsteady relationship between the word and sentence we describe both of these as *formless*, even when individual apparently finite verbal forms are to be recognized as closed [...]. Indisputably *formless* are then the *incorporating* or *sentence-word* languages [...] (Mexican, Greenlandic.) And the rest we can group together under the name *non-word* languages, but we have to recognize three subdivisions: the root-isolating, stem-isolating and concatenating languages, of 1. Chinese and Siamese (Burmese); 2. Malay and Dayak; 3. Egyptian-Coptic and the Bantu family.'

(p. 104): 'These [...] *non-word* languages of course lack proper form, since they have no true verb, and because of this the relationship of word and sentence is harmed. But they must not be considered *a priori* formless in the sense of misformed: unlike the incorporating languages that make an entire sentence into one word, they do not combine into whole elements that do not belong together, i.e. the two parts of the predicative, objective or attributive relation; they do not have such wholes or words. Neither do they attempt to create such wholes, as the pseudo-word, agglutinative languages do. For these reasons they stay preserved from all failures.'

(p. 226): 'Since forms that can at one moment consist of stacked up elements and then at the next be missing do not deserve this name.'

(p. 232): 'Since a sound structure that does not belong to a specific word class and does not bear a specific relationship to the whole of the sentence is not a *word*.'

(p. 241): 'There we encounter that *concern for mere comprehensibility* that characterizes the Ural-Altaic languages and which is not compatible with the sense for strict form.'

(p. 264): 'Corresponding to our *man* ("one" as a pronoun) [in Malay] is once again *oraṅ* (*orang*), actually "Men" (*Menschen*), and precisely this detail is suitable to shed light on the *difference between form and matter*: *man* separates itself so clearly from the substantive *Mann* ("man") through its being only singular and its lacking an article – just as much as French *on* distinguishes itself from *homme* through its phonetic form – that it is simply considered a pronoun, about whose actual meaning hardly anyone thinks any more. *Oraṅ* (*orang*), by contrast, is "men" [?] and not "one", if the different conceptions can be demonstrated on the word in question.'

(p. 268): 'Because of their [the formal elements'] greater number and phonetic slimness, Egyptian crosses over into *concatenation*;

361 oder auch da antreten, wo das Wort schon geschlossen erschien; das sind die agglutinirenden oder scheinwortigen Sprachen, Finnisch, Magyarisch, Jakutisch, Karnattisch; wegen des schwankenden Verhältnisses von Wort und Satz bezeichnen wir Beide als formlos, wenn gleich einzelne namentlich finite Verbalformen als geschlossene Wörter anzuerkennen sind ... Formlos ohne Frage sind dann die einverleibenden oder satzwortigen Sprachen ... (Mexicanisch, Grönländisch.) Und den Rest darf man mit dem Namen nichtwortig zusammenfassen, muss aber drei Abtheilungen unterscheiden: die Wurzel-isolirenden, Stamm-isolirenden und anreihenden Sprachen, oder 1. Chinesisch und Siamesisch (Barmanisch); 2. Malajisch und Dajakisch; 3. Ägyptisch-Koptisch und die Bantufamilie." Grönländisch.] 1901

(S. 104): „Diese ... nichtwortigen Sprachen ermangeln freilich der ||343|| richtigen Form, weil sie kein echtes Verbum besitzen, und damit auch das Verhältniss von Wort und Satz Schaden leidet, und dürfen doch nicht von vornherein als formlos im Sinne von missformig gelten; weder fassen sie in ungehöriger Weise, wie die satzwortigen, einverleibenden Sprachen, die zwei Glieder der prädicativen, objectiven, attributiven Beziehung in ein Ganzes zusammen (sie haben eben nicht solche Ganze oder Worte), noch machen sie den Versuch, solche Ganze zu schaffen, wie die scheinwortigen, agglutinirenden Sprachen, und bleiben daher vor jedem Misslingen bewahrt."

(S. 226): „Denn Formen, welche bald aus gehäuften Elementen bestehen bald wieder fehlen können, verdienen diesen Namen nicht."

(S. 232): „Denn ein Lautgebilde, das nicht einer bestimmten Wort-Kategorie angehört und ein bestimmtes Verhältniss zum Ganzen des Satzes an sich trägt, ist kein Wort."

(S. 241): „Da gewahrt man jene Rücksichtnahme auf blosse Verständlichkeit, die die uralaltaischen Sprachen charakterisirt und mit dem Sinne für strenge Form sich nicht vereint."

(S. 264): „Unserem „man" entspricht [im Malaischen] wieder *orań*, eigentl. „Menschen", und gerade diese Einzelheit ist geeignet, den Unterschied von Form und Stoff in's Licht zu setzen; „man" trennt sich durch den Singular und mangelnden Artikel so sehr vom Substantivum „Mann" ab, ebenso französisch *on* schon durch den Laut von *homme*, dass es füglich als Pronomen gelten darf, an dessen eigentliche Bedeutung man kaum mehr denkt; *orań* dagegen ist „Männer" [?] und nicht „man", wenn die verschiedene Auffassung am nämlichen Worte einleuchten soll."

(S. 268): „Wegen ihrer [der formalen Elemente] grösseren Zahl und lautlichen Schmächtigkeit tritt das Ägyptische in den Zustand der Anreihung;

362 since the formal elements and syllables only accompany the matter, assign themselves to it from outside, stand independently beside it.'

(p. 274): 'That *true form elements,* still so loosely appended, but confining and shaping the matter.'

(p. 275): 'That Egyptian, although it does not bind root and affix tightly, is *still, from a certain perspective, a form language*; the Egyptian thought formally, and insofar is his language formal.' [Cf. the quotation above, p. 337, from Steinthal's [1860] *Charakteristik,* p. 234.]

As in Steinthal's works, I have provided here the most extensive excerpts possible in order to be as fair as I can to all three researchers. I doubt, however, whether the matter has become much clearer because of this. A lot of what is reproduced above already has to do with the outer form, and all of this is directed towards the evaluation of languages, a topic to which I will devote a later section of this book. From my point of view, I see everywhere in this area differences of degree, more or less lively expressions of the drive to formation, and not actual oppositions that would justify my denying the inner or outer form of any particular language.

It seems to me rather awkward to want to apply the etymological standard everywhere, since this standard fails where the sources of language history do not deliver: its applicability depends very much on chance. An example from Kannada (Karnatta, Kannadi). Tense is indicated on the positive forms of the verb in this language by a suffix that is attached between the stem and the personal suffix. Stem *gey*, 'do': *geydam*, 'he did'; *geydaḷ*, 'she did'; *geyvam*, *geyvaḷ*, 'he, she will do'; *geydapam*, *geydapaḷ*, 'he, she does'. But the negative verb attaches the personal suffix directly to the stem: *geyyam*, *geyyaḷ*, 'he, she does not'. It is fortunate that the imperative, *do!*, has the simple form *gey* alongside the forms *geyya*, *geyvudu*, and that Tamil, Malayalam and Telugu exhibit traces of an old negative suffix *a*, otherwise we might be inclined to think erroneously that the Kannadigas nonsensically see the negative as primary and derive the positive from it. (Cf. R. Caldwell, *A comparative Grammar of the Dravidian or South-Indian Family of Languages*. 2nd ed. London 1875, pp. 360f, 442f.)

But we should also allow, on the one hand, for the mechanical sharpening of forms and, on the other, for the powerful innovations that can – often hand in hand with phonetic weathering – introduce a lively need for distinctness and clarity in language. Periphrastic constructions, emphatic reduplications or tautologies,

362 denn die formalen Elemente und Silben begleiten nur den Stoff, gesellen sich ihm von Aussen bei, treten ihm selbständig zur Seite."

(S. 274): „Dass wahre Formenelemente, noch so lose angefügt, doch den Stoff begrenzen und gestalten."

(S. 275): „Dass das Ägyptische, obwohl es Wurzel und Affix nicht fest miteinander verbindet, doch in gewisser Hinsicht eine Formsprache ist; der Ägypter hat formal gedacht, und insoweit ist seine Sprache formal." [Vergl. dazu das oben, S. 337 aus STEINTHAL's Charakteristik S. 234 Angeführte.]

{195}

Ich habe hier, wie bei STEINTHAL, möglichst ergiebige Auszüge mitgetheilt, um allen drei Forschern nach Kräften gerecht zu werden, zweifle aber, ob die Sache dadurch sehr an Klarheit gewonnen. Vieles des hier Wiedergegebenen betrifft schon die äussere Form, Alles zweckt auf die Werthabschätzung der Sprachen ab, der ich einen späteren Abschnitt widmen werde. Ich meinerseits ||344|| sehe hier überall Gradunterschiede, mehr oder minder lebhafte Äusserungen des Formungstriebes, nicht eigentliche Gegensätze, die mich berechtigten einer Sprache die innere oder äussere Form abzusprechen.

Steinthal 1889 beiden 1889 and 1891

Aeusserungen 1889

Den etymologischen Massstab überall anlegen zu wollen, scheint mir misslich. Denn dieser Massstab versagt da den Dienst, wo die Quellen der Sprachgeschichte versiegen: seine Anwendbarkeit hängt sehr vom Zufall ab. Ein Beispiel aus dem Canaresischen (Karnatta, Kannadi). Hier fügen die positiven Formen des Verbums zwischen den Stamm und die Personalsuffixe Tempuscharaktere. Stamm *gey*, thun: *geydam*, er that, *geydaḷ*, sie that, *geyvam*, *geyvaḷ*, er, sie wird thun, *geydapam*, *geydapaḷ*, er, sie thut. Dagegen fügt das negative Verbum die Personalendungen unmittelbar an den Stamm: *geyyam*, *geyyaḷ*, er, sie thut nicht. Ein Glück, dass der Imperativ: thue! neben den Formen *geyya*, *geyvudu* noch das einfache *gey* hat, und dass Tamulisch, Malayâlam und Telugu Spuren eines alten Negativsuffixes *a* aufweisen. Man wäre sonst versucht, dem Canaresen den Unsinn zuzumuthen, er betrachte das Negative als das Erste und leite erst davon das Positive ab. (Vergl. R. CALDWELL, A Comparative Grammar of the Dravidian or South-Indian Family of Languages. 2d. ed. London 1875, S. 360 flg. 442 flg.)

Dann aber sollte man auch einerseits dem mechanischen Formenschliffe und andrerseits jenen mächtigen Neuerungen Rechnung tragen, die, – oft Hand in Hand mit phonetischen Abnutzungen, – ein lebhaftes Deutlichkeits- und Anschaulichkeitsbedürfniss in die Sprache einführen kann. Periphrastische Gebilde, nachdrückliche Doppelungen oder Tautologien

363 and most figurative manners of speaking are certainly essentially sensual and material. But precisely because of this the mood and thought of the speaker are expressed all the more effectively in them, and even more securely will the listener be gripped and brought over in their soul to the speaker. Whatever develops and works in this way does not need to be ashamed of its origin or its destiny – and neither does it need to be ashamed after it has sunk, with faded features, from the heights of rhetoric and poetry into the grammar or dictionary of everyday usage and etymologists confront it with its birth certificate. By birth it was formal, perhaps in the highest, artistic sense, and formal it has remained, representing the 'intuition of an intuition'.

In order to establish the concept of inner form, I will take the genetic route. Every person has his inner world of a certain wider or narrower range, in other words, his circle of ideas. To this inner world, ruling it, there belongs also a subjective (inner) world order, which in a more or less fine and rich way groups things, puts them together or separates them. In other words: the circle of ideas is ruled by a particular point of view, which it of course conditions in turn. In both, in the circle of ideas and in the point of view, exists a certain community among the speakers of a language, thanks to the linguistic exchange of ideas, and this must find its expression in language. As far as this community is related to the circle of ideas, the type and amount of individual representations, it is material. By contrast, as far as it bears upon the point of view, it is formal, *inner form*. This inner form will show itself first of all in the vocabulary, in how it has been etymologically constructed and synonymically grouped; secondly, in the structure of the language, how, with more or less sharpness and liveliness, the representations are ordered into categories, their mutual relations in thought and the relations of the pronounced thought to the soul of the speaker are conceived and differentiated. It is quite common that formative elements, whether auxiliary particles or affixes, can be used or left out at will. Examples we could mention include German 'Das Haus, das er gebaut (hat)'; English: 'the house (which) he has built'; older German: 'es ist viel übrig (ge)blieben', and finally the augment in Homeric dialects. Such occurrences are much more usual in the isolating and agglutinative languages. Some have even thought they saw formlessness here, but it seems to me more correct to speak of 'form doublets' and to assume vague shadowing of the sense. When a Chinese speaker can say for 'to be in the house': *tsái kiā* (在家 *zài jiā*) = be-in house; *tsái iǖ kiā* (在於家 *zàiyú jiā*) be-in concerning house; *tsái iǖ kiā (čī) čū'n* (在於家（之）中 *zàiyú jiā [zhī] zhōng*) = be-in concerning house('s) centre: we notice how the intuition is made more distinct blow by blow.

363 und die meisten figürlichen Redeweisen sind freilich im Grunde sinnlich und stofflich. Aber um so wirksamer äussern sich eben darum in ihnen die Stimmung und der Gedanke des Redens, und um so sicherer wird durch sie der Hörer ergriffen und zum Redner seelisch hinübergezogen. Was so entsteht und so wirkt, das braucht sich weder seiner Herkunft noch seiner Bestimmung zu schämen, – auch dann nicht, wenn es aus den Höhen der Rhetorik und Poesie mit verblassten Zügen in die Grammatik oder das Wörterbuch des alltäglichen Gebrauchs hinabgestiegen ist, und ihm nun die Etymologen seinen Geburtsschein vorhalten. Formell, vielleicht im höchsten, künstlerischen Sinne, war es von Hause aus, und formell, die „Anschauung einer Anschauung" darstellend, ist es geblieben.

Um nun den Begriff der inneren Form festzustellen, schlage ich den genetischen Weg ein. Jeder Mensch hat seine innere Welt von einem gewissen engeren oder weiteren Umfange, mit anderen Worten seinen Ideenkreis. Zu dieser inneren Welt, sie beherrschend, gehört auch subjective (innere) Weltordnung, die in mehr oder minder feiner und reicher Weise die Dinge gruppirt, in Zusammenhang setzt oder trennt. Mit anderen Worten: der Ideenkreis wird beherrscht durch eine bestimmte Anschauungsweise, die er doch natürlich auch seinerseits wiederum bedingt. In Beidem nun, in jenem Ideenkreise und in ||345|| dieser Anschauungs|327|art, besteht, Dank dem sprachlichen Gedankenaustausche, eine gewisse Gemeinschaft unter den Sprachgenossen, die in der Sprache ihren Ausdruck finden muss. Soweit sie den Ideenkreis, die Art und Menge der einzelnen Vorstellungen betrifft, ist sie stofflich. Soweit sie dagegen in der Anschauungsweise beruht, ist sie formal, i n n e r e F o r m . Diese innere Form wird sich zeigen erstens im Wortschatze, wie er sich etymologisch aufgebaut hat und synonymisch gruppirt; zweitens im Sprachbaue, wie, mit mehr oder minderer Schärfe und Lebhaftigkeit, die Vorstellungen in Kategorien geordnet, ihre wechselseitigen Beziehungen im Gedanken und die Beziehungen des ausgesprochenen Gedankens zur Seele des Sprechenden erfasst und unterschieden werden.

gewissen 1889 gewissen, 1891
subjektive 1889
trennt; mit 1889
dank 1889
wie 1889
ansgesprochenen 1901

Der Fall ist überaus häufig, dass Formative, seien es Hülfswörter oder Affixe, beliebig gebraucht oder weggelassen werden können. Als Beispiele diene deutsch: „Das Haus, das er gebaut (hat)"; englisch: *the house* (*which*) *he has built*, älteres Deutsch: „es ist viel übrig (ge-)blieben", endlich das Augment im homerischen Dialekte. Viel gewöhnlicher ist Solches in den isolirenden und vielen agglutinirenden Sprachen. Man hat eben hierin Formlosigkeit erblicken wollen; mir aber scheint es richtiger, von Formdoubletten zu sprechen und leise Abschattungen des Sinnes anzunehmen. Wenn der Chinese für „im Hause sein" sagen kann: *tsái kiā* = sein-in Haus; *tsái iǖ kiā* = sein-in betreffs Haus; *tsái iǖ kiā* (*čī*) *čūṅ* = sein-in betreff Haus (-es) Mitte: merken wir ohne Weiteres, wie die Anschauung Schlag für Schlag schärfer ausgeprägt wird.

364 This too, the progression from indefinite to more definite, belongs in the area of the inner form. How the inner form conceives, divides up and combines the world of representation within the different language families and languages must be shown through the outer language form.

§ 4.2. Outer language form. Morphological classification

1. Let us also take a genetic approach to the examination of the outer linguistic form and begin with those utterances that still lack the decisive characteristic of human speech, namely articulation. Here we are thinking of the first utterances of children, those simple exclamations that do without any formal markings, and which express a thought through substitution: 'Ouch!' = that hurts; 'Woof woof!' = there is the dog; 'Num num!' = I want to eat, and so on. We might want to talk here of *unformed single-word sentences*; that is, single-word sentences in terms of their logical content, as opposed to those formed, articulated sentence-words like Latin *risit* 'he has laughed' and *lacrimabit* 'he will weep'.

2. Such exclamations, each standing for the expression of a complete thought, can now be stacked on top of one another without being bound into speech. They are spat out individually, each with its own value. But the person calling out knows how he moved from one thought to the next, and the listener perhaps guesses at it as well. A child is bitten by a dog; it hurts: 'Ouch!' The dog caused the pain: 'Woof woof!' The dog caused this pain by biting: 'Num!' Those are three thoughts, each one represented by an unformed sentence-word.

3. If we now eliminate the three exclamation marks – that is, those things that correspond to them in speech – 'Ouch woofwoof num', and translate this using the copula: the pain is (caused by the) dog, which bit. Here we have produced a three-part utterance that expresses the synthesis of the thought in the most unsophisticated way. A child that speaks like this is already speaking in human language, that is, in articulated language. But the synthesis consists in a mere concatenation; it is, like a dry-stone wall, without any binding mortar: the language is *isolating*. In such isolating languages children can make themselves understood very well. A little girl might tell of how she (*girl*) climbed onto a chair, fell down and as punishment was smacked (*smacksmack*) by her mother, but only moderately (*a little*): 'Girl chair, climb; bang! Mamma smacksmack, little!' If we understand the individual words in this childish babbling, then the construction is easy to discern: we only have the choice between

364 Auch dies, auch der Fortschritt vom Unbestimmten zum Bestimmteren, gehört in den Bereich der inneren Form. Wie diese innerhalb der verschiedenen Sprachfamilien und Sprachen die Welt der Vorstellungen auffasst, eintheilt und verknüpft, das muss die äussere Sprachform erweisen. Und alles dies muss 1889 and 1891

§ 4.2. Die äussere Sprachform. Die morphologische Classification.

1. Um auch bei der Betrachtung der äusseren Sprachform genetisch zu verfahren, beginnen wir bei jenen Sprachäusserungen, denen das entscheidende Merkmal der menschlichen Rede, die Gliederung noch fehlt. Wir denken an die ersten {196} Sprachäusserungen der Kinder, jene einfachen, aller Formenzeichen entbehrenden Ausrufe, die ersatzweise einen Gedanken ausdrücken: „Au!" = mir thut etwas weh; „Wauwau!" = da ist der Hund; Da 1889
„Happhapp!" = ich will essen, u. s. w. Man mag hier von u n g e f o r m t e n S a t z w ö r t e r n , von Satzwörtern im logisch-inhaltlichen Sinne reden, im Gegensatz zu jenen geformten, gegliederten Satzwörtern wie *risit, lacrimabit*.

2. Solche Ausrufe, jeder den Ausdruck eines vollständigen Gedankens vertretend, können nun auch gehäuft werden, ohne darum in der Rede verbunden zu sein. Sie werden einzeln hervorgestossen, jeder hat seinen Werth für sich. Aber der Rufende weiss, wie er von einem Gedanken auf den anderen gekommen ||346|| ist, und der Hörer ahnt es vielleicht auch. Ein Kind wird vom Hunde gebissen; das schmerzt: „Au!" Den Schmerz hat der Hund verursacht: „Wauwau!" Er hat ihn verursacht, indem er biss: „Happ!" Das sind drei Gedanken, jeder durch ein ungeformtes Satzwort vertreten. jedes 1889

3. Nun denken wir uns die drei Ausrufezeichen, das heisst das, was ihnen in der Rede entspricht, weg: „Au wauwau happ", und übersetzen |328| mittels der Copula: der Schmerz ist (verursacht durch den) Hund, welcher biss. Hier ist in rohester Weise die Synthese des Gedankens in einer dreigliedrigen Rede zum Ausdrucke gebracht; ein Kind, das so spricht, redet schon in menschlicher, das heisst gegliederter Sprache. Die Synthese aber besteht in einer blossen Aneinanderreihung, sie ist ohne bindenden Mörtel wie eine cyklopische Mauer: die Sprache ist i s o l i r e n d . In solcher isolirenden Sprache können Kinder sich sehr gut verständlich machen. Ein kleines Mädchen erzählte, wie sie (Mädel) auf den Stuhl geklettert, heruntergefallen und dafür von ihrer Mama Schläge (Puchpuch) erhalten habe, aber nur mässige (ein Bischen): „Mädel Tuhl, ketter; bum! Mama puchpuch, bissen!" Versteht man die einzelnen Wörter eines solchen Kindergelalles, so ist die Construction leicht zu deuten: man hat nur die Wahl zwischen

Wauwau 1889

isolirender 1889

dem 1889

herunter gefallen 1889

bischen): 1889 and 1891

365 the copula of being, of having or of cause. Some languages, and not only isolating ones, have not gone much further than this. But we must remember that a purely isolating language, namely Chinese, belongs among the most perfect carriers of thought: it knows how to articulate thought in a wonderfully fine way, connect thoughts wonderfully, and to mirror thought subtly.

4. As far as I know, only a few languages of South-East Asia are as purely isolating as Chinese: Siamese and Annamite along with their closest relatives. And in all of these languages isolation does not seem to be an original property, but tertiary or perhaps quaternary. In all of them there would seem to be a tendency for a part of the vocabulary to be phonetically and semantically volatile and, in comparison with the other components of speech, to be treated as less independent. Even in Old Chinese it is possible to see traces of this, and the modern Chinese dialects exhibit this to such a great extent that they hardly count as monosyllabic-isolating any more. Any form that loses its independence has to lean on another form, has to be bound with it. What was earlier an isolated form will have a part added to it, *compounded* with it, and now the compound looks like a part of the sentence of a higher order isolated to the outside. Up until this point, however, compounding is mainly an extension to the isolating structure, which it enriches.

5. Compounds can be of very different kinds, in respect to both the logical relations between their elements and the final product they form; we might think of such compounds as 'grindstone', 'pig-headed', 'dark brown', 'tally stick', 'oak tree', and so on. But all of them have one thing in common: that the parts are all in equal measure material. But even the oldest, still unarticulated human speech surely had expressions which, once they were brought into the complex sentence, were eminently suitable for formal use. The sounds of sensation expressed the subjective behaviour of the speaker, and we can also assume that deictic and interrogative calls were present at this earliest stage; even today our children quickly progress to saying 'da-da' when they reach for something or point to something. The call of negation will also be age-old; children learn very quickly to swap those unarticulated sounds that they use to express their dissatisfaction with the 'no' that they hear from their parents. Finally, we can be sure that the earliest urge to speak required designations for spatial positions, directions and movements, for above, below, in front, behind, towards, away, come, go, stay, being together, and furthermore for give, take, and also for thing, possession or accessory etc. As crudely sensual as that may be,

365 der Copula des Seins, des Habens oder der Ursache. Manche Völkersprachen, und nicht nur isolirende, haben es nicht viel weiter gebracht. Dafür wollen wir uns aber auch daran erinnern, dass eine rein isolirende Sprache, die chinesische, zu den vollkommensten Trägerinnen des Gedankens gehört, den sie wunderbar fein zu gliedern, wunderbar mannigfaltig zu verknüpfen und feinsinnig abzuschatten versteht.

mannichfaltig
1889 and 1891

{197}

4. So rein isolirend, wie das Chinesische, sind meines Wissens nur noch wenige Sprachen Hinterindiens: das Siamesische und das Annamitische sammt ihren nächsten Verwandten. Und bei allen Sprachen dieser Classe scheint die Isolation nicht ursprünglich, sondern tertiär, vielleicht quarternär zu sein. Wohl auch bei allen zeigt sie die Tendenz, einen Theil des Wortschatzes lautlich und inhaltlich zu verflüchtigen und im Vergleiche zu den übrigen Bestandtheilen der Rede als minder selbständig zu behandeln. Schon im Altchinesischen sind davon Spuren nachzuweisen, und neuchinesische Dialekte leisten hierin bereits soviel, dass sie kaum mehr als monosyllabisch-isolirend gelten können. Was seiner Selbständigkeit beraubt wird, muss sich an Anderes anlehnen, sich mit ihm verbinden. Was früher isolirt war, davon wird nun ein Theil zusammengefügt, componirt, und nun erscheint das Compositum als ein nach Aussen hin isolirter Satztheil höherer Ordnung. Bis jetzt jedoch ist die Zusammensetzung noch immer der Hauptsache nach eine bereichernde Erweiterung des isolirenden Baues.

Chinesische
1889

ursprünglich
1889

5. Zusammensetzungen können sehr verschiedener Art sein, in Rücksicht sowohl auf das logische Verhalten ihrer Glieder zu einander, als |329| auch auf ihr Gesammtergebniss; man denke an Gebilde wie Schleifstein, Dickkopf, schwarz||347||braun, Kerbholz, Eichbaum, schwarz-weiss-roth u. s. w. Eins aber ist ihnen gemein: dass die Glieder gleichermassen stofflich sind. Nun hatte aber sicher schon die älteste, noch ungegliederte menschliche Rede Ausdrücke, die, einmal in's Satzgefüge gebracht, sich vorzugsweise zur formalen Verwendung eigneten. Die Empfindungslaute drückten das subjective Verhalten des Redenden aus, und auch Deute- und Fragerufe dürfen wir jener frühesten Zeit zutrauen; wie bald sagen unsere Kinder „da-da", wenn sie nach etwas greifen oder auf etwas hinweisen. Auch der Verneinungsruf ist sicher uralt; lernen doch Kinder sehr schnell, jene unarticulirten Laute, durch die sie ihre Unzufriedenheit ausdrücken, mit dem Nein vertauschen, das sie von den Eltern hören. Endlich erforderte gewiss schon der früheste Redebedarf Bezeichnungen für örtliche Lage, Richtung und Bewegung, für oben, unten, vorn, hinten, herwärts, hinwärts, kommen, gehen, verweilen, beisammensein, ferner für geben, nehmen, wohl auch für Sache, Besitz oder Zubehör u. s. w. So grob sinnlich das an sich sein mag,

zueinander,
1889

gleichermaassen
1889

subjektive
1889

366 it offers ample material for the expression of purely formal relations. It is most difficult for us to grasp how the sounds of sensation could become suitable for such logical purposes. If we may be permitted to draw conclusions about prehistoric processes from historical facts then the East Asian languages provide us at least with examples of the same sound serving at once as a case marking and as an indicator of sensation. In Chinese *hû* 乎 (*hū*) can be a sound of questioning and doubt or a preposition indicating a general relation; similarly, *iǖ* (*ū*) 於 (*yú*) is both a preposition with a similar meaning to *hû* and an interjection. In Manchurian, the genitive marking *ni* is phonetically identical to the interrogative particle. And in Japanese *wo*, *ka* (*ga*), *mo* and *na* are at the same time sounds of sensation and auxiliary particles for logical relations. The object, taken in the widest sense, now appears as something that is sought by calls of questioning, complaining, desiring; now belonging, independence and authorship seem to be something doubtful, something to be deduced that is indicated by an interrogative. Even in our own language the interrogative sentence often represents the conditional sentence.[6] But let it be however it will: even isolating languages contain in them rich material for *auxiliary words*, which are felt to be formal as soon as they are used as formal means. A grammar has no less forming power just because it is purely syntactic, dependent on the means of word order and particles.

6. If formal words are perceived as such in contrast to material words, then it becomes natural to give this contrast expression in articulation. First of all in intonation, whether formal elements are treated as unstressed or, the other way round, that logical-psychological relations are strongly stressed. Both of these strategies occur in our unselfconscious speech, as is well known. In many languages, however, habit or rule has decided for one or the other of these strategies. In both cases the achievement here is that material and formal elements of speech bind together into acoustic units – it is as with waves, which are counted by their peaks. Once again we have compounds in the wider sense, but no longer compounds with equal material parts, but rather combinations of material and form. And this kind of compounding we call *appending* or *agglutination*. The auxiliary words become *affixes*; depending on their position, *prefixes*, *suffixes* or *infixes*. The last of these, infixes, come about when they, so to speak, push through the rind of the

6 In German it is possible to form a conditional clause by inverting the subject and finite verb; the same inversion is used to indicate yes/no questions. This conditional construction has a marginal existence in a much more restricted form in English with the auxiliary verbs *have* and *be* in the subjunctive, e.g. 'Had I done that…', 'Were I rich…'

366 so bietet es doch gefügigen Stoff zum Ausdrucke {198} rein formaler Beziehungen. Am Schwersten begreifen wir, wie Empfindungslaute zu solchen logischen Zwecken tauglich werden konnten. Dürfen wir jedoch aus geschichtlichen Thatsachen auf vorgeschichtliche Vorgänge schliessen, so liefern ostasiatische Sprachen wenigstens dafür Beispiele, dass dieselben Laute jetzt als Casuszeichen, jetzt als Empfindungsäusserungen dienen können. Im Chinesischen ist *hû* bald Laut der Frage und des Zweifels, bald Präposition der allgemeinen Beziehung, *iū̆* (*ū*) bald eine Präposition von ähnlicher Bedeutung, bald eine Interjection. Im Mandschuischen fällt das
Genetivzeichen *ni* lautlich mit der Partikel des Fragesatzes zusammen. Und zusammen; und 1889
im Japanischen sind *wo*, *ka* (*ga*), *mo* und *na* zugleich Empfindungslaute
und Hülfswörter für logische Beziehungen. Jetzt scheint das Object im Objekt 1889
weitesten Sinne, als ein Erstrebtes durch Rufe der Frage, der Klage, des
Begehrens, – jetzt Zugehörigkeit, Abhängigkeit und Urheberschaft als
ein Zweifelhaftes, nur zu Erschliessendes durch Fragelaute bezeichnet erschliessendes 1889
worden zu sein, – vertritt doch noch bei uns oft genug der Fragesatz den bedingenden. Doch sei dem wie ihm wolle: schon die isolirende Sprache trägt in sich reichen Stoff zu Hülfswörtern, die sie formal empfinden wird, sobald sie sie als Formenmittel anwendet. Eine Grammatik verliert dadurch nichts an formender Leistungskraft, dass sie |330| rein syntaktisch, auf die Mittel der Wortfolge und Formwörter angewiesen ist.

6. Werden nun die Formwörter als solche im Gegensatze zu den Stoffwörtern empfunden, so liegt es nahe, diesem Gegensatze auch in der Articulation Ausdruck zu geben. Zunächst in der Betonung, sei es, dass man das Formale leichthin behandelt, sei es, dass man umgekehrt die logisch-psychologischen Beziehungen scharf benachdruckt. In unserer
freien Rede kommt bekanntlich Beides vor; in vielen Sprachen aber haben hat 1889 and 1891
sich Gewohnheit und Regel für das Eine oder Andere entschieden. In beiden
Fällen ist nun aber der Erfolg der, dass sich Stoff- ||348|| und Formelemente nun ist der 1889
der Rede zu akustischen Einheiten verbinden, – es ist wie mit den Wellen, die man nach Gipfeln zählt. So haben wir wieder im weiteren Sinne Zusammensetzungen, nun aber nicht mehr solche von gleichwerthigen, stofflichen Gliedern, sondern Verbindungen von Stoff und Form. Und diese Art der Zusammensetzung nennen wir Anfügung oder Agglutination. Die Hülfswörter werden dann zu Affixen, je nach ihrer |199| Stellung zu Prä-,
Sub- oder Infixen. Letzteres, Infixe, werden sie dann, wenn sie so zu Infixe, 1889

word into its insides. This process is presumably always secondary: sonorant prefixes or suffixes, above all those containing liquids, have an impact, like a falling stone on soft ground, in virtue of a strong force of attraction. In Malay languages *m-kan* 'eat' gives rise to *kuman, koman* (instead of *km̥an*), and from the same stem *in-kan* or *ni-kan*:

sagen durch die Rinde des Stoffwortes hindurch in sein Inneres dringen. Dieser Vorgang ist vermuthlich immer secundär: leichtlautige Prä- oder Suffixe, zumal solche mit Liquidis, schlagen vermöge einer mächtigen Anziehungskraft ein, wie der fallende Stein in den weichen Boden. In malaischen Sprachen wird aus *m-kan*, essen: *kuman*, *koman* (statt *km̥an*), und von dem gleichen Stamme aus *in-kan* oder *ni-kan*:

sekundär: 1889

367 *kinan*. Latin *findo* and Sanskrit *bhinadmi*, which both mean 'split', replace **fidno*, **bhidnāmi*. In the languages of Kolar we find forms like *dal* 'hit': *dapal* 'hit each other', *dak'pal* 'hit each other hard', *hidžu* 'come': *hinidžu*, 'the coming'; and to the Hebrew *hit-phaʾel* corresponds the Arabic *i-f-ta-ʾala*: the *t* of the prefix has merged into the stem. For this reason we should distinguish between *pre-infixes* and *sub-infixes*.

7. As with everything in language, agglutination does not take place in one fell swoop, and what to one speaker may still feel like an independent auxiliary particle may be treated by another as a mere formative word element. But here, as everywhere where there are fluid boundaries, we can adopt the adage that controversial questions are trivial questions. The less the material and formal elements influence each other in sound and intonation, the freer the formative elements can be used or dispensed with, repositioned or, through the insertion of other elements, drawn away from the bearer of the material, the word stem. Finally, if the linguistic consciousness of speakers is reminded in a more lively way of the original identity of a formative with a material word that is still in use, then the auxiliary word will assert its independence longer in the feeling of the speaker. It seems as if the proclitic words are more secure here than the enclitics. For example, while in French and English many suffix forms of Latin and Anglo-Saxon have been replaced by formative elements inserted before them, these languages have not as a result moved in the direction of prefixing agglutination, but rather in the direction of a partial isolation. In contrast to this, the indivisible future forms of the modern Romance languages with *habere* – *je dirai* etc. – and the adverbial forms with *-mente*, *-ment* are instances of agglutination even though they encapsulate the infinitive and ablative. Indeed, they are just as much agglutination as our derived forms with the suffixes *-heit*, *-tum*, *-niss*, *-lich*, the material origin of which is known to linguistics, although no longer known to the linguistic consciousness of the people.

8. The progression from the isolating stage to the agglutinating stage occurs when a third means of formation, namely the affix, joins the previous two methods of word order and auxiliary words. But this still does not say quite enough, and the so-called *agglutinative language class* to which the vast majority of languages has been consigned is in the best case a conceptual embarrassment, from which we cannot get away, a kind of *cache-désordre*, a junk room, which we need in our house to be able to maintain order elsewhere. Only one thing should worry us: that in this case the junk room takes up the greater part of the building.

367 *kinan*. Lateinisch *findo* und das gleichbedeutende sanskritische *bhinadmi* stehen an Stelle von **fidno*, **bhidnāmi*. In den kolarischen Sprachen finden sich Gebilde wie *dal*, schlagen: *dapal*, einander schlagen, *dak'pal*, einander heftig schlagen, *hidžu*, kommen: *hinidžu*, das Kommen; und dem hebräischen *hit-phaʔel* entspricht im Arabischen *i-f-ta-ʔala*: das *t* des Präfixes ist in den Stamm gedrungen. So sollte man zwischen Prä-Infixen und Sub-Infixen unterscheiden.

kinan; lateinisch 1889

Kommen, 1889

7. Wie Alles in der Sprache, vollzieht sich auch die Agglutination nicht mit einem Schlage, und der Eine mag noch lange als selbständiges Hülfswort empfinden, was der Andere schon als blossen formativen Worttheil behandelt. Hier aber, wie überall, wo die Grenzen fliessen, gilt der Satz, dass streitige Fragen unerhebliche Fragen sind. Je weniger die stofflichen und formalen Elemente einander in Laut und Betonung beeinflussen, je freier die Formativen gebraucht oder weggelassen, um|331|gestellt oder durch Einschieben anderer Elemente von dem Träger des Stoffes, dem Wortstamme, abgedrängt worden können, je lebhafter endlich das Sprachbewusstsein an die ursprüngliche Identität des Formativs mit einem noch gebräuchlichen Stoffworte erinnert wird: desto länger behauptet das Hülfswort seine Selbständigkeit im Empfinden des Redenden, und es scheint, als wären die proklitischen Wörter hierin besser gesichert, als die enklitischen. Wenn z. B. das Französische und Englische viele Suffixformen des Lateinischen und Angelsächsischen durch vorgefügte Formenzeichen ersetzt haben, so sind sie damit noch nicht zu einer präfigirenden Agglutination, sondern nur zu einer theilweisen Isolation gelangt. Dagegen sind die unlöslichen Futurgebilde der neuromanischen Sprachen mit *habere*: *je dirai* u. s. w., die Adverbialformen durch *–mente*, *–ment*, trotz der darin eingekapselten Infinitive und Ablative ebensogute Agglutinationen, wie unsre Ableitungen mittels der {200} Suffixe –heit, –thum, –niss, –lich, deren stofflichen Ursprung wohl die Sprachwissenschaft, nicht aber mehr das Sprachbewusstsein des Volkes kennt.

überall 1889

Wortstamme 1889

Wörtchen 1889 and 1891

habere; 1901

||349||

8. Der Fortschritt von der isolirenden Stufe zu der agglutinirenden besteht darin, dass zu den beiden Formenmitteln der Wortstellung und der Hülfswörter noch ein drittes gekommen ist: das Affix. Dies besagt aber immerhin noch wenig genug, und die sogenannte agglutinirende Sprachenclasse, der man die ungeheure Mehrzahl der Sprachen eingereiht hat, ist im günstigsten Falle ein Verlegenheitsbegriff, dessen man nicht wohl entrathen kann, so eine Art *cache-désordre*, eine Rumpelkammer, wie man sie im Haushalte braucht, um anderwärts desto bessere Ordnung zu halten. Nur das sollte Bedenken einflössen, dass diesmal die Rumpelkammer den grössten Theil des Gebäudes einnimmt.

sogenannte 1889
so genannte 1891

368 *Subdivisions* are necessary, along the following dimensions.

a) The most obvious distinction is between those languages that have only suffixes, such as the Ural-Altaic, Dravidian and Australian languages, Greenlandic, Hottentott and many others, and those that also have prefixes, such as the Malayo-Polynesian, Kolarish and Bantu languages and so on. Infixes should be considered here as varieties of prefixes and suffixes.

b) The scale, the capability of expansion of agglutination, also demands to be considered. But here the possibilities and intermediate stages are innumerable, and there can be large differences between related languages. East Mongolian, for example, has no pronominal endings on the verb, while Kalmyk (Oirat) and Buryat dialects do have such endings. The same can be found within different members of the Tungus language family. To the Malay tribe belong the exceedingly versatile languages of the brown Philippine islanders (Tagalog, Bisaya, Pampanga, Iloca, Bicol, Zebu and so on) and on the other side Malay and Dayak, with their childish simplicity. A few examples should give us an idea of the productivity of agglutination. In Turkish the stem *sev* means 'love'. I will provide the following suffixes only in the form that they have due to vowel harmony with *-e-* stems. *-mek*, sign of the infinitive: *sevmek* 'to love'. *-iš* (*-iş*), sign of reciprocity: *sevišmek* (*sevişmek*), 'love one another'. *-dir*, sign of the causative: *sevdirmek* 'make [someone] love'. *sevišdirmek* (*seviştirmek*), 'make them love each other'. *-il*, passive suffix: *sevilmek*, 'to be loved', *sevišdirilmek* (*seviştirilmek*), 'to be caused to love one another'. *-me*, negation: *sevmemek*, 'to not love'; *sevišmemek* (*sevişmemek*), 'to not love one another', and so on. Finally, *sevišdirilememek* (*seviştirilmemek*) = 'to not be able to be caused to love one another'. Greenlandic goes to the extent of creating forms like this:

qasu-	*êr-*	*sar-*	*fi-*	*gssar-*	*sí-*	*ngit-*	*dluinar-*	*nar-*	*poq*
tired-	de-	make-	where	with which-	recuperate-	not-	completely-	v. impers.-	indicat.

= one did not find a place to rest at all, or one was completely unable to come to rest. The sentence 'Presumably he will try very hard to finish this as soon as possible' is expressed in this language in a single word: *inilertorniarpatdlãsarqôrpâ* (Kleinschmidt, *Gramm.*, p. 155). In Toumpakewa-Alifur from North Celebes, a language that is very close to the Philippine language in its structure, the stem

368 Unterabtheilungen sind nöthig, und zwar nach verschiedenen Eintheilungsgründen.

a) Am Nächsten liegt es, zu unterscheiden zwischen solchen Sprachen, die nur suffigiren, wie die uralaltaischen, die drâvidischen, die australischen, das Grönländische, das Hottentottische und viele andere, – und solchen, die auch Präfixe haben, wie die malaio-polynesischen, die kolarischen, die des Bantustammes u. s. w. Die Infixe kommen hierbei nur als Abarten der Prä- oder Suffixe in Frage.

a. 1889
nächsten 1889
ural-altaischen, 1889 and 1891

b) Auch der Umfang, die Erweiterungsfähigkeit der Agglutination verlangt Berücksichtigung. Hier aber sind die Möglichkeiten und |332| Zwischenstufen unzählig, und zwischen verwandten Sprachen können grosse Unterschiede bestehen. Das Ostmongolische z. B. kennt keine Pronominalendungen am Verbum, während das Kalmükische (Ölöt) und burjätische Dialekte solche besitzen. Das Gleiche findet sich innerhalb verschiedener Glieder der tungusischen Sprachfamilie. Zum malaischen Stamme gehören die so überaus bildsamen Sprachen der braunen Philippineninsulaner (Tagala, Bisaya, Pampanga, Iloca, Bicol, Zebu u. s. w.) und andrerseits das Malaische und das Dajak mit ihrer kindlichen Einfachheit. Von der Leistungsfähigkeit der Agglutination mögen ein paar Beispiele einen Begriff geben. Im Türkischen bedeutet der Stamm *sev*: lieben. Die nun folgenden Suffixe theile ich nur in der Form mit, die sie vermöge der Vocalharmonie mit dem –*e*–Stamme haben müssen. –*mek*, Zeichen des Infinitivs: *sevmek*, lieben. –*iš*, Zeichen der {201} Gegenseitigkeit: *sevišmek*, einander lieben. –*dir*, Zeichen des Causativums: *sevdirmek*, lieben machen; *sevišdirmek*, machen, dass sie einander lieben. –*il*, Passivsuffix: *sevilmek*, geliebt werden, *sevišdirilmek*, veranlasst werden einander zu lieben. –*me*, Negation: *sevmemek*, nicht lieben; *sevišmemek* einander nicht lieben u. s. w. Endlich *sevišdirilememek* = nicht veranlasst werden können, einander zu lieben. Das Grönländische versteigt sich zu Gebilden wie diesem:

b. 1889

sevišdirmek 1889 and 1891
machen 1889
sevĭsmemek 1889
sevičmemek 1901

qasu-	*êr-*	*sar-*	*fi-*	*gssar-*	*sí-*	*ngit-*	*dluinar-*	*nar-*	*poq*
müde –	ent –	machen –	wo –	womit –	erlangen –	nicht –	gänzlich –	v. impers.-	indicat.

||350||
= man hat durchaus keine Ausruhestelle gefunden, oder zur Ruhe kommen können. Den Satz: „Vermuthlich wird er sehr suchen, es baldigst fertig zu machen," drückt diese Sprache in einem Worte aus: *inilertorniarpatdlãsarqôrpâ* (Kleinschmidt, Gramm. S. 155). – Im Toumpakewa-Alifurischen von Nord-Celebes, einer in ihrem Baue den philippinischen nahestehenden Sprache, heisst der Stamm

369 *ilek* means 'see'. Active *milek*, stative *mailek*, passive of the object *ilekĕn*. Further *makailek*, 'get to see', 'realize', 'know'; *mapailek* 'make [someone] see', 'show'. To this we have the passive *pakailek*, 'become aware of', and *papailek* 'be shown'. Finally, from both of these combined: *mapakailek*, 'make someone know or recognize', and from this the passive *papakailek* and so on, since the means and combinations of the voice system are not exhausted by these.

c) Especially important, but of course allowing innumerable degrees, is the more or less close binding of formative and word stem. Malay, alongside the suffix *-kăn* (*-kan*), which shows the relation of the verb to an indirect object, also has an independent preposition with a related sense, *akăn* (*akan*). In the Fijian language there is, alongside the verbal prefix *faka*, the independent verb *faka*, 'to do'. These are signs of a particularly loose agglutination. It is better – that is indicating closer unity – when the Malay verbal prefixes *me-* and *pe-* (*meṅ-* [*meng-*], *peṅ-* [*peng-*]) the initial tenuis consonants in the stem *k*, *t*, *p* and *s* are turned into the corresponding nasals: *kības* (*kibas*), 'shake': *meṅības* (*mengibas*); *tāhan* (*tahan*), 'catch, detain': *menāhan* (*menahan*); *pādam* (*padam*) 'extinguish': *memādam* (*memadam*); *pemādam* (*pemadam*), *sāsal* (*sasal*), 'remorse': *meñāsal* (*menyasal*) etc. Similarly, in the same language, according to the rule that the penultimate syllable is lengthened, the quantity of the stem syllable will be influenced by the addition of suffixes, as will the quantity of the suffixes by the appending of additional ones: *krūboṅ* (*krubong*) 'to surround', with the active prefix and the transitive suffix: *meṅrubōṅi* (*mengrubongi*); *mūwat* (*muwat*) 'to load': *muwātan* (*muwatan*) 'load' (n.): *nanti* 'to wait': *nantīkan* (*nantikan*) 'to await'; *kāta* (*kata*; Sanskrit *kātha*) 'to say': with the transitive suffix *meṅatāi* (*mengatai*), passive: *dikatāi* (*dikatai*), with the possessive suffix *-ña* (*-nya*): *dikataīña* (*dikatainya*), 'it is said by him' = 'he says' – added to this the tense suffix *-lah*: *dikataiñālah* (*dikatainyalah*). The harmony rule in most Ural-Altaic languages demands that the suffix takes on a different vowel depending on the vowel in the stem. For example, Turkish *sevmék* 'to love', but from the stem *yaz* 'to write': *yazmáq* (*yazmak*); Magyar *kert* 'garden': *kerte* 'his garden', but *ház* 'house': *háza* 'his house'. Lappish does without vowel harmony, but has in its place in several of its dialects curious changes in the stem. In Norwegian dialects, for example: *gietta* 'hand', genitive: *gieđa*, but allative *gitti* – in Russian Lappish the present indicative of *poatte* 'to come' is conjugated: *poađam*, *poađah*, *poat*, *puättep*, *puätbetteđ*, *puätteb*; *puäđe* means 'come!' and *potme* 'coming'. Korean has five declensions depending on the final sound of the stem, with suffixes that are phonetically more or less different, but in its conjugations some word stems are additionally subject to noticeable modifications.

369 *ilek*: sehen: Activ: *milek*, zuständlich: *mailek*, Passiv des Objectes: *ilekĕn*, Aktiv 1889
mit causativer Schattirung *pailekĕn*. Weiter: *makailek*, zu sehen erlangen, Objektes: 1889
einsehen, wissen; *mapailek* sehen lassen, zeigen. Hierzu die Passiva: *pakailek*,
gewusst werden, und *papailek*, gezeigt werden. Endlich aus beiden combinirt:
mapakailek machen, dass man wisse oder erkenne, und davon wieder das
Passivum *papakailek* u. s. w., denn die Mittel und Combinationen der *genera*
verbi sind damit noch nicht erschöpft.

c) Besonders wichtig, aber natürlich zahllose Abstufungen zulassend, ist c. 1889
die mehr oder minder innige Verbindung von Formativ und Wortstamm.
Wenn das Malaische neben dem Suffixe *–kăn*, das die Beziehung des Ver-
bums auf ein indirektes Object anzeigt, noch eine selbständige Präposition
verwandten Sinnes *akăn*, besitzt, wenn in der Fidschi-Sprache neben dem Sinnes, 1889
Verbalpräfixe *faka* noch das selbständige Verbum *faka*, machen, besteht: so
sind das Anzeichen einer besonders losen Agglutination. Besser, auf engere
Vereinigung deutend, ist es schon, wenn die malaischen Verbalpräfixe
me– und *pe–* (*meṅ–*,*peṅ–*) die stamman|333|lautende tenuis *k*, *t*, *p* und pe 1889
s in den entsprechenden Nasal verwandeln: *kības*, schütteln: *meṅības*, tenuis 1889
tāhan, fangen: *menāhan*, *pādam*, auslöschen: *memādam*, *pemādam*, *sāsal*,
Reue: *meñāsal* u. s. w. Desgleichen, wenn in derselben Sprache nach dem
Gesetze, dass die offene Paenultima Dehnung erfährt, die Quantität der
Stammsylben durch antretende Suffixe, und die Quantität der Suffixe eintretende 1889
durch den Hinzutritt neuer beeinflusst wird: *krūboṅ* umringen, mit dem
Activpräfix und dem Transitivsuffix: *meṅrubōṅi*; *mūwat*, laden: *muwātan*, neṅrubōṅi; 1889
Ladung: *nanti*, warten: *nantīkan*, erwarten; *kāta* (sanskrit *kātha*) sagen: mantīkan, 1889
mit Transitivsuffix *meṅatāi*, Passiv: *dikatāi*, mit dem Possessivsuffix *–ña*:
dikataīña, es wird von ihm gesagt = er sagt, – hierzu das Zeitsuffix *–lah*:
dikataiñālah. {202} Das Harmoniegesetz der meisten uralaltaischen Sprachen
verlangt, dass die Suffixe je nach den Vocalen des Stammes verschiedene
Vocale annehmen, z. B. türkisch *sevmék* lieben, aber vom Stamme *yaz*,
schreiben: *yazmáq*; magyarisch *kert*, Garten: *kerte*, sein Garten, aber *ház*,
Haus: *háza*, sein Haus. Das Lappische entbehrt der Vocalharmonie, hat aber
statt deren in mehreren seiner Dialekte merkwürdige Stammveränder-
ungen. So im norwegischen Dialekte: *gietta*, Hand, Genitiv: *gieđa*, aber gieda, 1889
Allativ *gitti*; – im Russisch-Lappischen wird der Indicativ Präsentis von Ablativ 1889
poatte, kommen, conjugirt: *poađam*, *poađah*, *poat*, *puättep*, *puätbetteđ*, poadam, poadah, 1889
puätteb; *puäđe* heisst: komm! und *potme* kommend. Das Koreanische hat
je nach dem Stammauslaute fünf Declinationen mit lautlich mehr oder puätbetted, 1889
minder ||351|| verschiedenen Suffixen; in seinen Conjugationen aber sind puäde 1889
zudem auch manche Wortstämme auffälligen Veränderungen unterworfen.

370

d) Finally we have to ask: which word classes are formed agglutinatively, and which grammatical categories are served by the affix system? But here the outer form makes contact with the inner form, and the question as to the grammatical category is independent of the morphological class.

9. All agglutination is by nature composition; what distinguishes agglutination is only that the parts put together have different values in the linguistic consciousness. But any kind of composition, regardless of the types of its elements, can condense to the point that it is no longer perceived as such. This may be because speakers forget the independent value of one or both parts, or because elements meld together to the point where they are no longer phonetically recognizable. *Were* = 'man' in the compound *werewolf* is no longer understood but is still perceived as a particular part of the word; in *world*, from *were-old* = 'age of man, generation', nothing reminds us of the composition, neither the sound nor the sense. In γὰρ 'because' the Greeks would no longer have thought of γε ἄρα 'then'; certainly *nur* 'only' does not remind us of its origin: *ne ware* = *ni esset* 'were not'. The Roman may have still felt something of the old case ending *-bi* in *ibi* 'here', *ubi* 'where', but the Italian no longer feels anything in his *ivi* 'here', *ove* 'where', *dove* 'whence', to say nothing of the Frenchman in the monosyllabic *y* 'there', *où* 'where'. As different as these examples may be, they all have one thing in common, namely that the etymological consciousness is blunted and thus it is no longer a composite form that is available to it, but rather a simple one. To put it metaphorically, there is no longer a two-times-three but a six. The compound has turned into an amalgamation.

Of course the stronger the compounded parts influence one another phonetically, the more likely this process is to occur, since in this case one part merges into the other, in proportion to the degree of this effect. If the outer form should be decisive, and in a morphological classification this must be the case, then I do not know why the forms in Lappish should be considered less developed than the stem modifications in Indo-European: λείπω 'I leave', ἔλιπον 'I left', λέλοιπα 'I have left', *bind, band, gebunden*, bind, bound, are from their outer appearance no more developed than the forms of *poatte*; the *connection* is simply an especially inner appending, the *inflection* in the Indo-European sense is simply a more advanced agglutination. We have now lost another dualism; that is to say gained another scientific insight.

If we want to talk about a particularly inner combination of material and form in our languages, we might mention the dominance and diversity of our declensions and conjugations, of the genitives *Caesaris* 'Caesar's', *Pompeii* 'Pompey's', of the ablative plurals *pueris* 'from, by boys', *hominibus* 'from, by men', of the perfects *amavi* 'I loved',

370

d) Endlich wird auch zu fragen sein: welche Wortkategorien werden agglutinativ geformt, und welchen grammatischen Kategorien dient das Affixsystem? Hier berührt sich aber die äussere Form mit der inneren; und die Frage nach den grammatischen Kategorien gilt unabhängig von der morphologischen Classe. d. 1889

9. Alle Agglutination ist von Hause aus Composition; das Entscheidende liegt bei ihr nur darin, dass die zusammengefügten Bestandtheile dem Sprachbewusstsein als verschiedenwerthig gelten. Jederlei Composition, gleichviel welcher Art ihre Elemente seien, kann sich aber so verdichten, dass sie schliesslich gar nicht mehr als solche empfunden wird, sei es, dass man den selbständigen Werth eines oder beider Bestandtheile vergessen, sei es, dass sich die Elemente bis zur Unkenntlichkeit lautlich verwischt haben. „Wer" = Mann wird in der Zusammensetzung „Werwolf" nicht mehr verstanden, aber noch als besonderer Bestandtheil empfunden; in „Welt", aus Wêr-alt = Mannesalter, Generation, erinnert nichts mehr an die Zusammensetzung, weder der Klang noch der Sinn. Die Griechen werden bei γὰρ nicht mehr an γε ἄρα gedacht haben; sicher gemahnt uns „nur" nicht mehr an seinen Ursprung: *ne ware = ni esset*. Der Römer mochte bei *ibi, ubi* noch etwas von der alten Casusendung *–bi* empfinden; der Italiener thut dies bei seinem *ivi, ove, dove* nicht mehr, geschweige denn der Franzose bei den einsylbigen *y, où*. So verschieden: nun im Übrigen die angeführten Fälle sind: das Eine ist ihnen gemeinsam, dass das etymologische Bewusstsein abgestumpft und schliesslich in seinem Sinne nicht mehr ein Zusammengesetztes sondern ein Einfaches vorhanden ist, bildlich gesprochen, nicht mehr ein Zweimal drei, sondern eine Sechs, – die Zusammensetzung hat sich in Verschmelzung verwandelt.

Natürlich ist dies um so leichter möglich, je stärker die zusammenzusetzenden Theile einander lautlich beeinflussen; denn dann dringt ja, der Wirkung nach, der eine thatsächlich in den anderen ein. Soll nun die äussere Form entscheiden, und bei einer morphologischen Classification muss sie das, so wüsste ich nicht, wieso jene Erscheinungen im Lappischen hinter den indogermanischen Stammwandelungen zurückstünden: λείπω, ἔλιπον, λέλοιπα, binden, band, gebunden, haben der äusseren Erscheinung nach vor jenen Formen von *poatte* nichts voraus; die Anbildung ist nur eine besonders innige Anfügung, die Flexion im indogermanischen Sinne nur eine weit vorgeschrittene Agglutination. So |334| sind wir wieder um einen Dualismus ärmer, d. h. um eine wissenschaftliche Einsicht reicher. 9. Soll die 1889 and 1891 zurückständen: 1889 Anbilung [in den Berichtigungen, S. 502: Anbildung] 1891 Anbildung 1901

Will man nun doch von einer besonders innigen Verquickung von Stoff und Form in unseren Sprachen reden, so mag man an die Mehrheit und Verschiedenheit unserer Declinationen und Conjugationen erinnern, an die Genitive *Caesaris, Pompeii*, an die Ablative Pluralis *pueris, hominibus*, an die Perfecta ||352|| *amavi*,

371 *legi* 'I read', *scripsi* 'I wrote', of the plurals *Fuss – Füsse* 'foot – feet', *Mann – Männer* 'man – men', *Mensch – Menschen* 'person – persons' etc. But in this case the humble name 'defective system' is perhaps more appropriate than the beautiful sounding inflection. Since, as a matter of fact, we have here – just as in *sum* 'I am' – *fui* 'I was', λέγω 'I say' – εἶπον 'I said', among others – the situation where not every stem takes every form and not every form attaches to every stem. Similar phenomena occur elsewhere, without much being made of it; for example, in the plural forms of many African and American languages, and in the conjugations of many Caucasian languages. Clearly defective is also the alleged passive in Malay: *ku-īhat* (*kuihat*), *kau-līhat* (*kaulihat*), *līhat-ña* (*lihatnya*) = 'I see', 'you see', 'he sees'. But here it is only the third person, marked by the possessive suffix, that is undoubtedly passive, and **līhat-ku* (**lihatku*), **līhat-mu* (**lihatmu*) would correspond to it. The prefix *kau-*, on the contrary, and hence probably also the prefixed *ku-*, appear rather as subject forms, *ku* instead of *aku*, *kau*, from the evidence in related languages, perhaps not even as a contraction of *aṅkau* (*angkau*), but as a full but unexpanded personal pronoun. But this defective system is fundamentally different from the phono-mechanical interaction between stem and affixes that we encounter elsewhere in the world's languages. If we distinguish between the phonetic formal means and their values – the formal categories – we will be able to see that there were originally as many of the one as of the other, but that later the number of departments was reduced, so to speak, and the existing officials had to be redeployed to whatever remaining ministries they seemed to best fit.

With justice we praise the so-called agglutinative languages for their greater logical consistency in comparison to the Indo-European languages: in these languages every grammatical category has as a rule only a single phonetic expression, and the word order tends to be regulated by set rules. It is only in the subjective sense that such logical behaviour is less meritorious than it might seem to be, since it is in any case the most comfortable, being the simplest – whoever dispenses with what is superfluous can hardly pride himself on a feat of strength. But on the other hand what we praise in our 'inflecting' languages as opposed to the agglutinative ones belongs really to the inner language form rather than the outer form.

10. The agglutination theory has got us to this point: wherever we encountered a sound change in a word stem or affix, it was mechanically conditioned by the phonetic form, quantity and tone of the neighbouring syllables, and everywhere it varied within more or less narrowly measured boundaries. This is the explanation for the series of vowel strengthening and weakening in the Indo-European languages, as in Sanskrit *i-e-ai, u-o-au, r-ar-ār*. This applies also to the sharpening and lengthening in some languages of the Finno-Ugric family, as in Manchu

371 *legi, scripsi,* an die Plurale Fuss – Füsse, Mann – Männer, Mensch – Menschen u. s. w. Dann aber ist vielleicht der bescheidene Name des Defectivsystemes angebrachter, als der wohltönende der Flexion. Denn in der That liegt hier, ganz ähnlich wie bei *sum –fui*, λέγω – εἶπον u. a. m., der Fall so, dass {203} nicht jeder Stamm jede Form annimmt und sich nicht jede Form an jeden Stamm fügt. Ähnliches kommt auch anderwärts vor, ohne dass man viel Wesens davon machte, z. B. in den Pluralbildungen vieler afrikanischer und amerikanischer, in den Conjugationen kaukasischer Sprachen. Entschieden defectiv ist auch jenes angebliche Passivum des Malaischen: *ku-līhat, kau-līhat, līhat-ña* = ich sehe, du siehst, er sieht. Hier ist nur die dritte, das Possessivsuffix tragende Person zweifellos passivisch, und ihr würde **līhat-ku, *līhat-mu* entsprechen. Das Präfix *kau-* dagegen, und folglich wohl auch das vorgefügte *ku-* erscheinen eher als Subjectsformen, *ku* statt *aku, kau* nach dem Zeugnisse verwandter Sprachen vielleicht nicht einmal als Abkürzung von *aṅkau*, sondern als volles aber unerweitertes Personalpronomen. Grundsätzlich aber ist dieses Defectivwesen verschieden von jenen lautmechanischen Wechselwirkungen zwischen Stamm und Affixen, denen man auch sonst weithin in der Sprachenwelt begegnet. Unterscheiden wir zwischen den lautlichen Formenmitteln und deren Werthen, – den Formkategorien –: so werden wir annehmen dürfen, dass ursprünglich der einen soviele verschiedene waren, wie der anderen, dass aber nachmals sozusagen die Zahl der Ämter vermindert, und die einmal vorhandenen Beamten den Ressorts mit zugetheilt wurden, zu denen sie am Besten zu passen schienen.

annimmt, 1889
Aehnliches 1889
so zu sagen 1889
Aemter vermindert 1889

Mit Recht rühmt man den eigentlich sogenannten agglutinirenden Sprachen grössere logische Folgerichtigkeit nach, als den indogermanischen: jede grammatische Kategorie hat dort in der Regel nur einen einzigen lautlichen Ausdruck, und die Wortfolge pflegt durch feste Gesetze geregelt zu sein. Nur ist im subjectiven Sinne ein solches logisches Verhalten weniger verdienstlich, als es scheinen könnte; denn es ist jedenfalls das bequemste, weil einfachste, – wer Überflüssiges abwirft, darf sich darum kaum einer Kraftleistung rühmen. Was man aber andrerseits zum Lobe unserer „flectirenden" im Gegensatze zu jenen agglutinirenden hervorhebt, gehört im Grunde der inneren Sprachform, nicht der äusseren an.

10. Bis hierher sind wir mit der Agglutinationstheorie gelangt: wo wir einem Lautwandel der Wortstämme oder Affixe begegneten, da war er mechanisch bedingt durch Lautwesen, Quantität und Ton der benachbarten Sylben und bewegte sich überall in mehr oder minder eng bemessenen Grenzen. So haben wir Reihen der Vocalsteigerung und Schwächung in den indogermanischen Sprachen, z. B. im Sanskrit *i–e–ai, u–o–au, r–ar–ār*; so ferner jene Schärfungen und Zerdehnungen in einigen Sprachen des finnisch-ugrischen Stammes; so im Mandschu

einen 1891 and 1901
es 1901
Vocalsteigung 1889

372 in the suffixes with *a-e-o* and with *ô-u-o*. But we may assume that in the history of languages another force has also made itself felt from the very beginning: the drive to play, which whimsically remoulds words. This drive will act according to sense, the sounds symbolizing picturesquely, and it is conceivable that in time it will claim exclusive rights over part of the sounds and leave the others unmolested, conforming now within its limits to a strict formal system – since even the freest creative process demands order and rules as soon as it becomes itself the rule through dominance. If these principles are correct, if there are indeed such possibilities, then the Semitic languages occupy a very unique position among the languages of the world. Examples of symbolic sound changes are known from many other parts of the world, but it is only in one family where we find languages in which the consonants alone are constant, marking stems and roots, and the vowels serve formation according to a unified system. Perhaps we should speak of *symbolization* here.

To avoid any misunderstandings, let us make the following observation. The initial cause of these phenomena could lie in isolated phono-mechanical processes: it is possible that, for example, *ja* or *aj* change into *e* or *i*, *aw* or *wa* into *o* or *u*, and then that by a false analogy a corresponding vowel change has emerged in other cases. But what granted analogy this power here? And when the rule has become fixed: how does it further influence speakers' feeling for their language? It seems to me that the analogy not only requires a tendency towards symbolization of sounds, but also encourages it strongly for the future. I believe I will be able to show in one of the following sections (Section V., Manner of pronunciation or expression of mood) that symbolization is indeed one of the oldest means of formation in human language. But it is difficult to see how the Indo-European languages could be put together in a single class with the Semitic languages: in Indo-European we have inner agglutination, but in Semitic only loose agglutination; in Indo-European only suffixes, in Semitic also prefixes. In Indo-European there are no possessive suffixes and no objective conjugation, which are both used in Semitic; but there the forming of compounds is not allowed – in Indo-European it is almost unrestricted. And how fundamentally different are the two types of languages in their sentence construction!

Symbolization is compatible with each of the forms we have discussed so far. In the Hamito-Semitic language family, symbolization must have already been active at the isolating stage of development, since here the formative elements are, depending on the languages, at times prefixes and at other times suffixes. Even the jewel in the crown of the Semitic languages,

372 Suffixe mit *a–e–o* und solche mit *ô–u–o*. Allein ||353|| wir dürfen annehmen, es habe in der Geschichte der Sprachen von Anfang an noch eine andere Macht ihr Wesen getrieben: jener Spieltrieb, der die Wörter nach freier Laune umgestaltet. Er wird sinnig walten, malerisch die Laute symbolisirend, und es ist denkbar, dass er mit der Zeit einen Theil der Laute aus|335|schliesslich für sich in Ansprach nimmt, den andern unangetastet lässt und sich nun innerhalb seiner Schranken in ein strenges Formensystem fügt. Denn auch das freieste Schaffen verlangt Ordnung und Regel, sobald es selbst durch Überhandnehmen zur Regel geworden ist. Sind diese Sätze richtig, sind solche Möglichkeiten gegeben, dann nehmen unter den Sprachen der Erde die semitischen eine ganz vereinzelte Stellung ein. Zwar Beispiele von symbolischem Lautwandel sind auch sonst vieler Orten nachzuweisen. Dass aber in einer Sprache die Consonanten allein beständig, stamm- und wurzelhaft, die Vocale nach einem einheitlichen Systeme in den {204} Dienst der Formung genommen sind: das findet sich nur einmal. Vielleicht sollte man hier von Symbolisation sprechen.

anderen 1889

Ueberhandnehmen 1889

gewonnen 1901

Um nun Missverständnissen zu begegnen, sei gleich hier Folgendes bemerkt. Der erste Anlass kann auch bei diesen Erscheinungen in vereinzelten lautmechanischen Vorgängen gelegen, es mag sich z. B. *ja* oder *aj* in *e* oder *i*, *aw* oder *wa* in *o* oder *u* verwandelt, und dann nach falscher Analogie, auch in anderen Fällen ein entsprechender Vocalwechsel Platz gegriffen haben. Allein: was gab der Analogie gerade hier diese Macht? Und wenn die Regel sich gefestigt hatte: wie wirkte sie nun weiter auf das Sprachgefühl? Mir scheint, es werde von ihr eine Neigung zur Symbolisation des Lautes nicht nur vorausgesetzt, sondern auch für die Zukunft mächtig gefördert. Dass in der That die Symbolisation zu den allerältesten Formungsmitteln der menschlichen Sprache gehöre, glaube ich in einem der nächsten Abschnitte (V., Ausspracheweise oder Stimmungsmimik) wahrscheinlich machen zu können. Wie man aber die indogermanischen Sprachen mit den semitischen in eine Classe zusammensperren konnte, ist schwer einzusehen: dort sehr innige, hier sehr lose Agglutination; dort nur Suffixe, hier auch Präfixe. Dort fehlen die Possessivsuffixe und die Objectivconjugation, die hier im Gebrauch sind; dafür ist hier die Bildung von Compositis verwehrt, – dort fast unbeschränkt. Und wie grundverschieden sind die beiden Sprachtypen im Satzbaue.

Die Symbolisation ist mit jeder der bisher besprochenen Formen verträglich. Im hamito-semitischen Sprachstamme muss sie schon auf der isolirenden Entwickelungsstufe gewaltet haben; denn hier sind die Afformativa sprachenweise bald Prä-, bald Suffixe. Selbst das Pracht- und Schaustück der semitischen Sprachen,

373 the case suffixes *-a, -i, -u,* might find their opposites in the vocal initials of the substantives in the Berber languages.

11. Wherever we have fluid boundaries, we must observe the principle: *Denominatio fit a potiori* (the name is given from the more potent quality). And even this modest requirement is not always easily satisfied: Coptic is more or less a language with rather loose agglutination, just like its Ancient Egyptian mother. But alongside this it exhibits again and again in its form-building a change in the stem vowels that reminds us of the Semitic pattern and would have to be just as symbolic. We encounter a similar state of affairs in Tibetan. The Semitic languages, for their part, use vocalism more for stem-building than for form-building in the narrower sense – although it can be disputed whether the case endings, nominative *-u*, genitive *-i*, accusative *-a*, are symbolic or agglutinative. But otherwise the mutual relations of the sentence parts and sentences are shown partly through auxiliary words and partly through affixes. Organic, formative change of the final sound in the stem is, by the way, also found in other places, such as the conjugation of the Bantu languages and Japanese.

12. To allow ourselves to put the Indo-European and the Hamito-Semitic languages into a single common *inflecting* class, we say that it is only in these languages that the principle of word formation is carried through: root, stem, word and every material word is formed. But this is not true of our language family. We have the pronouns *me, te,* ἐμὲ, σὲ etc.,[7] which exhibit absolutely no sign of form and appear never to have done so. Furthermore we have formless nominatives in feminine words ending in *ā*, and in words like Sanskrit *manas* 'spirit', and Latin *arbor* 'tree', etc. And if we follow the further development of our languages then we see the increasing collapse of the case forms, above all of that much-praised hybrid between the subject and predicate case that we call the nominative. If our Indo-European ancestors did not consider the distinction between the nominative and accusative in the neuter necessary, their descendants have gone even further with this. In the modern Romance languages a mutilated case of the substantive or adjective replaces the word stem, case-marking has disappeared almost completely, and only number is still expressed through word forms.

13. The *incorporating* languages – also called *polysynthetic* languages – have been put forward as a special class. This class is principally represented by native American languages, and

7 ἐμὲ and σὲ here correct the forms ἐμε and σε in the critical edition.

373 die Casussuffixe *–a, –i, –u,* dürfte in den vocalischen Anlauten der Substantiva in den Berbersprachen sein Widerspiel finden.

11. Wo man es mit fliessenden Grenzen zu thun hat, da gilt der Satz: Denominatio fit a potiori. Und auch dieser bescheidenen Anforderung ist nicht ||354|| allemal leicht genügt: Das Koptische ist sonst, gleich seiner altägyptischen Mutter, eine Sprache von ziemlich loser Agglutination. Daneben aber zeigt es hin und wieder in seiner Formenbildung einen Wandel der Stammvocale, der an das semitische Muster erinnert und wohl ebenso symbolisirend ist. Ähnlichem begegnen wir im Tibetischen. Die semitischen Sprachen ihrerseits verwenden den Vocalismus doch mehr zur Stamm- als zur Formenbildung im engeren Sinne. Ob die Casusendungen: Nominativ *–u*, Genitiv *–i*, Accusativ *–a*, symbolisch oder agglutinativ seien, darüber mag man streiten. Sonst aber werden die gegenseitigen Beziehungen der Satztheile und Sätze theils durch Hülfswörter, theils durch Affixe angezeigt. Organischer, formativer Wandel des Stammauslautes findet sich übrigens auch sonst; so in der Conjugation der Bantusprachen und des Japanischen.

denominatio fit a potiori, und 1889

Aehnlichem 1889

Casusendungen, 1889

12. Um die indogermanischen Sprachen mit den hamito-semitischen einer gemeinsamen flectirenden Classe zuweisen zu können, beruft man sich darauf, es sei erst in diesen Sprachen das Prinzip der Wortformung durchgeführt: Wurzel, Stamm, Wort und jedes Stoffwort geformt. Für unsern Sprachstamm jedoch trifft das nicht zu. Wir haben die Pronomina *me, te,* ἐμε, σε u. s. w., die keinerlei Formzeichen tragen und anscheinend nie ein solches getragen haben. Wir haben weiter formlose Nominative bei den Femininen auf *ā*, bei Wörtern wie sanskrit *manas*, der Geist, lateinisch *arbor* u. s. w. Und verfolgen wir die fernere Entwickelung unserer Sprachen, so zeigt sich ein zunehmender Verfall der Casusformen, zumal jenes vielgepriesenen Zwitterdinges zwischen Sub|336|jects- und Prädicatscasus, das man den Nominativ nennt. Hatten schon unsere indogermanischen Urahnen eine Unterscheidung zwischen Nominativ und Accusativ im Neutrum nicht für nöthig gehalten, so durften ihre Nachkommen hierin noch weiter gehen. In den heutigen romanischen Sprachen ersetzt ein verstümmelter Casus des Substantivums oder Adjectivums den Wortstamm, die Casusbildung ist bis auf den letzten Rest geschwunden, und nur noch die Zahl durch Wortformen ausgedrückt.

Wort, 1889

Subjekts- 1889 das 1891 dass 1901

Geschlecht und Zahl 1889

{205}

13. Als eine besondere Classe hat man die einverleibende, incorporirende aufgestellt, die man wohl auch die polysynthetische nennt. Sie ist hauptsächlich durch Sprachen der amerikanischen Ureinwohner vertreten, und

374 Wilhelm von Humboldt (1999 [1836], *Über die Verschiedenheit des menschlichen Sprachbaues*, § 17) gives a characterization of the class using the example of Mexican (Nahuatl): 'To bring about the union of the simple sentence into a single sound-linked form, [Mexican] picks out the *verb* as the true centre, adjoins, so far as possible, the governing and governed parts of the sentence to it, and gives this conjunction the stamp of a bonded whole by means of sound formation: *ni-naca-qua* "I eat meat". This combination of the substantive with the verb might be regarded as a compound verb, like the Greek κρεωφαγέω "flesh-eating"; but the language obviously conceives it otherwise. For if, for any reason, the substantive itself is not incorporated, it is replaced by the third person pronoun, a clear proof that with the verb, and contained in it, the language demands to possess, at the same time, the schema of construction: *ni-c-qua in nacatl* "I-eat-it, the meat". In its form, the sentence is to appear already concluded in the verb, and is thereafter only determined more closely, as if by apposition. To the Mexican way of thinking, the verb simply cannot be conceived without these qualifiers to complete it. So if no particular object is present, the language combines with the verb a special indefinite pronoun, employed in two different forms for persons and things: *ni-tla-qua* "I-something-eat"; *ni-te-tla-maca* "I-somebody-something-give". The language betrays with the utmost clarity its intention of having these composite forms appear as a whole. For if such a verb, containing within it the sentence itself, or as it were the schema of this, is located in a past time, and thereby receives the augment *o*, this latter position itself at the beginning of the composite, which clearly shows that these qualifiers always and necessarily belong to the verb, whereas the augment attaches to it only contingently, as an indication of pastness. Thus from *ni-nemi* "I live", which as an intransitive verb can bring no other pronouns with it, we have the perfect, *o-ni-nen* "I have lived", and from *maca* "give", we get *o-ni-c-te-maca-c*, "I have given it to somebody". But it is still more important that in words used for incorporation the language very carefully distinguishes an absolute and an incorporative form, a precaution without which the whole method would become uncertain for the understanding, and which must therefore be regarded as basic to it. In incorporation, as in compound words, *nouns* drop the endings which always accompany them in the absolute state and characterize them as nouns. *Meat*, which we found incorporated above as *naca*, is called *nacatl* in its absolute form. Of the incorporated *pronouns*, none is used separately in the same form.

374 Wilhelm von Humoldt (Über die Verschiedenheit des menschlichen Sprachbaues § .17 schildert ihre Eigenart am Beispiele des Mexicanischen (Nahuatl) wie folgt: „Um die Verknüpfung des einfachen Satzes in eine lautverbundene Form hervorzubringen, hebt das Mexikanische das Verbum als den wahren Mittelpunkt desselben heraus, fügt, soviel es möglich ist, die regierenden und regierten Theile des Satzes an dasselbe an, und giebt dieser Verknüpfung durch Lautformung das Gepräge eines verbundenen Ganzen: *ni–naca–qua*, ich–Fleisch–esse. Man könnte hier das mit dem Verbum verbundene Substantiv als ein zusammengesetztes Verbum gleich dem griechischen κρεοφαγέω ansehen; ||355|| die Sprache aber nimmt es offenbar anders. Denn wenn aus irgend einem Grunde das Substantivum nicht selbst einverleibt wird, so ersetzt sie es durch das Pronomen der dritten Person, zum deutlichen Beweise, dass sie mit dem Verbum und in ihm enthalten, zugleich das Schema der Construction zu haben verlangt: *ni–c–qua in nacatl*, ich–es–esse, das Fleisch. Der Satz soll seiner Form nach schon im Verbum abgeschlossen erscheinen, und wird nur nachher, gleichsam durch Apposition, näher bestimmt. Das Verbum lässt sich gar nicht ohne diese vervollständigenden Nebenbestimmungen nach mexicanischer Vorstellungsweise denken. Wenn daher kein bestimmtes Object dasteht, so verbindet die Sprache mit dem Verbum ein eigenes, in doppelter Form für Person und Sachen gebrauchtes, unbestimmtes Pronomen: *ni–tla–qua*, ich–etwas–esse, *ni–te–tla–maca*, ich–jemandem–etwas–gebe. Ihre Absicht, diese Zusammenfügungen als ein Ganzes erscheinen zu lassen, bekundet die Sprache auf das Deutlichste. Denn wenn ein solches, den Satz selbst oder gleichsam sein Schema in sich fassendes Verbum in eine vergangene Zeit gestellt wird, und dadurch das Augment |337| *o–* erhält, so stellt sich dieses an den Anfang der Zusammenfügung, was klar anzeigt, dass jene Nebenbestimmungen dem Verbum immer und nothwendig angehören, das Augment ihm aber nur gelegentlich, als Vergangenheits-Andeutung, hinzutritt. So ist von *ni–nemi*, ich lebe, das als ein intransitives Verbum keine anderen Pronomina mit sich führen {206} kann, das Perfectum *o-ni-nen*, ich habe gelebt, von *maca*, geben, *o-ni-c-te-maca-c*, ich habe es jemandem gegeben. Noch wichtiger aber ist es, dass die Sprache für die zur Einverleibung gebrauchten Wörter sehr sorgfältig eine absolute und eine Einverleibungsform unterscheidet, eine Vorsicht, ohne welche diese ganze Methode misslich für das Verständniss werden würde, und die man daher als die Grundlage derselben anzusehen hat. Die Nomina legen in der Einverleibung ebenso wie in zusammengesetzten Wörtern, die Endungen ab, welche sie im absoluten Zustande immer begleiten und sie als Nomina charakterisiren. „Fleisch", das wir im Vorigen als *naca* fanden, heisst absolut *nacatl*. Von den einverleibten Pronominen wird keines in gleicher Form abgesondert gebraucht.

(Ueber 1889
§ 17 1889
Mexikanischen 1889
Eine 1889 and 1891
die mexikanische Sprache 1889
die mexicanische Sprache 1891
κρεωφαγέω 1889
nimmt es aber 1889 and 1891
mexikanischer 1889
Objekt 1889
deutlichste. 1889
Wörtern 1889
begleiten, 1889

375 The two indefinite pronouns simply do not occur in the language in the absolute state. Those referring to a specific object have a form more or less different from their independent form. But the method described already shows automatically that the incorporative form must have two different forms, one for the governing and one for the governed pronoun. The independent personal pronouns can, indeed, be placed for special emphasis ahead of the forms here depicted, but the incorporated pronouns relating to them are not therefore absent. The subject of the sentence expressed in a particular word is not incorporated; but its presence is evinced in the form by the fact that in the third person a governing pronoun indicating it is always missing.

'If we survey the different ways in which even simple sentences may be represented in the understanding, we readily perceive that the strict system of incorporation cannot be carried through all the different cases. Concepts must often be extricated in single words from the form, which cannot encompass everything. But in this the language always pursues the path once chosen, and invents new artificial aids where it runs into difficulties. Thus if, for example, one thing is said to happen in relation to another, either for or against it, and the definite governed pronoun would now produce obscurity, since it would have to relate to two objects, then the language creates a special class of such verbs, by means of an accreted ending, and otherwise proceeds as usual. The schema of the sentence now lies entirely once more in the coupled form, the indication of a thing done in the governed pronoun, and the by-relation to another in the ending; and now, with certainty of understanding, the language can let these two objects follow on outside the verb, without equipping them with marks of their relationship: *chihua*, "to make", *chihui-lia*, "to make for or against someone", with change of *a* into *i* by the law of assimilation, *ni-c-chihui-lia in no-piltzin ce calli*,[8] "I-make-it-for the my-son a house".

'The Mexican method of incorporation demonstrates a correct sense of sentence-formation in that it attaches the designation of its relations precisely to the verb, and thus to the point at which the sentence ties itself together into unity. It thereby differs essentially, and to its advantage, from the want of indication in Chinese, in which the verb cannot even be known with certainty from its position, but is often recognizable only materially, from its meaning.[9] But in the parts standing outside the verb in more complicated sentences,

8 *Calli* here corrects *catli* in the critical edition.

9 An error, which has since been corrected.

375 Die beiden unbestimmten kommen im absoluten Zustande gar nicht in der Sprache vor. Die auf ein bestimmtes Object gehenden haben eine von ihrer selbständigen mehr oder weniger verschiedene Form. Die beschriebene Methode zeigt aber schon von selbst, dass die Einverleibungsform eine doppelte haben müsse, eine für das regierende und eine für das regierte Pronomen. sein 1889 and 1891
Die selbständigen persönlichen Pronomina können zwar den hier geschilderten Formen zu besonderem Nachdruck vorgesetzt werden, die sich auf sie beziehenden einverleibten bleiben aber darum nicht weg. Das in einem eigenen Worte ausgedrückte Subject des Satzes wird nicht einverleibt, sein Subjekt 1889
Vorhandensein zeigt sich aber an der Form dadurch, dass in dieser allemal bei der dritten Person ein sie andeutendes regierendes Pronomen fehlt.

||356||

„Wenn man die Verschiedenheit der Art überschlägt, in welcher sich auch der einfache Satz dem Verstande darstellen kann, so sieht man leicht ein, dass das strenge Einverleibungssystem nicht durch alle verschiedenen Fälle durchgeführt werden kann. Es müssen daher oft Begriffe in einzelnen Wörtern aus der Form, welche sie nicht alle umschliessen kann, herausgestellt werden. Die Sprache verfolgt aber hierbei immer die einmal gewählte Bahn, und ersinnt, wo sie auf Schwierigkeiten stösst, neue künstliche Abhelfungsmittel. Wenn also z. B. eine Sache in Beziehung auf einen Anderen, für oder wider ihn, geschehen |338| soll, und nun das bestimmte regierte Pronomen, da es sich auf zwei Objecte beziehen müsste, Undeutlichkeit erregen würde, so bildet sie, vermittelst einer zuwachsenden Endung, eine eigene Gattung solcher Verben, und verfährt übrigens wie {207} gewöhnlich. Das Schema des Satzes liegt nun wieder vollständig in der verknüpften Form, die Andeutung einer verrichteten Sache im regierten Pronomen, die Nebenbeziehung auf einen Anderen in der Endung, und sie kann jetzt mit Sicherheit das Verständniss dieser beiden Objecte, ohne sie mit Kennzeichen ihrer Beziehung auszustatten, ausserhalb Objekte, 1889
nachfolgen lassen: *chihua*, machen, *chihui-lia*, für oder wider jemand machen, mit Veränderung des *a* in *i* nach dem Assimilationsgesetz: *ni-c-chihui-lia in no-piltzin ce catli* ich–es–mache–für der mein–Sohn ein Haus.

„Die mexicanische Einverleibungsmethode zeugt darin von einem richtigen Gefühle der Bildung des Satzes, dass sie die Bezeichnung seiner Beziehungen gerade an das Verbum anknüpft, also den Punkt, in welchem sich derselbe zur Einheit zusammenschlingt. Sie unterscheidet sich dadurch wesentlich und vortheilhaft von der chinesischen Andeutungslosigkeit, in welcher das Verbum nicht einmal sicher durch seine Stellung, sondern oft nur materiell an seiner Bedeutung kenntlich ist.[10] In den bei verwickelteren Sätzen ausserhalb des Verbums stehenden Theilen aber

10 Ein Irrthum, der seitdem berichtigt worden ist.

376 it is again entirely on a par with the latter. For in projecting the whole of its indicative activity upon the verb, it leaves the noun completely uninflected. It approximates, indeed, to the Sanskrit procedure, insofar as it truly provides the threads that link up the parts of the sentence; but for the rest, it stands in a remarkable opposition to the latter. In a wholly simple and natural way, Sanskrit designates every word as a constitutive part of the sentence. The incorporative method does not do this; where it cannot lump everything into one, it has markers emerging as arrows from the midpoint of the sentence, to show the directions in which particular parts must be looked for, in accordance with their relation to the sentence. We are not exempted from looking and guessing, being thrown back, rather, by this particular type of indication, into the opposite system of non-indication. But although this procedure thus has something in common with the other two, we would be mistaking its nature if we wished to regard it as a mixture of them, or to suppose that the inner linguistic sense had just not had the power to carry the system of indication throughout all parts of the language. This Mexican sentence structure obviously harbours, on the contrary, a peculiar mode of conception. The sentence is not to be constructed, not to be built up gradually out of parts, but must be given all at once as a form stamped into unity.'

I break off the quotation at this point. Humboldt now extends, with justice, the concept of incorporation to all cases where, in addition to the subject, the verb formally marks the object, and logically he could have extended it further to also cover the case where the subject alone is expressed in the conjugation; that is, where the finite verb also depicts a logical sentence. He includes here also the case where pronominal possessive affixes combine with substantives; he even mentions the Bahuvrîhi compounds of Sanskrit. Obviously this has more to do with the inner form than the outer form. A further statement from Humboldt is relevant to this point (ibid., p. 182 [141]): 'Strictly speaking, the *method of incorporation* is by nature itself in true opposition to inflection, in that the latter proceeds from the particular, while the former sets out from the whole. Only in part can it return again to inflection, through the victorious influence of the inner linguistic sense. But incorporation always betrays the fact that, owing to its lesser power, the objects are not presented to intuition with clarity and distinctness matching the points in those objects which impinge individually upon feeling.'

376 kommt sie der Letzteren wieder vollkommen gleich. Denn indem sie ihre ganze Andeutungs-Geschäftigkeit auf das Verbum wirft, lässt sie das Nomen durchaus beugungslos. Dem sanskritischen Verfahren nähert sie sich zwar insofern, als sie den die Theile des Satzes verknüpfenden Faden wirklich angiebt; übrigens aber steht sie mit demselben in einem merkwürdigen Gegensatz. Das Sanskrit bezeichnet auf ganz einfache und natürliche Weise jedes Wort als constitutiven Theil des Satzes. Die Einverleibungsmethode thut dies nicht, sondern lässt, wo sie nicht Alles in Eins zusammenschlagen kann, aus dem Mittelpunkte des Satzes Kennzeichen, gleichsam wie Spitzen, ausgehen, die Richtungen anzeigen, in welchen die einzelnen Theile, ihrem Verhältniss zum Satze gemäss, gesucht werden müssen. Des Suchens und Rathens wird man nicht überhoben, vielmehr durch die bestimmte Art der Andeutung in das ||357|| entgegengesetzte System der Andeutungslosigkeit zurückgeworfen. Wenn aber auch dies Verfahren auf diese Weise etwas mit den beiden übrigen |339| gemein hat, so würde man seine Natur dennoch verkennen, wenn man es als eine Mischung von Beiden ansehen, oder es so auffassen {208} wollte, als hätte der innere Sprachsinn nicht die Kraft besessen, das Andeutungssystem durch alle Theile der Sprache durchzuführen. Es liegt vielmehr offenbar in dieser mexicanischen Satzbildung eine eigenthümliche Vorstellungsweise. Der Satz soll nicht construirt, nicht aus Theilen allmählig aufgebaut, sondern als zur Einheit geprägte Form auf Einmal hingegeben werden."

letzten 1889 letzteren 1891
kann 1889
welchem 1889
beiden 1889 and 1891
mexikanischen 1889
allmählich 1889 and 1891

Ich breche hiermit das Citat ab. HUMBOLDT dehnt nun mit Recht den Begriff der Einverleibung auf alle die Fälle aus, wo das Verbum nächst der Subjects- auch die Objectsbezeichnung formal andeutet, und folgerichtig hätte er ihn wohl weiterhin auch auf den Fall erstrecken mögen, wo bloss das Subject in der Conjugation ausgedrückt wird, wo also doch auch das Verbum finitum einen logischen Satz darstellt. Auch den Fall zieht er hierher, wo pronominale Possessiv-Affixe sich mit Substantiven verbinden, ja selbst die Bahuvrîhi-Composita des Sanskrit erwähnt er. Offenbar handelt es sich dabei mehr um die innere Form als um die äussere. Und hierzu stimmt auch eine weitere Äusserung (S. 182:) „Die Einverleibungsmethode befindet sich, streng genommen, in ihrem Wesen selbst in wahrem Gegensatze mit der Flexion, indem diese vom Einzelnen, sie aber vom Ganzen ausgeht. Nur theilweise kann sie durch den siegreichen Einfluss des inneren Sprachsinnes wieder zu ihr zurückkehren. Immer aber verräth sich in ihr, dass durch seine geringere Stärke die Gegenstände sich nicht in gleicher Klarheit und Sonderung der in ihnen das Gefühl einzeln berührenden Punkte vor der Anschauung darlegen."

Subjekts- 1889
Objektsbeziehung 1889
blos das Subjekt 1889

377

If we confine ourselves to 'incorporation' in the sense of this process that results in sentence-words, as should perhaps be recommended in the interests of a more solid terminology, then the following questions arise:

a) What cases are incorporated in the verb? Just the subject case, as in the Indo-European languages, or also the object case? Just the accusative, as in the Semitic verb, or also the dative? Finally, other relations: kind, instrument, genitive to the object, etc. Cherokee creates verbal forms like the following:

galölisananihiha, I come in order to bind it repeatedly;
galŏ̈lisanega, I go in order to bind it repeatedly;
galŏ̈lidolihiha, I come in order to bind here and there;
galŏ̈lidolega, I go in order to bind here and there;
galŏ̈stanihiha, I come in order to bind with it;
galŏ̈stanega, I go in order to bind with it;
galŏ̈stisotiha, I bind with it repeatedly;
galŏ̈stisotanihiha, I come in order to bind with it repeatedly;
galŏ̈stisotanega, I go in order to bind with it repeatedly;
galŏ̈stanidoha, I bind here and there with it;
galŏ̈stanidoliliha, I come in order to bind with it here and there;
galŏ̈stanidolega, I go in order to bind with it here and there;
galŏ̈stisanidoha, I bind with it here and there repeatedly;
galŏ̈stisanidolihiha, I come in order to bind with it here and there repeatedly;
galŏ̈stisanidolega, I go in order to bind with it here and there repeatedly;
galŏ̈onihiha, I come in order to be finished with binding;
galŏ̈onega, I go in order to be finished with binding;
galŏ̈onisiha, I will be finished with or stop repeatedly binding, etc.; (cf. H.C. v. d. Gabelentz [1852], 'Kurze Grammatik der tscherokesischen Sprache', in Höfer's *Zeitschrift* III, p. 298).

In Cree, a language of the Algonkian family, not only the object, but also the genitive or dative that belongs to it is shown in the verb. And the Santhali language (Munda family) exhibits such forms as *dal-e-a-e* 'he will hit him', *dal-ae-a-e* 'he will hit for him', *dal-t-ae-a-e* 'he will hit his (one)', *dal-t-ae-t-iñ-a-e* 'he will hit my one of his', *dal-ae-t-ae-a-e* 'he will hit for his', *dal-ae-t-ae-t-iñ-a-e* 'he will hit for my one of his', etc.

377

Hält man sich, wie wohl im Interesse einer festeren Terminologie zu empfehlen wäre, lediglich an die satzwortbildende Einverleibung, so ergeben sich folgende Fragen:

a) Welche Casus werden dem Verbum einverleibt? Bloss der Subjectscasus, wie bei den indogermanischen Sprachen, – oder überdies der Objectscasus? bloss der accusativische, wie im semitischen Verbum, oder auch der dativische? – endlich noch andere Beziehungen: Art, Werkzeug, Genitiv zum Objecte u. s. w. Das Tscheroki bildet Verbalformen wie folgende:

Blos der Subjektscasus, 1889
Objektscasus? blos der accusativische 1889
Werkzeug 1889
Beziehungen, 1889
Objekte 1889
galŏ̈lisanihiha 1891
Tscheroki drückt in einem Worte aus: „ich gehe um wiederholt damit zu binden". Im […] 1889

galölisananihiha, ich komme um es wiederholt zu binden;
galŏ̈lisanega, ich gehe um es wiederholt zu binden;
galŏ̈lidolihiha, ich komme um hier und da zu binden:
galŏ̈lidolega, ich gehe um hier und da zu binden;
galŏ̈stanihiha, ich komme um damit zu binden;
galŏ̈stanega, ich gehe um damit zu binden;
galŏ̈stisotiha, ich binde wiederholt damit;
||358||
galŏ̈stisotanihiha, ich komme um wiederholt damit zu binden;
galŏ̈stisotanega, ich gehe um wiederholt damit zu binden;
galŏ̈stanidoha, ich binde hier und da damit;
|340|
galŏ̈stanidoliliha, ich komme um hier und da damit zu binden;
galŏ̈stanidolega, ich gehe um hier und da damit zu binden;
galŏ̈stisanidoha, ich binde wiederholt hier und da damit;
galŏ̈stisanidolihiha, ich komme um wiederholt hier und da damit zu binden;
galŏ̈stisanidolega, ich gehe um wiederholt hier und da damit zu binden;
galŏ̈onihiha, ich komme um mit Binden fertig zu werden;
galŏ̈onega, ich gehe um mit Binden fertig zu werden;
galŏ̈onisiha, ich werde fertig oder höre auf wiederholt damit zu binden u. s. w.; (vgl. H.C. v. d. Gabelentz, Kurze Grammatik der tscherokesischen Sprache, in Höfer's Zeitschrift III, S. 298).

galölidolihiha 1901
galöstanidolega 1901

Im Kri, einer Sprache der Algonkin-Familie, wird nicht nur das Object, sondern auch ein ihm zugehöriger Genitiv oder Dativus commodi im Verbum angezeigt. Und die (kolarische) Santalsprache weist Formen auf wie: *dal-e-a-e*, er wird ihn schlagen, *dal-ae-a-e*, er wird für ihn schlagen, *dal-t-ae-a-e*, er wird Seinen schlagen, *dal-t-ae-t-iñ-a-e*, er wird meinen Seinigen schlagen, *dal-ae-t-ae-a-e*, er wird für Seinen schlagen, *dal-ae-t-ae-t-iñ-a-e*, er wird für meinen Seinigen schlagen u. s. w.

Objekt, 1889
dativus 1889

378

b) Secondly we must ask whether only formal elements can be incorporated, or whether foreign, above all substantive, elements can also be incorporated. We may ignore fixed compounds, such as *handhaben, beratschlagen, verschlimmbessern* and similar,[11] which leaves us as representatives of this kind only those incorporating connections like those we find in some American languages. The term *polysynthesis* fits these well. In other – and by far the most – cases the verbal stem combines only with verbal, pronominal and perhaps prepositional auxiliary elements; that is, with formal additions.

c) We must also consider the negative side. In the vast majority of American languages the verb shoots its arrows at the substantive parts of the sentence, and these act simply as targets, and do not react through case marking; it is only through the topography of the sentence that we can see which arrows were intended for what targets. Exceptions, languages with better developed case forms, such as Yakama, Choctaw (Chahta) and its relatives, and Mutsun, are rare in America. By contrast, in the incorporation manifested in the languages of the Old World it is the rule that the substantive parts of the sentence – perhaps with the exception of the subject – answer with corresponding case-markings to the verbal forms that point to them. While Santhali has no accusative suffix, its close relative Kolh does.

However diverse the possibilities of incorporation and polysynthesis may be, all relevant phenomena belong, in terms of outer form, to the class of appending and composition. Symbolizing forms are perhaps compatible with this. In this way Cree symbolizes the subjunctive through the lengthening of the vowel in the stem: *ā* becomes *iā*, *ă*: *ē*, *e*: *iē*, *ī* in the initial position *ayi*, in the nucleus *ā*; *ĭ* becomes *e*, *u*: *wa*, *o* in the initial position *wē*, in the nucleus *io*.

14. A widespread type of formation is one that I would like to call *syntactic composition*. It is an apparent hybrid between word formation and syntax, between the isolating and agglutinative procedure. The essence of the so-called true compounds of the Indo-European languages consists in the fact that two or more word stems are combined with one other in such a way that

11 There are no etymologically transparent, precisely parallel English examples of Gabelentz's fixed incorporated compounds. *Handhaben* means to handle or manipulate, and is made up of the noun 'hand' plus the verb 'have'. *Beratschlagen* is to deliberate, made up of the noun *Ratschlag* 'suggestion' verbalized with the transitivizing prefix *be-* and the infinitive ending *-en*. *Verschlimmbessern* means to try to improve something but in fact make it worse; it is a blend of *verschlimmern*, 'make worse', and *verbessern*, 'make better', each of which contains an adjective – *schlimmer* 'worse' and *besser* 'better' – verbalized with the prefix *ver-* and the infinitive ending *-(e)n*.

378

b) Zweitens fragt es sich, ob nur Formenelemente, oder auch fremde, zumal substantivische Stoffelemente incorporirt werden. Von festen Compositis wie handhaben, berathschlagen, verschlimmbessern u. dgl. dürfen wir absehen, und so bleiben als Vertreter der zweiten Art wohl nur jene frei einverleiben{209}den Verbindungen, wie wir sie in einigen amerikanischen Sprachen antreffen. Für diese passt der Ausdruck Polysynthetismus. In anderen und wohl weitaus den meisten Fällen verbindet sich der Verbalstamm nur mit verbalen, pronominalen und etwa präpositionalen Hülfselementen, das heisst mit formalen Zusätzen.

incoporirt 1889
verschlimmbes-sern, 1889

c) Auch die negative Seite verdient Beachtung. In der grossen Mehrzahl der amerikanischen Sprachen schiesst das Verbum seine Pfeile nach den sustantivischen Satztheilen, und diese stehen lediglich Scheibe, ohne durch Casuszeichen zu reagiren; nur aus der Topographie des Satzes ist zu entnehmen, welcher Pfeil diesem und welcher jenem Ziele gilt. Ausnahmen, Sprachen mit besser entwickelten Casusformen, wie das Yakama, das Choctaw (Tschahta) mit seinen Verwandten und das Mutsun, gehören in Amerika zu den Seltenheiten. Dagegen ist es bei den Ein|341|verleibungen in den Sprachen der alten Welt geradezu die Regel, dass die substantivischen Satztheile – das Subject etwa ausgenommen, – auf die hindeutenden Verbalformen congruenzmässig mit entsprechenden Casuszeichen antworten. Hat das Santal kein Accusativsuffix, so besitzt dafür das ihm verschwisterte Kolh ein solches.

Substantivi-schen 1889
Subjekt 1889

||359||

So mannigfaltig demnach die Möglichkeiten der Incorporation und des Polysynthetismus sind, so gehören doch alle einschlägigen Erscheinungen der äusseren Form nach in die Classen der Anfügung und Zusammensetzung. Symbolisirende Formen sind damit wohl vereinbar. So kennzeichnet das Kri den Conjunctiv durch Zerdehnung des Stammvocales: *ā* wird *iā*, *ă*: *ē*, *e*: *iē*, *ī* im Anlaute *ayi*, im Inlaute *ā*; *ĭ* wird *e*, *u*: *wa*, *o* im Anlaute *wē*, im Inlaute *io*.

mannichfaltig 1889 and 1891
Stammvocals: 1889

14. Eine weit verbreitete Formungsart ist die, welche ich als syntaktische Composition bezeichnen möchte. Es ist dies ein scheinbar bedenkliches Zwitterding zwischen Wortbildung und Syntax, zwischen isolirendem und agglutinativem Verfahren. Das Wesen der sogenannten echten Composita der indogermanischen Sprachen besteht darin, dass zwei oder mehrere Wortstämme so miteinander verbunden sind, dass nur der letzte mit Formenzeichen versehen ist, während die grammatisch-logischen Beziehungen

ē, 1889
weitverbreitete 1889
ächten 1889 and 1891
mit einander 1889

only the last of them carries formal elements, while the grammatical-logical relations of the remaining stems are left unexpressed and can only be worked out from their material meaning and their position relative to one another. Sanskrit *putradârau* (dual) = 'child and wife', Latin *suovetaurilia* = 'sacrifice of pigs, sheep and bulls' are so-called *dvandva* with co-ordinated parts.

der übrigen unausgedrückt bleiben und nur aus dem Verhältnisse ihrer materiellen Bedeutungen und aus ihrer gegenseitigen Stellung zu erschliessen {210} sind. Sanskrit *putradârau* (Dual) = Kind und Gattin, lateinisch *suovetaurilia* = Opfer von Schweinen, Schafen und Stieren, sind sogenannte Dvandva mit coordinirten Gliedern.

379 In other cases the first part seems to have an attributive relation to the following one: *family home*, *triangle*, *high-rise*, *snow-white*, Latin *purificare*, Sanskrit *açvavid* (*aśvavid*) = 'horse-expert'. Similarly in Japanese we find: *asi-fayaku*, 'foot-fast' = 'fleet-footed', *ama-midu* (*ame-mizu*) 'rainwater', *nige-asi* 'run-feet', that is feet in such a position as if they were ready to flee, etc. Here we have constructions of the isolating type of languages encapsulated in word units. Similar forms can be found in languages in which the formation of compounds does not occur or occurs only to a small degree. There words that in their relations to one other express a unified representation remain independent, but are arranged alongside one other without any phonetic signs indicating their relations. In Manchurian *aniya biya* (pronounced *aña bya*) 'year-month' means first month of the year, rather than *aniya-i biya*; *gala obombi* 'to wash hands', perhaps more precisely 'hand-washing', *nure omimbi* 'wine-drinking'; as opposed to *nure be omimbi* = 'to drink the wine'. The verb forms a syntactic compound with its preceding object; if an accusative suffix *be* or a genitive marker *-i* were to stand in between them, the unity of the representation would be interrupted. Similarly in Modern Persian: *maj nūšīdam* = 'I drank wine', as opposed to the form with the object suffix: *maj-rā nūšīdam* = 'I drank the wine'. In this way in a form-poor language the suppression of the formative sounds at the right moment can be meaningful; it has in the truest sense a formal effect, precisely because of the contrast it creates.

15. Important for both the outer and inner form of languages are the phenomena of *word position*. To what extent is the order of the parts of the sentence free or rigid? Does the grammatical subject stand before or after the verb? Do the genitive, adjectival and adverbial attributes stand before or after their heads? Does the object stand before or after the verb, and is it distinguished from the adverbial attribute or not? And what about those phenomena that are referred to as nested structures? Do they occur? Are they required or only permitted? If we want to classify languages morphologically, it is one-sided to look only at word formation, since it is not the word but the sentence that is the organic unit of human speech; it is, to put it metaphorically, the living body. And not only the nature of its parts but also their topology is decisive for the organism. We will now, proceeding from the whole to the parts in an analytic fashion, begin with the examination of the phenomena of position.

379 In anderen Fällen verhält sich das erste Glied zu dem folgenden irgendwie attributiv: Vaterhaus, Dreieck, Hochbau, schneeweiss, *purificare*, sanskrit *açvavid* = pferdekundig. Ähnlich z. B. im Japanischen: *asi-fayaku*, fussgeschwind = schnellfüssig, *ama-midu*, Regenwasser, *nige-asi*, Rennfüsse, d. h. Füsse in der Stellung, als wären sie zum Fliehen bereit u. s. w. Wir haben da in Worteinheiten verkapselte Gebilde isolirender Sprachweise. Ähnliches findet sich auch in Sprachen, denen die Bildung zusammengesetzter Wörter entweder gänzlich versagt, oder doch nur in beschränktem Masse gestattet ist. Da bleiben die Wörter, die in ihrem Zusammenwirken eine einheitliche Vorstellung ausdrücken sollen, zwar selbständig, werden aber ohne lautliche Beziehungszeichen aneinander gereiht. Im Mandschu heisst *aniya biya* (spr. *aña bya*) „Jahresmonat" = erster Monat des Jahres, statt *aniya-i biya*; *gala obombi* die Hände waschen, genauer etwa „handwaschen", *nure omimbi*, Wein trinken; dagegen: *nure be omimbi* = den Wein trinken. Das Verbum bildet mit seinem vorausgehenden Objecte ein syntaktisches Compositum; stände hier das Accusativsuffix *be*, dort das Genitivzeichen *–i* dazwischen, so wäre die Einheit der Vorstellung unterbrochen. Ähnlich im Neupersischen: *maj nūšīdam* = ich trank Wein; dagegen mit dem Objectssuffixe: *maj-rā nūšīdam* = ich trank den Wein. So kann auch in einer formenarmen Sprache die rechtzeitige Unterdrückung der Formativlaute sinnig bedeutsam werden; sie wirkt, gerade durch den Gegensatz, im wahrsten Sinne formell.

15. Wichtig für die äussere wie für die innere Form der Sprachen sind die Erscheinungen der W o r t s t e l l u n g. Inwieweit ist die Ordnung der Satzglieder freibeweglich oder starr? Steht das grammatische Subject vor oder hinter ||360|| dem Verbum? Stehen die genitivischen, adjectivischen, adverbialen Attribute voran oder nach? Steht das Object vor oder hinter dem Verbum, und wird es vom adverbialen Attribute unterschieden oder nicht? Und wie steht es mit jenen Erscheinungen, die man als Einschachtelungen bezeichnet? Kommen sie vor? Sind sie geboten oder nur erlaubt? Will man die Sprachen morphologisch classificiren, so ist es einseitig, sich bloss an die Wortbildung zu halten, denn nicht das Wort, sondern der Satz ist die organische Einheit der menschlichen Rede, ist, bildlich gesprochen, der lebende Körper. Und nicht nur die Beschaffenheit, sondern auch die Topik seiner Glieder ist für den Organismus entscheidend. Wir werden hernach, analytisch vom Ganzen zu den Theilen fortschreitend, in erster Reihe die Stellungserscheinungen in's Auge fassen.

purificare, sanskrit açvavid = pferdekundig, schneeweiss. Aehnlich 1889
Regenwasser 1889
Aehnliches 1889
sich nun auch 1889 and 1891
trinken. 1889 and 1891
Objekte 1889
dsr 1901
werden. 1889 and 1891
Subjekt 1889
adjektivischen, 1889
sind 1889
blos 1889

380
§ 5. C. The Drive to Formation
The usual purpose of speech is communication. In many cases this purpose can be satisfied even with completely formless means, such as facial expressions, gestures or calls, which are not on a higher level. A 'Hey!' or 'Psst!' is a command, a 'Tut!' is a completely adequate expression of a moral or aesthetic judgement. Language returns, wherever possible, always to these shortest and most comfortable means, which are almost certainly among its most original means. Words, even whole sentences, change into formless exclamations and calls; the *bhōs, bhō* = 'Hey! You!' of the ancient Indians is a contraction of *bhavas*, 'one who is'; French *hélas*, English *alas* correspond to Italian *ah, mi lasso!*

We never say all that we think, but rather almost always let the rational listener fill out this or that detail. But many languages, when we take them by their word and letter, say much more than is necessary for understanding, certainly also more than the speaker thought and intended. Those conventional expressions which we use without thinking, but which we and others never omit from our speech: they have equal companions in linguistic forms. Those intuitions that, for example, are the basis of the grammatical genders have long since become strange to our thinking and sensation, but the language clings to the corresponding forms, and the German child will laugh at the foreigner who says *die Mond* (instead of *der Mond*) or *der Bein* (instead of *das Bein*). So strongly rooted is this useless weed in the soil of our sense of language.

But it not surprising that we cling to old relics through the force of habit; we cannot do otherwise. The only puzzling thing is what might have driven our ancestors to burden their speech with so much useless bric-a-brac. We have double reason to ask them this question: first, because they bear the responsibility, and second, because nowhere else do we enjoy such a wide-reaching view into the prehistory of language.

Much in their language seems useless to us, although it must certainly have had a purpose for them, even if it was not the purpose of communication. For the purpose of communication it is mere decoration and ballast, and we must explain where this came from and what it was for. Why did the ancestors of the Veda-singers keep things that the English have long since thrown on the scrap heap, such as genders, agreement and countless forms that sound different but mean the same thing, each of which was applicable only to a restricted subset of the vocabulary?

380

§ .5. C. Der Formungstrieb.

Der regelmässige Zweck der Rede ist Mittheilung. Diesem Zwecke kann in vielen Fällen auch durch ganz formlose Mittel genügt werden, durch Mienen, Gesten oder Zurufe, die auf nicht höherer Stufe stehen. Ein He! oder St! ist ein Befehl, ein Pfui! ist der völlig zureichende Ausdruck eines sittlichen oder ästhetischen Urtheils. Die Sprache kehrt, wo es angeht, immer und immer wieder zu diesen kürzesten und bequemsten Mitteln zurück, die ja gewiss auch zu ihren ursprünglichsten Mitteln gehören. Wörter, ja ganze Sätze verwandelt sie in formlose Aus- und Zurufe; das *bhōs*, *bhō* = He! Sie! der alten Inder gilt für eine Zusammenziehung von *bhavas*, Seiender; das französische *hélas*, englisch *alas* entspricht dem italienischen *ah, mi lasso!*

ist ist 1901
italiänischen 1889

Wir sagen wohl nie Alles, was wir denken, lassen fast immer den |343| verständigen Hörer noch dies und jenes hinzuergänzen. Und doch sagen viele Sprachen, wenn man sie beim Worte und beim Buchstaben nimmt, weit mehr, als zur Verständigung nöthig ist, gewiss auch mehr, als der Redner gedacht und beabsichtigt hat. Jene conventionellen Redensarten, bei denen man nichts denkt, und die man doch weder selbst vernachlässigen noch bei Anderen vermissen mag: sie haben in den sprachlichen Formen ihre ebenbürtigen Seitenstücke. Jene Anschauungen z. B., die dem grammatischen Geschlechte zu Grunde liegen, sind längst unserm Denken und Empfinden fremd geworden, und doch hält die Sprache an den entsprechenden Formen fest, und das deutsche Kind lacht über den Ausländer, der etwa „die Mond" oder „der Bein" sagt. So fest wurzelt noch das nutzlose Gewächs im Boden unseres Sprachgefühls.

Alles 1889
hinzu ergänzen. 1889
man sich nichts 1889 and 1891
unserem 1889
Ausländer 1889

Doch das ist kein Wunder, dass wir mit der Kraft der Gewohnheit am Altererbten festhalten; das kann ja nicht anders sein. Räthselhaft ist es nur, was unsere Urahnen dazu getrieben haben mag, ihre Rede mit so vielem zweck||361||losen Tand zu belasten. Denn an sie müssen wir die Frage richten aus doppeltem Grunde: einmal, weil sie die Verantwortlichkeit trifft, und zweitens, weil uns nirgends sonst ein so weitreichender Blick in die Vorgeschichte der Sprache vergönnt ist.

einmal 1889

Für zwecklos aber gilt uns in ihrer Sprache Vieles, was doch in ihrem, der Alten, Sinne gewiss auch seinen Zweck haben musste, wenn das auch nicht der Zweck der Mittheilung war. {212} Dem gegenüber ist es Tand und Ballast, und es gilt zu erklären, woher und wozu dieser? Warum hielten die Ahnen der Vedensänger auf Dinge, die die Engländer längst in's alte Eisen geworfen haben: auf das Genuswesen, auf die Congruenz und auf jene Menge verschieden lautender und gleichwerthiger Formen, deren jede nur für einen beschränkten Kreis des Wortschatzes brauchbar ist?

Jener, 1889
haben, 1889

381 That is, where did this luxury of the different conjugations and declensions come from?

Here, as always, general linguistics must seek contact with the history of human civilization. But if in our case linguistics wants to compare, for instance, a European steam engine or one of those gigantic railway bridges of our engineers with those richly and tastefully decorated instruments of a savage people that serve them as unsophisticated tools, and if it then places our languages alongside those agglutinating languages of brown and black people, then it is as if the roles were suddenly reversed, since it is on our side that the waste of form is to be found, and over there the judicious, purposeful economy. The parallel remains, however, even when the lines seem to run in opposite directions. And precisely that which those youthful barbarian peoples practise on their tools, that is what our race also did with language in its youth.

It is just a higher level of the drive to play, that joy at free, artistic formation, which in fresh fancy puts the stamp of its own individuality and mood on every creation. Even if the artistic achievement may have be slight, that extra effort I made on my work over and above bare utility was already a piece of love and gave for all time the dead matter a breath of the personal. And precisely the same thing happened with language. The soul demands something more than that simple business-like style that says in objective clarity all that is necessary and nothing more. It wants to identify itself with the thing, how it relates to its world, temperamentally, fancifully, moodily. I repeat the expression from earlier: the soul wants not only to express something but also itself, and wants not only to compel the listener to share a thought but also to share a feeling. Here the soul gives freely from its abundance – the soul is indeed so rich that it adds some of its own ingredient even to the smallest bit, at first according to the inspiration of the moment, apparently without rule but always meaningfully; then as time progresses ever more under the force of habitual norms. Insofar as the form doublets of our languages date from prehistoric times, they may be considered monuments of that period of unbridled desire for moulding and shaping.

We do not know whether those sober languages that we describe as agglutinative formerly went through such a stage, and I see no means to prove whether they did or not. If their youth was similar to that of Indo-European then we must admit that they 'put away childish things' much earlier.[12] And even this cannot be considered either a merit or a fortune on their part. The same drive to mould and shape,

12 'Put away childish things' is an allusion to Corinthians 13:11.

381 Woher also der Luxus der verschiedenen Conjugationen und Declinationen?

Die allgemeine Sprachwissenschaft muss hier wie immer Fühlung suchen mit der Gesittungsgeschichte der Menschheit. Will sie aber in unserem Falle etwa eine europäische Dampfmaschine oder eine jener riesigen Eisenbahnbrücken unserer Ingenieure mit den als Werkzeuge rohen und doch so reich und geschmackvoll verzierten Geräthen eines wilden Volkes vergleichen, und stellt sie dann unseren Sprachen jene agglutinirenden der braunen oder schwarzen Menschen gegenüber, so ist es, als wären plötzlich die Rollen vertauscht. Denn nun ist auf unserer |344| Seite die Formverschwendung, und drüben die verständige, zielbewusste Ökonomie. Die Parallele bleibt aber doch, auch wenn die Linien in entgegengesetzten Richtungen zu laufen scheinen. Und eben das, was jene jugendlichen Barbarenvölker an ihren Geräthen üben, das hat unsere Rasse in ihrer Jugendzeit auch mit der Sprache gethan.

es 1889

Oekonomie. 1889 and 1891

Raçe 1889

Es ist doch nur eine höhere Stufe des Spieltriebes, jenes Gefallen an freier, künstlerischer Formung, das in frischer Laune jeder Schöpfung den Stempel der eigenen Individualität und Stimmung aufdrücken muss. Es sei die künstlerische Leistung noch so gering: schon jener Überschuss von Arbeit, die ich meinem Werke über den blossen Nützlichkeitsbedarf hinaus zugewendet habe, ist ein Stück Liebe gewesen und hat dem todten Stoffe für alle Zeiten einen Hauch des Persönlichen gegeben. Und ebenso geschah es mit der Sprache. Die Seele verlangte ein Mehreres als jenen Geschäftsstil, der in objectiver Klarheit alles Nothwendige sagt und weiter nichts. Sie will in der Sache sich selbst wiederfinden, wie sie sich ihrer Welt gegenüber gemüthvoll, phantastisch, launenhaft verhält, will, – dass ich den Ausdruck wiederhole, – nicht nur etwas, sondern auch sich selbst aussprechen, und wird um so sicherer den Hörer nicht nur zum Mitdenken, sondern auch zum Mitfühlen zwingen. Da wirthschaftet sie aus dem Vollen, – sie ist ja so reich; da {213} wird auch dem Kleinsten etwas von eigener Zuthat angeheftet, erst nach der Eingebung des Augenblickes, scheinbar ||362|| regellos und doch immer bedeutsam; dann je länger je mehr unter dem Zwange gewohnheitsmässiger Normen. Soweit die Formdoubletten unserer Sprachen aus Urzeiten herrühren, mögen sie als Denkmäler jener Periode ungezügelter Gestaltungslust gelten.

Ueberschuss 1889

gemüthsvoll, 1889

erst, 1889 and 1891

Ob jene nüchternen Sprachen, die wir agglutinirende nennen, vormals einen ähnlichen Zustand durchlebt haben, wissen wir nicht, und ich sehe kein Mittel, einen Beweis für oder wider zu führen. War ihr Jugendleben dem indogermanischen ähnlich, so müssen wir zugestehen, dass sie viel früher „abgelegt haben, was kindisch war“. Und auch dies ist ihnen weder als Verdienst, noch als Glück anzurechnen. Derselbe Gestaltungstrieb,

382 whose half-wilted fruits still hang from our languages, created in later periods the wonders of European and Indian art; form was made a purpose in itself, and in just this way the purposes were elevated, the technique more rigorously trained. Where Gothic cathedrals rose up, those bold and yet sober iron bridges and the prosaic machines that roll over them were eventually invented. And where the Vedic hymns and the Homeric songs were composed, philosophy later could use languages for which the highest tasks were not too high.

It is strange, however, that the same praise is due not only to the mature modern Indo-European languages, which strive with all their might to rid themselves of formal pomp, but also to Chinese, which has been trying to get rid of it for at least four thousand years. A deep, thoughtful lyric, a rich, finely carved prose, serviceable and appropriate to every flight of thought, blossomed in the soil of an isolating language. And the drive to formation is no weaker here than it was among our prehistoric Indo-European ancestors; it has simply taken a different course and worked on different matter: not word composition and word formation, but syntax.

I believe that wherever poetry and rhetoric bloom we may speak with confidence of the linguistic drive to formation. And where do poetry and rhetoric not bloom, or where does not at least one of them bloom? We read about the artful and powerful speeches of American Indians, about the delicate, eloquent pleas in the court procedures of the Bantu negroes, we hear ever more about the song repertoire of the Ural-Altaic and Malay peoples. Indeed, even the poor, degenerate bushmen use their chortling language for impressive tales that play with sound, artworks which are the equal of their masterful animal silhouettes, and just like them have flowed out of the double spring of fine observation of nature and a creative drive to mould and shape.

If we now judge languages according their suitability for what I will call 'business' purposes, for want of a better word, we may ask: How much do they require for the purpose of sufficient, completely clear communication of thought? It is immediately apparent that this is dependent on the type of the thought to be communicated and its scope. A certain number of individual representations and a certain number of logical relations demand expression and find it. And the extent of this requirement will be very different depending on the intellectual life of each people. Now we can ask empirically: How does the formal apparatus of languages relate to their 'business' requirements? There we will find a more or less significant surplus. The more we get to know a language, the richer its repertoire of grammatical synonyms will unfold before us.

382 dessen halbwelke Früchte noch unsern Sprachen anhaften, hat nachmals die Wunderwerke europäischer und indischer Kunst geschaffen; die Form wurde zum Zwecke erhoben, und eben damit wurden die Zwecke höher gesetzt, die Technik härter geschult. Wo sich gothische Dome erheben, da erfand man schliesslich auch jene küh|345|nen und doch so nüchternen eisernen Brücken und die prosaischen Maschinen, die darüber rollen. Und wo die Vedenlieder und die homerischen Gesänge gedichtet wurden, da fand nachmals die Philosophie Sprachen vor, denen die höchsten Aufgaben nicht zu hoch waren.

Seltsam aber, das gleiche Lob gilt nicht nur von den gealterten neueren indogermanischen Sprachen, die sich des Formengepränges nach Kräften entledigen, sondern auch vom Chinesischen, das seiner seit mindestens viertausend Jahren entbehrt. Eine innige, sinnige Lyrik, eine reiche fein ausgemeisselte Prosa, jedem Fluge des Gedankens dienstbar und gerecht, sind auf dem Boden einer isolirenden Sprache erblüht. Und schwerlich ist der Formungstrieb hier schwächer, als er bei unsern indogermanischen Urahnen war; er hat nur andere Bahnen eingeschlagen und einen anderen Stoff bearbeitet: nicht das Wortbildungs- und Wortformungswesen, sondern die Syntax.

gealterten, 1889
reiche, 1889 and 1891

Ich glaube nun, wo Poesie oder Rhetorik blüht, da dürfe man getrost auch von sprachlichem Formungstriebe reden. Und wo blühen nicht die Beiden, oder wo blüht nicht wenigstens die eine oder andere von ihnen? Wir lesen von den kunst- und kraftvollen Reden amerikanischer Indianer, von den zierlich gewandten Plaidoyers in den Gerichtsverhandlungen der Bantu{214}neger, erfahren immer mehr von den Liederschätzen der ural-altaischen und der malaischen Völker. Ja selbst die armen, verkommenen Buschmänner verwenden ihre glucksende Sprache zu ausdrucksvoll klangspielenden Erzählungen, Kunstwerken, ebenbürdig ihren meisterhaften Thiersilhouetten, gleich diesen entflossen der doppelten Quelle feinfühliger Naturbeobachtung und schöpferischer Gestaltungslust.

Beidens 1901
Plaidoyés 1889
kluksende 1889
Kunstwerken 1889
ebenbürtig 1889 and 1891

Legen wir an die Sprachen den Massstab, den ich in Ermangelung eines passenderen Wortes den geschäftlichen nennen will, fragen wir: Wieviel haben ||363|| sie nöthig zum Zwecke der zureichenden, völlig deutlichen Gedankenmittheilung? so ergiebt sich erstens ohne Weiteres, dass dies abhängig ist von Art und Umfang der mitzutheilenden Gedanken. So und soviele Einzelvorstellungen, so und soviele logische Beziehungen verlangen Ausdruck und finden ihn. Und dies Bedürfnissmass wird je nach dem Gedankenleben der einzelnen Völker ein sehr verschiedenes sein. Nun stellen wir zweitens an die Erfahrung die Frage: Wie verhält sich der Formenvorrath der Sprachen zu diesen ihren geschäftlichen Erfordernissen? Da wird sich ein mehr oder minder bedeutender Über|346|schuss herausstellen. Je genauer wir Sprachen kennen lernen, desto reicher pflegt sich vor unseren Blicken ihre grammatische Synonymik zu entfalten.

eine 1901
wieviel 1889
Bedürfniss-maass 1889
Ueberschuss 1889

383

From this 'business' perspective, everything that transmits the same thought content to the listener counts as synonymous. Perhaps it would be good to present three of the most obvious examples here:

1. Alongside the active voice, many languages also possess the passive; Tagalog and its relatives can even turn the instrument and place of an event into the subject of the sentence. No matter whether I say: 'I'm searching for the book in the room with the light' or 'The book is being searched for by me with the light in the room', or 'The room is being searched by me with the light for the book', or 'The light is being used by me in the room to search for the book', the listener is always being told about the same fact.

2. The same is true of cases where we use a predicative or attributive expression, whether we say: 'high mountain' or 'the mountain is high'. Both arouse the same complete representation. And to climb into higher regions of syntax, the period is of the same value as the individual sentences that are lined up through conjunctions, into which we divide it. Such doublets are common in the world's languages, even if the construction of the period might resemble a train made up of carriages coupled together, as it does in the Ural-Altaic languages, and even if the repertoire of conjunctions might be quite meagre.

3. In the great majority of languages word position is bound to narrow and strict rules. But, as far as I know, nowhere are these rules so inviolable and unavoidable that they do not permit certain freedoms. The human mind, in its inertia, would no doubt like to make the habitual into a binding rule. But it is not always willing to walk along a prescribed marching route; it reserves certain freedoms for itself, it even seems to acquire them surreptitiously, making use of the laws of language by submitting itself to them. French is well known for delivering masterpieces of this art; Chinese too. But even many of those languages that are normally considered 'formless' manifest precisely at this point an unmistakable urge towards the free shaping of the sentence.

That in all of this and in many other things languages allow themselves a certain degree of luxury is undeniable. But even so this apparent luxury must correspond to a need, since whatever does not answer to a need has no long-term existence. In our case this driving or preserving need cannot be objective, demanded by the thing. Consequently, it must be justified subjectively, that is, in the soul of the speaker; it is not directed at the thing but at the form itself. In other words:

383

Es hat aber von dem gewählten rein geschäftlichen Standpunkte aus alles das als synonym zu gelten, wodurch dem Hörer derselbe Gedankeninhalt übermittelt wird. Vielleicht ist es gut, hierfür drei der nächstliegenden. Beispiele anzuführen.

1. Neben der activen Ausdrucksweise besitzen viele Sprachen noch die passive, das Tagalische und seine Verwandten können sogar das Werkzeug und den Ort der Handlung zum Subjecte des Satzes erheben. Ob ich nun sage: Ich suche das Buch mit dem Lichte in der Kammer, – oder: Das Buch wird von mir mit dem Lichte in der Kammer gesucht, – oder: Die Kammer wird von mir mit dem Lichte nach dem Buche durchsucht, – oder: Das Licht wird von mir in der Kammer zum Suchen nach dem Buche gebraucht: – immer erfährt der Hörer genau dieselbe Thatsache.

Subjekte 1889
das 1889
die 1889
das 1889

2. Das Gleiche gilt, ob man den prädicativen oder den attributiven Ausdruck wählt, ob man etwa sagt: Hoher Berg, oder der Berg ist hoch. Beides erweckt dieselbe Gesammtvorstellung. Und um in höhere Kreise der Syntax zu steigen, {215} auch die Periode ist gleichwerthig den durch Conjunctionen aneinandergereihten Einzelsätzen, in die wir sie auflösen. Auch solche Doubletten sind in der Sprachenwelt häufig, mag auch der Periodenbau, wie der ural-altaische, einem aneinandergekoppelten Eisenbahnzuge gleichen, und der Vorrath an Conjunctionen ärmlich genug sein.

oder: 1889 and 1891
uralaltaische, 1889 and 1891

3. Wohl in weitaus den meisten Sprachen ist die Wortstellung an enge und strenge Regeln gebunden. So unverbrüchlich und unumgänglich in allen Stücken sind aber diese Regeln meines Wissens nirgends, dass sie nicht doch noch gewisse Freiheiten zuliessen. Der menschliche Geist lässt wohl gern in seiner Trägheit das Gewöhnliche zur bindenden Regel werden. Aber nicht immer ist er gewillt, mit Zwangspass auf gebundener Marschroute zu wandern; gewisse Freiheiten behält er sich vor, scheint wohl gar sie zu erschleichen, wenn er sich die Gesetze der Sprache dienstbar macht, indem er sich ihnen unterwirft. Das Französische liefert bekanntlich Meisterstücke dieser Kunst; das Chinesische desgleichen. Doch auch viele jener Sprachen, die man sonst wohl ||364|| als formlose schilt, äussern gerade in diesem Punkte einen unverkennbaren Drang nach freierer Gestaltung des Satzes.

|347|

Dass in Alledem und noch in vielen anderen Dingen die Sprachen sich einen gewissen Luxus gestatten, ist doch unleugbar. Und gleichwohl muss auch dieser scheinbare Luxus einem Bedürfnisse entsprechen; denn was keinem Bedürfnisse entspricht, hat auf die Dauer keinen Bestand. Nun kann in unserem Falle das treibende oder erhaltende Bedürfniss ein objectiv, durch die Sache gebotenes nicht sein. Folglich muss es subjectiv, das heisst in der Seele des Redenden begründet sein und nicht der Sache, sondern der Form selbst gelten. Mit anderen Worten:

objektiv, 1889
subjektiv, 1889

384 I want not only to say *this*, but I want also to say it *like this*, and to be able to say it differently as I please, depending on my mood and the circumstances. Obviously a drive to formation manifests itself here as well, even if it is modest in its demands.

We must surely acknowledge this drive from the beginning in every human language. I would like to deny it perhaps only to those impoverished newly born hybrid languages, those bedraggled languages that are created for use in international intercourse and exist often only briefly, and indeed really for business use. Depending on its degree and direction, the drive manifests itself differently, and for this reason only the most in-depth language competence is capable of judging it justly. An exhaustive synonymy, of both the lexical and grammatical aspects of the language, would have to lead to knowledge of how and where the language-creating mind has actuated the urge for subjective shaping of the world. And then etymology would have to demonstrate what means the mind has made use of for this shaping work. Here we will stray in the case of most languages into the shadowy kingdom of hypotheses. In the end, it only the deepest knowledge of the living use of the language that can decide the important points: how far the formal consciousness is still alive in the language or whether the form is simply maintained out of inertial habit, and how far an originally material means is felt to be formal. For this reason it seems to me that the questions that belong here are rather delicate, and that science can only hesitantly approach them, after long historical-critical groundwork. Only that which is available to us right now permits an immediate judgement.

It is something else to refute a claim through the proof of the opposite, something else to shift the burden of proof onto the shoulders of the opponent who made the claim. But I cannot deliver the positive, inductive proof of my opinion in accordance with my fundamental requirements. I hope now to be able to claim an *a priori* probability for my view, however, and in this case the burden of proof falls on those who charge the great majority of languages with formlessness, since the controversial point is precisely whether the linguistic drive to formation is something generally human or is only a prerogative of certain, especially gifted races. In trying to discuss this point, it was inevitable that I jumped ahead into that part of our science that concerns itself with the evaluation of languages.

But before we move onto this, we need to examine some of the most important means and reasons of language formation.

384 ich will nicht nur *das* sagen, sondern ich will es auch *so* sagen und es nach Laune und Umständen auch noch ganz anders sagen können. Offenbar bekundet sich auch hier ein Formungstrieb, mag er auch noch so bescheiden in seinen Ansprüchen sein.

Diesen Trieb müssen wir wohl von vornherein jeder menschlichen Sprache zuerkennen. Ganz absprechen möchte ich ihn höchstens etwa jenen armseligen neugeborenen Blendlingssprachen, jenen verwahrlosten, die sich der internationale Verkehr oft nur zum flüchtigen Gebrauche erzeugt, und zwar {216} recht eigentlich zum geschäftlichen Gebrauche. Nach Mass und Richtung äussert der Trieb sich verschieden, und insoweit vermag nur die eingehendste Sprachkenntniss ihn gerecht zu beurtheilen. Eine erschöpfende Synonymik, sowohl eine lexikalische wie eine grammatische, müsste zu der Erkenntniss führen, inwieweit und wo der sprachbildende Geist den Drang nach subjectiver Gestaltung der Welt bethätigt. Und dann müsste die Etymologie nachweisen, welcher Mittel er sich zu dieser gestaltenden Arbeit bedient habe. Da wird man denn bei den meisten Sprachen sehr bald in das Nebelreich der Hypothesen gerathen. Endlich kann nur die gründlichste Kenntniss des lebendigen Sprachgebrauches über jene wichtigen Punkte entscheiden: inwieweit das Formbewusstsein noch in der Sprache lebendig, oder die Form nur aus träger Gewohnheit beibehalten sei, und inwieweit ein ursprünglich stoffliches Mittel formal empfunden werde? Deshalb dünken mir die hierher gehörigen Fragen so heikler Art, dass die Wissenschaft nur zögernd, nach langer historisch-kritischer Vorarbeit, an sie herantreten sollte. Nur das zu Tage Liegende gestattet eine sofortige Beurtheilung.

Bländlingssprachen, 1889

Maass 1889

Richtung aber äussert 1889 and 1891

subjektiver 1889

Ein Anderes ist es, eine Behauptung durch den Beweis des Gegentheiles widerlegen, ein Anderes, dem Gegner, der die Behauptung aufgestellt hat, die Beweislast auf die Schultern wälzen. Jenes, den positiven, inductiven Beweis für meine Meinung, durfte ich eben nach |348| meinen grundsätzlichen Anforderungen nicht unternehmen. Eine apriorische Wahrscheinlichkeit aber hoffe ich nunmehr für meine Ansicht beanspruchen zu dürfen, und damit fiele die Beweislast auf Jene, die die grosse Mehrzahl der Sprachen der Formlosigkeit anklagen. Denn das eben ist streitig, ob der sprachliche Formungstrieb allgemein menschlich ||365|| oder nur ein Vorrecht gewisser, besonders begnadeter Rassen sei. Indem ich dies zu erörtern versuchte, war es unausbleiblich, dass ich ein Stück vor- und hineingriff in jenen Theil unserer Wissenschaft, der sich mit der Werthbestimmung der Sprachen beschäftigt.

Racen 1889

Ehe wir aber zu dieser übergehen, müssen wir eine Reihe der wichtigsten Mittel und Gründe der Sprachformung untersuchen.

385
III. Word order[13]

§ 1. Psychological subject and predicate

According to the agglutination theory, affixes are the remnants of former independent words: the word, which has been formed by appending or accretion, therefore exhibits prehistoric laws of word order in fossilized form. But if we trace the phenomena of word order back to their ultimate cause, nothing will be achieved with a palaeontological investigation. We have to seek out language at the place where it still lacks its defining feature of articulation; we must look at the pre-synthetic stage of language, where it still does its work in isolated verbal expressions. For this is the beginning, or rather the predecessor, of all human speech: of the speech of the child, this we can observe daily; and of the speech of primitive man, this we can confidently assume.

There is not a great deal that can be expressed in this simplest of ways, but that is why this expression is sufficient for its modest purposes. Let us try to bring these purposes under a single unified category.

In this investigation we do not need to confine ourselves to the small repertoire of so-called natural sounds, which continue to live on in our languages as humble outsiders that are no more than tolerated. We may assume that the linguistic means available to our ancestors satisfied their requirements for communication. Whenever new purposes demanded new means, our ancestors, in those times of youthful creativity, will not have had any difficulty; and the living conditions were still so simple that they could be sure they would be easily understood by the listener. If pain, worry, surprise or joy squeezed a scream from them, this would have been expressive enough; and the same would have been true for those onomatopoeic imitations of sounds with which they would have designated the objects and events of the outer world. Calls, commands, expressions of agreement and disagreement, deictic sounds, names for the closest companions and the most immediate needs, perhaps even for 'me' and 'here' and a questioning call would soon become common property;

13 Cf. my essays in Lazarus and Steinthal's *Zeitschrift für Völkerpsyhologie und Sprachwissenschaft* vols. VI and VIII and in Techmer's *Internationale Zeitschrift für allgemeine Sprachwissenschaft* vol. III, p. 100 ff. [Gabelentz (1869; 1875; 1887)]

385
III. Die Wortstellung.[14]

§ .1. Psychologisches Subject und Prädicat.

Nach der Agglutinationstheorie haben wir in den Affixen Überbleibsel vormals selbständiger Wörter zu erblicken: das durch Anfügung oder Anbildung gestaltete Wort zeigt mithin vorgeschichtliche Wortstellungsgesetze gleichsam in fossiler Erstarrung. Wollen wir aber den Erscheinungen der Wortfolge bis auf ihren letzten Grund nachgehen, so ist es mit der paläontologischen Untersuchung nicht gethan. Wir haben die Sprache da aufzusuchen, wo ihr das ausschliessliche Merkmal der Gliederung noch fehlt: auf ihrem vorsynthetischen Standpunkte, da wo sie noch ihre Arbeit in vereinzelten Stimmäusserungen leistet. Denn dies ist der Anfang, oder richtiger der Vorläufer aller menschlichen Rede: der Rede des Kindes, – das können wir täglich beobachten, – und der Rede der Urmenschen, – das dürfen wir getrost annehmen.

Was auf diese einfachste Weise ausgedrückt werden kann, ist nicht viel. Aber eben darum ist die Ausdrucksweise für ihre bescheidenen Zwecke genügend. Versuchen wir, diese Zwecke unter eine einheitliche Kategorie zu bringen.

|349|

Bei diesem Versuche brauchen wir uns nicht auf den kleinen Vorrath sogenannter Naturlaute zu beschränken, die in unseren Sprachen noch als bescheidene Geduldete fortleben. Wir dürfen voraussetzen, dass der Sprachschatz unserer Urahnen ihrem Mittheilungsbedürfnisse reichlich entsprochen habe. Wo neue Zwecke neue Mittel erheischten, da wird man in jenen Zeiten jugendlicher Schöpfungskraft nicht arg in Verlegenheit gewesen sein; und die Lebensbedingungen waren noch so einfach, dass man auf das Verständniss der Hörer rechnen durfte. Presste der Schmerz, die Angst, das Erstaunen, die Freude einen Schrei aus, so war dieser an sich ausdrucksvoll genug; und das Gleiche ||366|| gilt erst recht von jenen Schallnachahmungen, womit man Gegenstände und Ereignisse der Aussenwelt bezeichnen mochte. Rufe, Befehle, Ausdrücke der Zustimmung oder Ablehnung, Deutelaute, Namen für die nächststehenden Personen und die nächstliegenden Bedürfnisse, vielleicht auch für das Ich oder das Hier und ein Frageruf wurden bald gemeinüblich;

so genannter 1891

14 Vergl. meine Aufsätze in LAZARUS und STEINTHAL's Ztschr. f. Völkerpsych. und Sprachwiss. Bd. VI und VIII und in TECHMER's Internat. Ztschr. f. allgem. Sprachwissensch. Bd. III, S. 100 flg.

386 and in this way the outlines of that oldest language would have been sketched.

The task was then to translate this language into articulated human speech. There are two possibilities here. Either the call contained subject and predicate together, still wrapped in one another, or the call was still one-sided in terms of its content, so that of the two necessary parts of the (logical) sentence one remained unexpressed and therefore had to be added. But this shows that the need for completion is self-evident, otherwise the calls would not have been sufficient to their purpose.

In those sounds of sensation, the first person is obviously the subject: *Ouch!* = I am in pain. The second person is the obvious subject of commands: *Hey!* = Come here! *Psst!* = Be quiet! and so on.

Even today we are satisfied with such fragmentary speech in innumerable other cases, and now we want to observe a series of such cases.

In the forest I see something move, point to it with my finger and call out: 'There!' Through this action the attention of the listener is aroused and directed towards a specific point. The other says, 'A squirrel', since he assumes that I have not recognized the animal; and now I know what it is.

In a horse race all eyes are focused on a galloper that suddenly charges out of the back of the pack to the front with redoubled effort. Someone calls out, 'The blue one!', someone else, 'The chestnut mare!' The former is indicating the rider, the latter the horse. Both mean the same thing, and everyone understands them.

When talking to servants, we only have to name what we want, and the servant knows that he should bring it. In like manner, the servant announces briefly: 'Mr So-and-so' and we understand that the person named is outside and would like to make his visit.

The most common case, however, is that of answering and of continuing a conversation. Who is that? – So-and-so – And the other? – X. – From where? – From A. – What are they doing here? – They're having a look around. – In order to move here? – Maybe.

These examples will suffice. They show that wherever speech is satisfied with bare (logical) parts of the sentence the missing parts are provided from somewhere else, whether it is through the words of the interlocutor or from the shared condition of those speaking. The closer the relationship between the people, the more shared and limited their mental horizon, and the clearer the circumstances reflect their common interest, then the more certainly will this elliptical half-speech fulfil its purpose. Complemented by lively facial expression and gestures, such speech will have satisfied the simple needs of primitive man for a long time.

386 und somit dürften die Umrisse jener ältesten Sprache gezeichnet sein.

Jetzt gelte es, diese Sprache in gegliederte menschliche Rede zu übertragen, so sind zwei Fälle möglich. Entweder der Ruf enthält noch unentfaltet Subject und Prädicat zugleich, oder er ist seinem Inhalte nach dergestalt einseitig, dass von den zwei nothwendigen Gliedern des (logischen) Satzes immer eins unausgedrückt bleibt, also zu ergänzen ist. Aber auch das leuchtet ein, dass die Ergänzung sich als selbstverständlich bieten muss; denn sonst würden die Rufe ihren Zweck verfehlen.

Bei jenen Empfindungslauten gilt ohne Weiteres die erste Person als Subject: Au! = ich leide Schmerz. Die zweite Person ist selbstverständliches Shmerz. 1901
Subject der Befehle: He! = Komm her! St! = Sei still! u. s. w.

Noch heute jedoch begnügen wir uns in unzähligen anderen Fällen mit der bruchstückweisen Rede, und eine Reihe solcher Fälle wollen wir betrachten.

Ich sehe im Walde sich etwas bewegen, weise mit dem Finger hin und rufe: „Dort!" Die Aufmerksamkeit des Hörers wird dadurch erregt und auf einen bestimmten Punkt gerichtet. Der Andere sagt: „Ein Eichhorn;" denn er nimmt Eichhorn"; 1901
an, dass ich das Thier nicht erkannt habe; und nun bin ich verständigt.

Bei einem Wettrennen richten sich Aller Blicke auf einen Renner, der eben mit plötzlich verdoppelter Anstrengung aus dem hintersten Gliede in's Vordertreffen strebt. Man ruft: „Der Blaue!" „Die Fuchs|350|stute!" Der Eine deutet auf das Pferd, der Andere auf den Reiter, Beide meinen dasselbe, und Jeder versteht sie.

Dem Diener braucht man bloss die gewünschte Sache zu nennen, und er weiss, dass er sie bringen soll. Dafür meldet er ebenso kurz: „Herr N. N." und die Herrschaft versteht, dass der Genannte draussen ist und seinen Besuch machen will.

Der gewöhnlichste Fall aber ist der der Antwort und des fortgesetzten Gespräches. Wer ist das? – N. N. – Und der Andere? – X. – Woher? – Aus A. – Was machen sie hier? – Sie sehen sich um. – Um sich hier niederzulassen? – Vielleicht.

Diese Beispiele werden genügen. Sie zeigen, dass überall da, wo die Rede sich mit blossen (logischen) Satztheilen begnügt, das ihr Fehlende anderwärtsher gegeben ist, sei es durch die Worte des Gesprächsgenossen, sei es durch die sonstige gemeinsame oder gegenseitige Lebenslage der beiden Redenden. Je enger also die Beziehungen zwischen den Menschen sind, je gemeinsamer und ||367|| beschränkter ihr geistiger Gesichtskreis ist, je deutlicher sich den Umständen nach ihr gemeinschaftliches Interesse ergiebt: desto sicherer wird jene halbe, elliptische Rede ihren Zweck erfüllen. Durch lebhafte Mienen und Gesten ergänzt, mochte sie lange Zeit hindurch den einfachen Bedürfnissen der Urmenschen genügen.

387
In addition, we want to consider a related process in the creation of words. Even today man proceeds in the linguistic designation of his individual representations just as he did with the expression of his thoughts in former times: suggestively, giving fragments rather than the whole. Among all the features of the bird, he grabs onto one or the other and calls it the 'feathered one', the 'flying one', the 'one who lays eggs', and he knows that the listener will fill out this image. The only advantage we have is that such abbreviations have become unambiguous through habit, just like the commands: 'Halt!', 'Fire!'

Of course there will have also been occasions when someone will have made such one-word utterances one after the other. He turned to A – 'Come!' – and then to B – 'Go!' Then he drew C's attention to two things, and called out perhaps, pointing with his finger: 'A hog! ... A wolf!' Those are two utterances (calls) that cohere just as little in terms of logic as in terms grammar.

But it sometimes happened, and still happens today, that such grammatically isolated – that is, in fact, ungrammatical – calls were connected to one another by a logical thread. The process in such a case is always then that the first call did not satisfy the caller himself, and then he made a second call, or perhaps even more additional calls. There are two possible cases here.

Either the calls that follow one after the other represent the same part of the thought, such as the predicate, 'Wonderful! ... Marvellous!', or the vocative, 'Karl! ... Brother!' – or something else.

Or the calls complement one another in such a way that they produce both parts of the sentence. Linguistically they are still not bound to one another, in the soul of the caller each call was the product of a unique impulse; these impulses are related to each other in terms of their causes, however, and the individual calls form themselves in the soul of the caller, as in the soul of the listener, to a higher unit, which is no longer one-sided. We must bring in a few examples to demonstrate this. Someone falls and breaks his arm. His call 'Ouch!' says only that he is in pain, but the others around him do not know why he is in pain. Then he calls: 'My arm!' – and then, since he has recognized the damage, 'Broken!' – A mother calls to her husband: 'Our child!' She has suddenly noticed alarming changes in the child. Then she notices the danger it is in and shouts: 'Cramps!' – A and B are walking through the forest. A sees deer in the distance, points to them and whispers: 'There!' B seems not to understand what he means, so A adds: 'Deer!' The result is the same everywhere. The individual representations accumulate in the soul of the listener to produce a complete image,

387

Nebenbei wollen wir eines verwandten Herganges in der Wortschöpfung gedenken. Denn der Mensch verfährt in der sprachlichen Bezeichnung seiner Einzelvorstellungen noch heute genau so, wie er es vor Alters mit dem Ausdrucke seiner Gedanken gethan hat: andeutend, Bruchstücke statt des Ganzen gebend. Unter allen Merkmalen des Vogels greift er die eine oder andere heraus, nennt ihn jetzt den Gefiederten, jetzt den Fliegenden, jetzt den Eierlegenden, und weiss, dass der Hörer das Bild ergänzen wird. Der Vorzug liegt nur darin, dass solche Abkürzungen durch die Gewohnheit eindeutig geworden sind, etwa wie die Commandorufe: Halt! Feuer!

Natürlich geschah es auch zuweilen, dass Einer mehrere solcher einwortigen Äusserungen nacheinander that. Jetzt wendete er sich an den A: „Komm!“ und dann an den B: „Geh!“ Jetzt machte er den C auf Zweierlei aufmerksam, rief etwa, mit dem Finger deutend: „Ein Eber! ... Ein Wolf!“ Das sind je zwei Reden (Rufe), die logisch ebensowenig zusammenhängen, wie grammatisch.

|351|

Aber auch das kam und kommt heute noch vor, dass solche grammatisch isolirte, also eigentlich ungrammatische Rufe durch ein logisches Band miteinander verknüpft sind. Der Hergang ist dann allemal der, dass der erste Ruf dem Rufenden selbst nicht genügt hat, und er nun einen zweiten, vielleicht noch mehrere weitere folgen lässt. Hier sind zwei Fälle möglich.

Entweder die aufeinanderfolgenden Rufe vertreten denselben Theil des Gedankens, etwa das Prädicat: „Herrlich! ... Prächtig!“ oder den Vocativ: „Karl! ... Bruder!“ – oder was sonst.

Oder die Rufe ergänzen einander dergestalt, dass sie beide Seiten des Satzes ergeben. Sprachlich sind sie noch unverbunden, in der Seele des Rufenden war jeder Ruf Erzeugniss eines besonderen Impulses; diese Impulse aber hängen unter sich ursächlich zusammen, und die einzelnen Rufe gestalten sich in der Seele des Rufers wie des Hörers zu einer höheren Einheit, die nur nicht mehr einseitig ist. Wir müssen dies an ein paar Beispielen erläutern. Jemand fällt und bricht sich den Arm. Sein Ruf „Au!“ besagt nur, dass er Schmerz empfindet; was ihm weh thut, wissen die Anderen nicht. Nun ruft er: „Mein Arm!“ – und dann, da er den Schaden erkannt hat: „Gebrochen!“ – Eine Mutter ruft ihrem Manne zu: „Unser Kind!“ Sie hat plötzlich an dem Kinde schreckhafte Veränderungen wahrgenommen. Jetzt merkt sie, in welcher Gefahr es schwebt, und schreit auf: „Krämpfe!“ – A. und B. gehen durch den Wald. A. sieht in der Ferne Rehe, weist hin und flüstert: „Dort!“ B. scheint nicht ||368|| zu bemerken, um was es sich handelt, und nun fügt A. hinzu: „Rehe!“ Das Ergebniss ist überall dasselbe. Es ballen sich in der Seele des Hörenden die Einzelvorstellungen zu einem Gesammtbilde zusammen,

nun, [in den Berichtigungen, S. 502: nur,] 1891

388 and this of course occurs in the order in which the individual representations are received. Whether the caller had the same complete image from the very beginning, or whether it was only later that he felt the need to complete it and make it comprehensible through additions, or whether he gained the new representations that his calls arouse only through further observation: these are all questions that we cannot answer yet.

It is easy to see that something similar must have occurred very early and very frequently in the prehistory of human language. It was the preliminary stage of articulated – that is, actual – human speech, it was the preliminary stage of syntax. The next step to be taken was clearly vanishingly small, and yet decisive. But the decisive aspect was essentially in the inner facts of the soul. Where earlier the parts of the total representation broke out jerkily in calls, by drives that came in fits and bursts, now from the beginning the mind broke down the thought that floated before it into its parts, in order to reconstitute it from these parts in front of the listener in continuous speech. The outer form will not have added much here, and when I speak of continuous, coherent speech, this must be understood as a euphemism. In reality this new manifestation of language will not have stood in any better relation to the old one than for instance the call 'There, deer!' is better than the two calls one after the other 'There! ... Deer!'

The inner connections between these parts of the sentence of course remained indefinite. It was just like the language of children today, where 'Daddy hat' can mean both 'The father's hat' and 'The father has put the hat on'. With our formed language and our trained thinking we distinguish sharply between these different conceptions of the same total representation. At one moment we talk in sentences: 'The fox catches the hare' or 'The hare is caught by the fox'. And then we construct simple sentence fragments and allow each one of the three individual representation to be determined by the other two: 'The fox that catches the hare', 'The hare caught by the fox', and even 'The hare-catch of the fox'. It is essentially the same image that appears before us. At this most naive level of human speech there were at most as many possibilities of expression as possibilities of order: *abc*, *acb*, *bac*, *bca*, *cab*, *cba*: 'Fox catch hare', 'Fox hare catch', 'Catch fox hare' etc. The image broke itself down into three parts – the two animals and the catching – and the speaker could choose how he wanted to order these parts.

388 und dies geschieht natürlich in der Ordnung, in welcher die Einzelvorstellungen empfangen werden. Ob dem Rufenden von Anfang an dasselbe Gesammtbild vorgeschwebt, und er nur hinterdrein das Bedürfniss empfunden, es durch Zusätze zu ergänzen und verständlich zu machen, oder ob er die neuen Vorstellungen, die seine Rufe erwecken, selbst erst durch weitere Beobachtung gewonnen hat: das bleibt vorerst dahingestellt.

Es ist leicht einzusehen, dass sich in der Vorgeschichte der menschlichen Sprache dergleichen schon sehr früh und sehr oft ereignet haben muss. Es war die Vorstufe der gegliederten, das heisst der eigentlich menschlichen Rede, es war die Vorstufe der Syntax. Der nächste Schritt, |352| den es zu thun galt, war anscheinend verschwindend klein, und doch geradezu entscheidend. Das Entscheidende aber lag wesentlich im inneren, seelischen Thatbestande. Waren früher die Bestandtheile der Gesammt-Vorstellungen ruckweise, auf ruckweise sich einfindende Antriebe im Rufen hervorgebrochen: so zerlegte nunmehr der Geist vom Beginn der Rede an den ihm vorschwebenden Gedanken in seine Theile, um ihn aus jenen Theilen in fortlaufender Rede vor dem Hörer wieder aufzubauen. Das Äussere dabei will nicht viel besagen, und wenn ich von fortlaufender, zusammenhängender Rede spreche, so lasse man dies als einen Euphemismus gelten. In der That wird sich die neue Erscheinungsform der Sprache zu der alten nicht besser verhalten haben, als etwa der Ruf: „Da Rehe!" sich zu den zwei aufeinanderfolgenden Rufen: „Da! … Rehe!" verhält.

ein findende [in den Berichtigungen, S. 502: einfindende] 1891

Die Art des inneren Zusammenhanges zwischen den Redegliedern blieb natürlich noch unbestimmt. Es war wie noch heute mit der Sprache der Kinder, wo „Papa Hut" Beides bedeuten kann, sowohl: Des Vaters Hut, als auch: Der Vater hat den Hut aufgesetzt. Wir mit unserer gebildeten Sprache und mit unserm geschulten Denken unterscheiden wohl scharf zwischen den verschiedenen Auffassungen derselben Gesammtvorstellung. Jetzt reden wir in Sätzen: „Der Fuchs fängt den Hasen", oder: „Der Hase wird vom Fuchse gefangen". Jetzt wieder bilden wir blosse Satztheile, lassen je eine der drei Einzelvorstellungen durch die beiden anderen näher bestimmt werden: „Der den Hasen fangende Fuchs, – der vom Fuchse gefangene Hase", wohl auch: „des Fuchses Hasenfang". Im Grunde ist es immer dasselbe Bild, das uns vorschwebt. Auf jenem naivsten Standpunkte menschlicher Rede gab es höchstens soviele Ausdrucksmöglichkeiten als Möglichkeiten der Stellung: *abc*, *acb*, *bac*, *bca*, *cab*, *cba*: „Fuchs fangen Hase, Fuchs Hase fangen, Fangen Fuchs Hase" u. s. w. Das Bild zerlegte sich in drei Theile: die beiden Thiere und das Fangen, und der Redende hatte die Wahl, wie er diese Theile ordnen wollte.

389
This choice may have been free, but it was certainly not random, but rather meaningful. We have to ask how it was determined; that is, what was expressed by it.

This investigation becomes easier if we first put ourselves into the soul of the listener. This soul receives with the first word that it perceives a representation *a*. It asks expectantly: What about *a*? Then *b* is given to it. Then it adds *a* and *b* to make a unit $(a + b)$ and asks further: What is $(a + b)$? In answer it receives the new representation *c*. In the sentence 'Hare fox catch', the first word tells it that it should imagine a hare, the second word shows it that the hare has entered into some sort of relationship to the fox, and the third word explains to it what kind of relationship this is. The fact that the hare is the one suffering in this relationship can be concluded only from the nature of the situation. Precisely the same process occurs in us when we listen in on a child's babbling attempts at speech: what we hear accumulates from word to word to form a unit, and if we were to express our expectation in the form of a question, it would be: 'What about it?' And if the child loses the thread and pauses, then in this case we also ask: 'What about it?' *That is, what we hear has the same relation to what we expect as a subject to its predicate.* It is like the two rolls in a telegraph machine: on the one side the roll being written on, which swells up more and more, and on the other the smooth strip of paper that is still to be filled out and in being filled out slides over to the other side.

And to stay with this metaphor: the speaker knows both what is on the written roll and what is yet to be put on the blank part of the strip. Apart from this, however, the process in his soul and in that of the listener is parallel. Those experiences that he prompts in the listener he has himself experienced innumerable times. He speaks – that is, he demands that the listener follow him in thinking what he thinks himself and how he thinks himself. In this way he has already put himself halfway into the soul of the listener. He leads the thinking of the listener to a certain representation, and then further and ever further, arousing ever more expectations and then satisfying them immediately after that. In this sense there is a natural sympathy between the 'I' and the 'you', and it is involuntary and unconscious. The speaker has only to follow his own drive and he will have the corresponding effect on the listener.

But what is this drive? The image of the whole appears before me; I hold the parts in my hands in order to build up their replicas. What is it that determines what I present first, then second, then third, and so on? Obviously the principle is

389
||369||
Diese Wahl mochte sehr frei sein, aber gewiss war sie nicht zufällig, sondern bedeutsam. Wir müssen fragen, wodurch sie bestimmt, was also durch sie ausgedrückt wurde.

Die Untersuchung wird erleichtert, wenn wir uns zunächst in die Seele des Hörenden versetzen. Die empfängt mit dem ersten Worte, das sie vernimmt, eine Vorstellung *a*. Sie fragt erwartungsvoll: Was ist |353| mit *a*? Da wird ihr *b* gegeben. Nun summirt sie *a* und *b* zu einer Einheit ($a + b$) und fragt weiter: Was ist ($a + b$)? Als Antwort empfängt sie die neue Vorstellung *c*. Der Satz laute: „Hase Fuchs fangen", so sagt ihr das erste Wort, dass sie sich einen Hasen vorzustellen habe. Das zweite Wort zeigt ihr an, dass der Hase in irgendwelche Beziehung zum Fuchse getreten sei. Und das dritte Wort erklärt ihr die Art dieser Beziehung. Dass der Hase dabei der leidende Theil ist, ergiebt sich derweile nur aus der Natur der Sache. Ganz dasselbe geht in uns vor, wenn wir dem lallenden Redeversuche eines Kindes lauschen: von Wort zu Wort ballt sich das Gehörte zu einer Einheit zusammen, und sollen wir unsere Erwartung in Form einer Frage ausdrücken, so lautet die: „Was ist damit?" Und wenn das Kind den Faden verloren hat und innehält, so fragen wir es auch: „Was ist damit." Das heisst: Das Gehörte verhält sich zu dem weiter Erwarteten, wie ein Subject zu seinem Prädicate. Es ist wie mit den zwei Papierrollen im Telegraphenapparate: hier die beschriebene, die immer stärker anschwillt, dort der glatte Papierstreifen, der noch vollgeschrieben werden soll und, indem er es wird, zur anderen Rolle hinübergleitet.

Um nun im Bilde zu bleiben: der Redende weiss Beides, sowohl was auf der beschriebenen Rolle steht, als auch was auf den leeren Theil des Streifens noch kommen soll. Im Ubrigen jedoch wird der seelische Hergang bei ihm und beim Hörenden parallel sein. Was er eben den Andern erfahren lässt, das hat er selbst unzählige Male erlebt. Er redet, das heisst er verlangt, dass der Andere ihm nachdenke, was und wie er ihm vordenkt. So hat er sich nachdenke 1891
schon zur Hälfte in die Seele des Hörers hineinversetzt. Nun leitet er mit dem ersten Worte des Anderen Denken auf eine gewisse Vorstellung und dann weiter und immer weiter, immer neue Erwartungen jetzt erweckend, jetzt, gleich darauf, befriedigend. Insoweit und in diesem Sinne waltet zwischen dem Ich und dem Du eine natürliche Sympathie. Und sie waltet auch ungewollt und unbewusst. Der Redner braucht nur seinem eigenen Triebe zu folgen, so wird er auf den Hörer die entsprechende Wirkung üben.

Welches ist aber dieser Antrieb? Das Bild des Ganzen schwebt mir vor; die Theile halte ich in Händen, um sie nachschaffend aufzubauen. Was bestimmt mich, erst diesen aufzustellen, dann den, dann jenen? Offenbar ist es dies,

390 that I first name what stimulates my thinking, what I am thinking about, my *psychological subject*, and then I say what I think about it, my *psychological predicate*, and then, when necessary, I can make both of these the object of further thinking and speaking.

This can all be concluded simply from the nature of the speech. If, however, we want to make an inductive proof, we have to be selective with examples, since the phenomena of word order in languages are not all of the same value.

First of all, in most languages word order is to some extent fixed: the attribute or object always comes either first or last, the verb must be either before or after its subject, etc. Here the thought typically follows a predefined marching route. If the thought leaves this route, if the thought is able to get around it through inversion or modified sentence construction, then it simply proves in this way that another force in the soul has for the moment overcome the omnipotence of habit. Among the most closely related languages to ours, French offers us especially instructive examples of this kind.

Secondly, the need in the soul that we are talking about here can only develop itself fully in cases where other influences do not interfere. Most clearly of course in communicative speech. But it too has to be as free and isolated as possible. An answer is dependent on the question that prompts it; the continuation of speech is under the influence of what has gone before. Simple judgements, actual communications or aphorisms will offer the most suitable material for examination. Let us choose our examples according to this.

I can say: 'The 16th of March is my birthday', or 'My birthday is the 16th of March.' Here the difference between the two sentences is immediately obvious, since subject and predicate are in equal measure substantive parts of the sentence, and everyone feels how these parts of the sentence swap roles in the different positions.

But if, instead of 'the 16th of March', I say 'yesterday' or 'three days ago …' then the problem seems to become more difficult, since now one of the two parts of the sentence is adverbial and so cannot be the grammatical subject. But this does not make it any less of a psychological subject, since I am still talking about a particular day and am saying about it that it was my birthday.

In the proverb *Mit Speck fängt man Mäuse* 'With bacon one catches mice' the grammatical subject is *man* 'one'. But this is most certainly not the psychological subject, not what the speech is about. Rather the speech is about the means – *mit Speck* 'with bacon' – and the sense is: enticements

390 dass ich erst dasjenige nenne, was mein Denken anregt, worüber ich nachdenke, mein psychologisches Subject, und |354| dann das, was ich darüber denke, ||370|| mein psychologisches Prädicat, und dann wo nöthig wieder Beides zum Gegenstande weiteren Denkens und Redens mache.

Dies Alles dürfte sich aus der Natur der Sache ergeben. Gilt es aber, den inductiven Beweis zu führen, so müssen wir mit den Beispielen wählerisch sein; denn die Stellungserscheinungen der Sprachen sind nicht gleichwerthig. aber 1891

Erstens ist in den meisten Sprachen die Wortstellung einigermassen gebunden: das Attribut oder Object steht ein- für allemale voran oder nach, das Verbum muss entweder hinter oder vor seinem Subjecte stehen u. s. w. Hier wandert der Gedanke gewohnheitsmässig auf einer vorgezeichneten Marschroute. Verlässt er diese, gelingt es ihm, sie durch Inversionen oder veränderte Satzbildung zu umgehen: so beweist er eben dadurch, dass eine andere seelische Macht zeitweilig die Allgewalt der Gewohnheit überwunden hat. Unter den uns nächstliegenden Sprachen bietet die französische besonders lehrreiche Beispiele dieser Art.

Zweitens kann sich das seelische Bedürfniss, um das es sich hier handelt, nur da voll entfalten, wo nicht noch andere Einflüsse mitwirken. Am Klarsten natürlich in der mittheilenden Rede. Aber auch die muss möglichst frei und isolirt sein. Die Antwort ist abhängig von der Frage; die Fortsetzung der eigenen Rede steht unter dem Einflusse des Vorhergegangenen. Also einfache Urtheile, thatsächliche Mittheilungen oder Sinnsprüche werden den brauchbarsten Untersuchungsstoff bieten. Wählen wir darnach unsere Beispiele.

Ich kann sagen: „Der 16. März ist mein Geburtstag“, oder: „Mein Geburtstag ist der 16. März.“ Hier leuchtet der Unterschied beider Sätze ohne Weiteres ein; denn Subject und Prädicat sind gleichermassen substantivische Satztheile, und jeder empfindet, wie diese Satztheile mit der Stellung auch die Rollen wechseln.

Setze ich statt: der 16. März: gestern, oder vor drei Tagen war …, so wird anscheinend die Sache schon schwieriger, denn nun ist der eine der beiden Satztheile adverbial, kann also nicht grammatisches Subject sein. Psychologisches Subject ist er aber darum nicht minder; denn nach wie vor rede ich von einem gewissen Tage und sage von ihm aus, dass er mein Geburtstag war.

In dem Sprichworte: „Mit Speck fängt man Mäuse“ ist das grammatische Subject „man“. Ganz gewiss ist dies aber nicht das psycho|355|logische Subject, nicht Dasjenige, wovon die Rede ist. Vielmehr ist die Rede vom Mittel, und der Sinn ist: Lockungen und Schmeicheleien sind das Mittel, um Mittel 1891

and flattery are the means for catching frivolous people. If the phrase were *Mäuse fängt man mit Speck* 'Mice, one catches them with bacon', then the sentence would be about frivolous people and would say how they are to be caught. The remaining three words, *fängt man Mäuse* 'one catches mice', do not of course allow any change in order and so offer no material for observation.

Leichtsinnige zu fangen. Hiesse es umgekehrt: „Mäuse fängt man mit Speck", so handelte der Satz von den leichtsinnigen Leuten und sagte von ihnen aus, wie sie zu fangen seien. Die letzten drei Wörter „fängt man Mäuse" dulden bekanntlich keine Umstellung, bieten daher zu keiner Beobachtung Stoff.

391

In the sentence *Manus manum lavat* 'One hand washes the other' the ordinary Latin word order is observed: subject, object, verb. In principle, every other ordering would be allowed, but only this order corresponds to the relevant sense, that those who are alike help one another.

I can give the distance between Leipzig and Dresden in the following four forms, among others:

1. Leipzig is 115 kilometres from Dresden.
2. Dresden is 115 kilometres from Leipzig.
3. From Leipzig, Dresden is 115 kilometres.
4. From Dresden, Leipzig is 115 kilometres.

We see that in (1) and (3) we are talking about Leipzig, and in (2) and (4) about Dresden, but in (1) and (2) both cities are thought of silently as the destination, and in (3) and (4) explicitly as the point of departure. Such paradigms illustrate the principle quite clearly.

Cicero's first *Cataline Oration* (Chapter I, § 3 [1919, pp. 280-281]) contains the sentence: 'For we have a resolution of the senate, a formidable and authoritative decree against you, O Catiline.' If we were to ask a schoolboy to translate this into Latin, the result would presumably be something like: *Vehemens et grave in te, Catalina, senatus consultum habemus*. That starts blusteringly. But Cicero does it differently, with something like the cold cruelty of an executioner from past centuries, who shows his victim the instruments of torture before he begins: 'Here I have something – *habemus* –. Look here, it is a pair of pincers – *senatus consultum* –. I will pinch you with it – *in te, Catalina* –. It will hurt you quite a lot – *vehemens et grave!*' How differently does this affect our nerves!

But this is of course not to deny the fact that in languages with free word order rhetoric and poetics can enlist the order of the parts of speech for the purposes of euphony. This has long been known to grammarians and may have even contributed to misleading them.

Exceedingly effective and relevant for our purposes are such French turns of phrase as: *Votre frère, j'ai de ses nouvelles* 'Your brother, I have some news of him'. *Cette lettre, je l'ai lue* 'This letter, I have read it'. Here the psychological subject is isolated, and not put at the beginning of the grammatical sentence, but rather before it, and then repeated associatively from within the sentence using a deictic. Exactly this process is also common in Chinese.

Languages in which adverbial expressions have a fixed position, whether before or after the verb, tend in the case of specifications of time, place, reason or circumstances to make exceptions that virtually become the rule. It is, for example, characteristic of the chronicle genre in Chinese that the regular order for relating an event is to first name the time, then the place,

391

In dem Satze: „Manus manum lavat“ herrscht die gewöhnliche lateinische ||371|| Wortfolge: Subject, Object, Verbum. An sich wäre nun jede andere Stellung auch erlaubt; dem Sinne aber, dass Gleich und Gleich einander helfen, entspricht doch nur die eine Wortordnung.

Die Entfernung zwischen Leipzig und Dresden kann ich u. A. in folgenden vier Formen angeben:

Leipzig ist von Dresden	Dresden ist von Leipzig …
Von Leipzig ist Dresden	Von Dresden ist Leipzig …

– 115 Kilometer entfernt.

Man sieht: auf der linken Seite ist von Leipzig, auf der rechten von Dresden die Rede, aber beide Städte werden in der ersten Zeile stillschweigend als Ziel, in der zweiten ausdrücklich als Ausgangspunkt gedacht. Solche Paradigmen dürften besonders einleuchtend wirken.

In Ciceros erster catilinarischer Rede (cap. I § . 3), steht der Satz: „Wir haben gegen Dich, Catilina, einen gewaltigen und strengen Senatsbeschluss.“ Lässt man dies einen Schüler in's Lateinische übersetzen, so wird vermuthlich das Ergebniss sein: Vehemens et grave in te, Catilina, senatusconsultum habemus. Das fängt gleich polternd an. Cicero aber macht es anders, so etwa mit der kalten Grausamkeit eines Henkers vergangener Jahrhunderte, der der Folter die Territion vorausgehen liess: „Da habe ich etwas – *habemus* –. Siehe her, es ist eine Zange – *senatusconsultum* –. Damit werde ich Dich zwicken – *in te, Catilina* –. Es wird dir aber sehr weh thun – *vehemens et grave!*“ Wie anders wirkt das auf die Nerven!

Damit ist nun natürlich nicht verneint, dass die Rhetorik und Poetik in Sprachen mit freier Wortstellung die Ordnung der Redetheile auch den Wohllautszwecken dienstbar machen können. Dies war den Grammatikern längst bekannt und mag mit dazu beigetragen haben, sie irrezuleiten.

Überaus wirksam und für unsern Zweck zutreffend sind französische Wendungen wie: „Votre frère, j'ai de ses nouvelles.“ „Cette lettre, je l'ai lue.“ Hier wird das psychologische Subject isolirt, nicht an, sondern vor den Anfang des grammmatischen Satzes gestellt und dann im Satzinnern durch Deutewörter gedanklich wiederholt. Genau dasselbe Verfahren ist im Chinesischen üblich.

Sprachen, in denen sonst die adverbiale Bestimmung ihren festen Platz, sei es vor, sei es hinter dem Verbum hat, pflegen wohl für Angaben der Zeit, des Ortes, des Grundes, der Umstände Ausnahmen zu gestatten, die geradezu zur Regel werden können. So ist es z. B. für die chronikalische Erzählungsweise der Chinesen bezeichnend, dass regelmässig erst die Zeit, dann der Ort,

392 and then the subject of the event recounted. Those are the three headings, so to speak, that are represented by a psychological subject that progressively narrows itself down: What happened at that time? What happened there at that time? What happened there at that time with so-and-so?

It can also often be observed that languages that otherwise put the verb after the subject will allow the reversed order when the verb shows a coming into being or a becoming perceptible. The sensory impression is in this case the psychological subject, and something will be said about its cause, the grammatical and logical subject. We see the following in Chinese: *hiá iù* (下雨 *xià yǔ*), *descendit pluvia* 'falls rain' = it's raining, *hîng lûi* (行雷 *xíng léi*), *it tonitrus* 'goes thunder' = it's thundering. Semitic grammar distinguishes sharply between verbal sentences and nominal sentences. In the former a verbal predicate precedes its subject; in the latter the substantive or pronominal subject precedes the predicate, regardless of whether it is nominal or verbal.[15] According to this, therefore, the difference lies not in the grammatical predicate, but in the psychological subject, and as such the verbal predicate will, as a rule, also be treated in this way: *Māta Zaidun* = 'died has Said'. But here the Semitic feeling for language puts a complete sentence in the first word: 'Died has he' – and the name *Said* comes to this virtual subject as an apposition. The most important point is that only special circumstances license the placing of the subject noun before the verb.

Those cases in which languages tend towards a fixed word order or even prefer it are typical – and both cases can be found everywhere. For what else could it have been that put limits on personal arbitrariness other than a national habit of thought, according to which the representations prefer to roll out in a certain order? Here language shows itself as simultaneously the product and producer of the national spirit, a topic to which we will return later.

I believe that we have just become acquainted with one of the oldest – embryonic, so to speak – grammatical categories to be found in all human languages. For a long time it may have been, alongside the various kinds of emphasis, the only one

15 Here I follow Wright, Arabic Grammar II, p. 272 and Caspari, Gramm. arabe trad. par E. Uricoechea, § 478, which also treat *Zaidun māta* = 'Said has died' as nominal sentences. S. de Sacy, Gramm. arabe, IIe éd. II, p. 511 demands that a verbal sentence has a verbal predicate and that a nominal sentence has a nominal predicate. On this point, cf. M.S. Howell, A Practical Gramm. of the Class. Arab. Lang. I, Introd. p.IV ff

392 dann das Subject der Begebenheit genannt wird. Es sind das so zu sagen drei Überschriften, die ein sich stufenweise verengendes psychologisches Subject darstellen: Was geschah damals? Was geschah damals dort? Was geschah damals dort mit Dem und Dem?

|356| ||372||

Auch das ist öfter zu beobachten, dass Sprachen, die sonst das Verbum hinter das Subject setzen, die umgekehrte Reihenfolge gestatten, wenn das Verbum ein Inerscheinungtreten oder Sinnfälligwerden anzeigt. Der Sinneseindruck ist dann psychologisches Subject, und der Urheber, das grammatische und logische Subject, wird davon ausgesagt. Im Chinesischen ist *hiá iù̀, descendit pluvia* = es regnet, *hîng lûi, it tonitrus* = es donnert. Die semitische Grammatik unterscheidet scharf zwischen Verbalsätzen und Nominalsätzen. In jenen steht ein verbales Prädicat vor seinem Subjecte; in diesen steht das substantivische oder pronominale Subject vor dem Prädicate, sei dies nun nominal oder verbal[16]. Darnach beruht also der Unterschied nicht im grammatischen Prädicate, sondern im psychologischen Subjecte, und als solches wird in der Regel auch das verbale Prädicat behandelt: *Māta Zaidun* = gestorben ist Said. Hier aber legt das semitische Sprachgefühl schon in das erste Wort einen vollständigen Satz: „Gestorben ist er," – und zu diesem gedachten Subjecte „er" tritt der Name Said als Apposition. Das Wichtigste ist, dass nur besondere Umstände die Stellung des Subjectsnomens vor dem Verbum gestatten.

Bezeichnend sind überhaupt jene Fälle, wo die Sprachen einer festen Wortfolge huldigen oder doch eine solche bevorzugen, – und das Eine oder das Andere mag wohl überall stattfinden. Denn was anderes war es, was der Willkür Schranken setzte, als eine nationale Denkgewohnheit, derzufolge die Vorstellungen sich am Liebsten in einer bestimmten Reihenfolge abrollen? Hier zeigt sich die Sprache zugleich als Gebilde und Bildnerin des Volksgeistes, und darauf werden wir an einer späteren Stelle zurückkommen.

Ich glaube, wir haben hiermit eine der ältesten, so zu sagen embryonalen grammatischen Kategorien aller menschlichen Sprache kennen gelernt. Lange Zeit hindurch mochte sie neben den verschiedenen Arten der Emphase die einzige

16 Ich folge hier WRIGHT, Arabic Grammar II, pg. 272 und CASPARI, Gramm. arabe trad. par E. URICOECHEA, § . 478, die auch *Zaidun māta* = Said ist gestorben, als Nominalsatz bezeichnen. S. DE SACY, Gramm. arabe, IIe éd. II, p. 511 dagegen verlangt zum Verbalsatz verbales, zum Nominalsatz nominales Prädicat. Vergl. dazu M. S. HOWELL, A Practical Gramm. of the Class. Arab. Lang. I, Introd. p. IV flg. Laidun 1901

393 of its kind, and because of this it seems to us just as indefinite as it was common. Certainly it was logically indefinite inasmuch as it still did not accommodate the huge diversity of logical relations. But for all that it was no less a category, and indeed a grammatical category. If my deduction is correct, then this category must be present in all languages, in however many different ways its effects may manifest themselves.

But if following the lead of the agglutination theory we go back into the prehistory of the language families that are known to us, and look behind the agglutination for old syntactic structures, then we will make some surprising observations. In ancient times the Indo-Europeans always put the verb before its logical subject; that is, in the manner of the Semites and Malays, the Indo-Europeans made the verbal predicate into the psychological subject: **ed-mi* = 'I eat'. By contrast, the Semites put the complete action before the subject, and the incomplete action after it, as in Arabic قَتَلْتَ *qatal-ta*, 'you have killed', contrasted to تَقْتُلُ *ta-qtulu*, 'you will kill'. In other words, for the *fait accompli* they said who the actor was, but for the active subject they said what he was doing. And so everywhere composition and agglutination present fossilized laws of word order. It would perhaps be worthwhile to examine the peculiar conjugational structures of American languages from this perspective.

IV. Stress

From Latin and Greek grammars to descriptions of languages of the most diverse types a view has spread that we need to look at more closely here. This is the view that word order, especially the rarer sequences, can be explained in terms of stronger or weaker stress that speakers place on parts of a sentence. The more important or significant a speaker considers a part of the utterance to be – that is, the sharper he stresses it – the closer he brings it to the beginning of the sentence. And then he saves up an especially important part of the sentence until the end, in order to keep up the suspense for the listener. The beginning and end of the sentence are therefore the preferred positions for stress.

As far as I know, that is more or less the traditional doctrine contained in our Latin and Greek school books, and we cannot deny that it seems on the surface to be justified. First of all, there is of course a certain emphasis on that

393 ihrer Art sein, und eben darum erscheint sie uns als ebenso unbestimmt, wie sie allgemein war. Logisch unbestimmt war sie allerdings insofern, als sie der Mannigfaltigkeit der logischen Beziehungen noch nicht Rechnung trug. Aber eine Kategorie war sie darum nicht weniger, und zwar eine grammatische Kategorie. Wenn anders meine Deduction richtig war, so muss diese Kategorie in allen Sprachen gegenwärtig sein, so mannigfaltig sich auch ihre Wirkungen äussern mögen.

Mannichfaltigkeit 1891

mannichfaltig 1891

||373||

Gehen wir aber nun nach Anleitung der Agglutinationstheorie auf die Vorzeit der uns bekannten Sprachstämme zurück, suchen wir hinter den Agglutinationen alte syntaktische Gebilde: so ergeben sich manche überraschende Beobachtungen. Der Indogermane hat vor Alters das Verbum stets vor seinem logischen Subjecte genannt, also nach Art der Semiten und Malaien das verbale Prädicat zum psychologischen Subjecte gemacht; **ed-mi* = ich esse. Der Semite dagegen nannte die vollendete That vor, die noch unvollendete dagegen nach ihrem Subjecte: arabisch: قَتَلْتَ *qatal-ta*, du hast getödtet, dagegen: تَقْتُلُ *ta-qtulu* du wirst tödten; mit anderen Worten: von der fertigen Thatsache sagte er aus, |357| wer ihr Urheber war, vom thätigen Subjecte aber sagte er aus, was es im Werke hatte. Und so bieten überall die Composition und die Agglutination fossile Stellungsgesetze. Es lohnte sich wohl, unter diesem Gesichtspunkte die wunderlichen Conjugationsgebilde amerikanischer Sprachen zu betrachten.

IV. Die Betonung.

Von den lateinischen und griechischen Grammatiken aus hat sich durch die Darstellungen der verschiedenartigsten Sprachen eine Ansicht verbreitet, die wir hier näher erörtern müssen. Man hat die Stellungserscheinungen, zumal die selteneren, aus dem stärkeren oder schwächeren Nachdrucke erklären wollen, den der Redende auf einen Satztheil legt. Je wichtiger, bedeutsamer ihm ein Glied der Rede sei, je schärfer er es betone, desto weiter rücke er es nach vorn. Und dann spare er wohl noch einen besonders wichtigen Satztheil, um den Hörer in Spannung zu erhalten, bis an's Ende auf. Anfang und Ende des Satzes seien also die vorzugsweise betonten Stellen.

So etwa lautet meines Wissens die herkömmliche Lehre unserer lateinischen und griechischen Schulbücher; und ein Schein der Berechtigung ist ihr nicht abzusprechen. Denn erstens ruht natürlich ein gewisser Nachdruck auf demjenigen

394 part of an utterance that is put at the front as its topic – that is, on the psychological subject. Secondly, the listener's attention is held in suspense, and is only satisfied with the last, concluding psychological predicate. And thirdly, and most importantly, there are very often cases in speech where the first element of the sentence carries without a doubt the main accent. We have to examine these cases more closely, and to this end we must clarify the concept of stress. In the interests of a secure theory of word order, we must clear the ground for induction, we must judge the forces that seem to interfere and clear them out of the way.

It is well known that every word and every syllable can be stressed in one way or another. But here we are talking about preferential stress, that is about the emphatic accentuation of a part of an utterance through greater exertion of the voice. We pronounce whatever we stress louder and probably also with a higher pitch.

It sometimes happens that entire utterances are delivered with such an increased vocal effort. When does this happen?

When I speak to someone I want him to listen to me and understand me. If I am worried that he might not hear me, if he is, say, far away from me or is hard of hearing, then I will talk loudly. Uneducated people would probably also shout when talking to foreigners, who they think will not understand them, since they ascribe this lack of understanding to a problem with hearing. When children and common people argue, their speech can very easily degenerate into shouting. It is not only an angry mood that stimulates them to expend greater energy, but also the fact that naive people involuntarily treat others who have a different view from themselves as people who do not understand them. The fact that it is often easier to make speech clearer using the speech organs of the mouth – that is, through sharper articulation – than with the lungs – through shouting – is an insight that raw and excited people tend not to grasp.

The reason why we sometimes pronounce part of our utterances with particular effort of the speech organs will be analogous: we wish that just this part is not missed and not misunderstood. What we stress for the ear, what we underline or typographically emphasize for the eye is that thing which is the point for us, the most important for us. But this importance comes from a contrast that is either present or introduced into the discourse by us. This is the motivation for lively stress in a polemical speech against an opponent, who can be either standing before us or imagined. Every emphatic stressing of a segment of an utterance is antithetical, regardless of whether it is a specific opposition that is supposed to be excluded or everything apart from what is stressed.

394 Theile der Rede, der als ihr Thema vorangestellt wird, also auf dem psychologischen Subjecte. Zweitens bleibt die Aufmerksamkeit des Hörers bis an's Ende gespannt, wird erst mit dem letzten psychologischen Prädicate, dem abschliessenden, befriedigt Und drittens und hauptsächlich giebt es in der Rede sehr häufige Fälle, wo das erste Satzglied zweifellos den Hauptton trägt. Diese Fälle müssen wir näher betrachten und zu dem Ende uns den Begriff der Betonung klar machen. Es gilt, im Interesse einer sicheren Wortstellungstheorie den Boden für die Induction zu säubern, die scheinbar störenden Mächte zu beurtheilen und wegzuräumen.

Irgendwie betont werden bekanntlich jedes Wort und jede Sylbe. Hier aber handelt es sich um die vorzugsweise Betonung, das ist um die nachdrückliche ||374|| Hervorhebung eines Theiles der Rede durch stärkere Anstrengung der Stimmmittel. Was wir so betonen, das sprechen wir lauter, wohl auch in gesteigerter Stimmhöhe aus.

wird 1891

Es geschieht, dass ganze Reden so mit erhöhter Stimmkraft geäussert werden. Wann geschieht dies?

Indem ich zu Jemand rede, will ich, dass er mich höre und verstehe. Fürchte ich, dass er mich überhören könne, steht er etwa fern |358| von mir oder ist er schwerhörig, so spreche ich laut. Ungebildete schreien wohl auch im Gespräche mit Ausländern, von denen sie nicht verstanden zu werden glauben; denn sie schreiben diesen Mangel an Verständniss einem Gehörfehler zu. Wenn Kinder oder gemeine Leute sich streiten, so verfallen sie leicht in's Schreien. Es ist nicht nur die zornige Gemüthserregung, die sie zu einer erhöhten Kraftäusserung anreizt, sondern auch das, dass der naive Mensch unwillkürlich den, der anders denkt als er, gleich Einem behandelt, der ihn nicht versteht. Dass man die Deutlichkeit der Rede oft besser mit den Sprachorganen des Mundes, durch schärfere Articulation, als mit der Lunge, durch Schreien erzielt, darauf pflegen rohe und erregte Menschen nicht zu verfallen.

Wenn wir nun einen Theil unserer Rede mit besonderer Anstrengung der Stimmorgane aussprechen, so wird der Grund ein analoger sein: wir wünschen, dass gerade dieser Theil nicht überhört und nicht missverstanden werde. Was wir für's Ohr betonen, für's Auge unterstreichen oder typographisch auszeichnen lassen, ist also dasjenige, worauf es uns besonders ankommt, was uns das Wichtigste ist. Wichtig aber ist es uns in Rücksicht auf einen vorhandenen oder vorgestellten Gegensatz. Daher das lebhafte Betonen bei polemischer Rede wider einen vor uns stehenden oder fingirten Gegner. Jede nachdrückliche Betonung eines Redegliedes ist antithetisch, gleichviel ob ein bestimmter einzelner Gegensatz oder alles Andere als das Betonte ausgeschlossen sein soll.

395
If stress determined word position, if we were able to explain one by the other, then the initial part of the sentence would always be the stressed part, and it would not be possible to put the main emphasis on words in the middle or at the end of the sentence. In order to test this, let us stay with sayings and aphorisms, since in these the stress tends to be bound to the meaning. Here we will find that every part of the sentence can be stressed. The first part is stressed, for example, in the sentences: ὕδωρ μὲν ἄριστον 'water is best'; *Actor sequitur forum rei* 'The plaintiff follows the forum of the thing in dispute'; *Negans non excipit* 'Those who refuse are granted no exceptions'; *Eines Mannes Rede ist keines Mannes Rede* 'The word of (only) one man is the word of no man'. Often it is the last part of the first clause that is stressed: *On commence par être dúpe, on finit par être fripon* 'One begins as a fóol, one ends as a rascal'; *Qualis páter, talis filius* 'Like fáther, like son'; *Wie der Hérr, so 's Geschirr* 'Like the master, so the servants'; *Chi ha térra, ha guerra* 'He who has land, has war', *Chi va piáno, va sano* 'He who goes slowly, goes safely'. Just so our description is complete, let us observe here how in parallel sentences the main or high accent tends to be in opposition to a low accent. Other sentences show how elements in the middle of a sentence, and even multiple elements of a single sentence can all have emphasis: *Wér den Gróschen nicht ehrt, ist des Thàlers nicht werth* 'He who does not respect the penny, does not deserve the guinea'; *Qui tácet, consentíre videtur* 'He who remains silent, agrees'. In the dictum Τωὐτόν δ'ἐστι νοεῖν τε καὶ οὕνεκέν ἐστι νόημα 'The same thing is for conceiving as is cause of the thought conceived', the words νόημα and οὕνεκέν are stressed to the same degree, but the main accent is on τωὐτόν, that is on the negation of an apparent difference.

To find counterexamples, let us select sentences in which every element can be stressed, depending on the circumstances. We change the order of words – for example, reverse them: 'Today is my birthday' and 'My birthday is today' – and ask ourselves when we would stress this or that word among the four in the sentence. The best approach would be to say the two sentences out loud alternately and accentuate the same word in both sentences. In this way we can convince ourselves that the stress always has the same effect, independently of the word order. It always expresses an opposition; it is always polemical, so to speak: *today* – not yesterday or tomorrow; *is* – not will be or was etc. The feeling for this is so deeply ingrained in us that, in order to emphasize the oppositions, we change even the stress on the syllables of the words. At one moment we might talk of a 'nátional rally' and of 'internátional trade', but then of 'nátional and ínternational rights'.

395

Wäre nun die Betonung für die Wortstellung massgebend, wäre die eine aus der anderen zu erklären, so müsste der voranstehende Satztheil immer der betonte sein, es dürfte nie der Hauptnachdruck auf einem in der Mitte oder am Ende des Satzes stehenden Worte ruhen. Um zu erproben, wie es damit stehe, halten wir uns an Sprichwörter und Sentenzen; denn in diesen pflegt die Betonung an den Sinn gebunden zu sein. Da werden wir denn finden, dass jedes Satzglied das Betonte sein kann. Das erste ist es z. B. in den Sätzen: ὕδωρ μὲν ἄριστον. Actor sequitur forum rei. Negans non excipit. Eines Mannes Rede ist keines Mannes Rede. Oft ist das letzte Glied des Vordersatzes betont: On commence par être dúpe, on finit par être fripon. Qualis páter, talis filius. Wie der Hérr, so 's Geschirr. Chi a térra, a guerra. Chi va piáno, va sano. Nur der Vollständigkeit halber sei hier bemerkt, wie in Parallelsätzen dem Haupt- oder Hochtone ein Tiefton gegenüber zu stehen pflegt. Andere Sätze werden ||375|| nun zeigen, wie auch mittlere Glieder, auch mehrere |359| Glieder desselben Satzes den Nachdruck haben können: Wer den Gróschen nicht ehrt, ist des Thàlers nicht werth. Qui tácet, consentíre videtur. In dem Lehrspruche: Τωὐτόν δ'ἐστι νοεῖν τε καὶ οὕνεκέν ἐστι νόημα sind νοεῖν und οὕνεκεν gleich stark betont, der Hauptton aber fällt auf τωὐτόν, das heisst auf die Verneinung τωὐτον, 1891
eines scheinbaren Unterschiedes.

Zur Gegenprobe wählen wir Sätze, in denen je nach den Umständen jedes Glied betont werden kann. Wir verändern die Wortfolge, stellen z. B. einander gegenüber: „Heute ist mein Geburtstag" und: „Mein Geburtstag ist heute", und fragen uns nun, wann wir das eine oder das andere der vier Wörter betonen werden. Am Besten sprechen wir die beiden Sätze viermal abwechselnd aus und heben dabei in beiden dasselbe Wort durch die Stimme hervor. Da überzeugen wir uns, dass diese Betonung immer dieselbe Wirkung hat, unabhängig von der Wortstellung. Immer nämlich drückt sie einen Gegensatz aus, ist so zu sagen polemisch: h e u t e , – nicht gestern oder morgen; i s t , – nicht wird sein oder war u. s. w. Das Gefühl hierfür steckt so tief in uns, dass wir, um Gegensätze hervorzuheben, sogar die Sylbenbetonung der Wörter ändern. Jetzt reden wir von einer nationálen Kundgebung, jetzt wieder vom internationálen Handel, dann aber von nátionalem und ínternationalem Rechte.

Jetzt wird man mir zugeben: wenn es Kennzeichen falscher Theorien giebt, so lässt jene Betonungstheorie deren keines vermissen. Und doch ist es wahr, dass sehr oft der Hauptton auf das erste Satzglied fällt. Wie ist das zu erklären? Ich glaube aus drei Ursachen.

If there are any indications that a theory is wrong then this theory of stress has them all. And yet it is true that the main accent very often falls on the first element in the sentence. How can we explain this? I believe there are three reasons for this.

First of all it is a feature of an excited mood that at the very beginning of our utterance we attack the opposite of our thought, whether this occurs through the positive emphasizing of what we mean or

Erstens entspricht es einer erregten Stimmung, dass wir gleich zu Anfang unserer Rede den Gegensatz unsres Gedankens feindlich anfallen, geschehe dies nun mittelbar durch positive Hervorhebung des von uns Gemeinten, oder

396 indirectly in negative form: 'It's *you* who I'm talking to. C'est à *vous* que je parle. Not *tomorrow*, but today it has to be finished.'

Secondly, our own utterance can stimulate us reciprocally as it progresses. I take an example from Tacitus' (1999) *Germania*, chap. XV: 'Quotiens bella non ineunt, non multum venatibus, plus per otium transigunt, dediti somno ciboque. Fortissimus quisque ac bellicosissimus nihil agens. *delegata* domus et pantium et agrorum cura feminis senibusque [...] *ipsi* hebent.' (Whenever not engaged in war, they spend a little time hunting but much more relaxing, devoting themselves to sleep and food. All the bravest and most bellicose men do nothing: care of hearth and home and fields is *left* to the women and old men, [...] while they *themselves* loll about in a stupor.)

A third reason is that with the first element of the sentence the speaker has often already said everything that seems necessary to him under the circumstances, and after that simply turns the expression of the thought into a complete sentence as a way of making it clear, in a kind of apposition. It is like this in answers, especially in sullen or peevish answers: 'Where were you last night?' – '*At home* is where I was.' Further in the continuation of the speaker's own utterance: 'I wanted to reward him; decorate him is what I wanted to do.' This occurs also in all those other cases where external circumstances seem to complete the sentence, but we still decide to bring it to a conclusion explicitly. If both sides, the speaker and the listener, are perceiving the same object – if they are, for example, both looking at the same picture – then a predicate alone is enough for understanding: 'A masterpiece!' The addition '...is what this picture is' is unnecessary, and therefore unstressed. But the result is and remains this: it is not the stress, but rather the psychological subject and predicate relations that determine the preferred order of the elements of the sentence, and the mental behaviour that is expressed in the stress has nothing to do with that relation.

But it is just as certain that stress – free stress as much as bound stress, word accent as much as sentence accent – is a most important indicator of the psychological state that lies at the foundation of languages and individual utterances. It is inevitably significant whether a people puts the main emphasis on the material stem syllable or on the formal suffixes, or whether a people stresses in all cases the first, last or penultimate syllable according to a mechanical law, whether a people prefers sentence accent or word accent, and what shape it gives to that accent. Our judgement of languages will have to take all of this into account and try to discover the *Ratio legis* – the reason why – for each law.

396 unmittelbar in verneinender Form: „Dir gilt das. C'est à vous que je parle. Nicht morgen, sondern heute noch muss es fertig werden."

Zweitens kann auch die eigene Rede in ihrem Fortgange uns gegensätzlich anregen. Ein Beispiel wähle ich aus des Tacitus Germania, cap. XV: Quotiens bella non ineunt, non multum venatibus, plus per otium transigunt, dediti somno ciboque. Fortissimus quisque ac bellicosissimus nihil agens. delegata domus et penatium et agrorum cura feminis senibusque ... ipsi hebent.

Ein dritter Grund liegt darin, dass der Redner oft schon mit dem |360| ersten Satzgliede Alles gesagt hat, was ihm den Umständen nach nöthig schien, und nur noch verdeutlichend, in einer Art Apposition, den Gedankenausdruck zum Satze vervollständigt. So in Antworten, zumal in trotzigen oder verdriesslichen: Wo warst du gestern Abend? – Zu Hause war ich. So ferner im Fortführen der eigenen Rede: „Ich wollte ihn belohnen; auszeichnen wollte ich ihn." So aber auch in allen jenen anderen Fällen, wo äussere Umstände den Satz zu er||376||gänzen schienen, und man dann doch vorzieht, ihn ausdrücklich zu vollenden. Ist beiden Theilen, dem Redner und dem Hörer, derselbe Gegenstand im Sinne, betrachten z. B. Beide dasselbe Bild, so genügt ein blosses Prädicat zur Verständigung: „Ein Meisterstück!" Der Zusatz: „ist dies Bild" ist unerheblich, daher unbetont. Das Ergebniss ist und bleibt aber dies: Nicht die Betonung, sondern die psychologischen Subjects- und Prädicatsverhältnisse entscheiden über die bevorzugte Stellung der Satzglieder, und das seelische Verhalten, das sich in der Betonung äussert, hat mit jenem Verhältnisse nichts zu thun.

Ebenso sicher ist aber auch die Betonung, die freie wie die gebundene, der Wort- wie der Satzaccent, ein höchst wichtiges Anzeichen für die den Sprachen und den einzelnen Reden zu Grunde liegende seelische Verfassung. Es kann nicht gleichgültig sein, ob ein Volk auf die stofflichen Stammsylben oder auf die formenden Suffixe den Hauptnachdruck legt, oder ob es nach einem mechanischen Gesetze ein- für allemal die erste, letzte oder vorletzte Sylbe betont, ob es den Satzaccent, oder den Wortaccent bevorzugt, und wie es jenen gestaltet hat. Die Beurtheilung der Sprachen wird auf alles dies Rücksicht zu nehmen haben und versuchen müssen, für jedes Gesetz die Ratio legis zu entdecken.

ein 1891

ratio 1891

V. Manner of pronunciation or expression of mood

What I called stress or emphasis above is really only one kind of those multiply meaningful nuances we produce with our organs of speech, which we must now examine further. In them

V. Die Ausspracheweise oder Stimmungsmimik.

Was ich im Vorigen als Betonung oder Nachdruck bezeichnete, ist doch nur eine Art jener vielerlei bedeutsamen Abschattungen im Gebrauche der Sprachorgane, die wir nun weiterhin betrachten müssen. In ihnen

397 I see the second original means of linguistic formation, alongside the order of the parts of an utterance. It must be just as original as word order, since it represents immediately, it appeals directly to the senses, is vivid, if we may say that about aural impressions. The order of representations corresponded to the order of parts of the sentence in the way they come forward in the soul of the speaker and the listener. But what the speaking voice expresses in its modulations are those moods in the soul that accompany individual representations or entire thoughts. Insofar as the moods are dependent on the world of objects, their expression can *indirectly* achieve objective meaning at the same time: what scares me a lot will be very scary; what makes me hesitate will still be floating in uncertainty. But if the tone of my voice is questioning, requesting or commanding then the tone will have to be recognized as the *direct* expression of syntactic categories.

What we have here is an object which science is still very poorly prepared to deal with. Every language allows a certain amount of freedom in its sound and tonal structure, within the limits of which its speakers must stay, but whose freedoms they must judiciously use in order to speak properly. Even freedom is not without rules: it is a concession that the language makes to the idiosyncrasies and current mood of the speaker, whether it has to do with style or with sound and tone production. In this latter sense I call the use of this *manner of pronunciation* or *expression of mood*. I use this latter term in contrast to the kind of expression that imitates objects or occurrences and which has its linguistic representative in onomatopoeia. The first name is supposed to indicate the outer phenomenon and the second the content and effect of this phenomenon.

The limits and the permissible expressions of this freedom are very different depending on the language and dialect, and the relevant phenomena are often easier to imitate orally and to perceive aurally than to describe with our graphic means. Everything that falls under the French term *accent* belongs here: pitch and inflection of the tone, rhythm, the way in which sound is produced. But also the negative counterpart, the pause, is significant.

We might have wanted to express the tone of speech in musical notes – we speak of singing speech, after all. But hardly any language or dialect could be singing in the sense that musical notation would completely fulfil its purpose. If it were possible to precisely reproduce the tone of speech of a Thuringian, a Swiss or an Italian using a string instrument, then there would be no end to the unnameable impure mid-tones, and it would be difficult to speak of the melodic language of the Italian. And it is certainly no coincidence

397 erblicke ich neben der Reihenfolge der Redetheile das zweite ursprüngliche Mittel sprachlicher Formung. Gleich jener ist es ursprünglich, denn es ist unmittelbar darstellend, es ist ohne Weiteres sinnfällig, |361| anschaulich, wenn man das von Gehörseindrücken sagen dürfte. Der Ordnung der Satzglieder entsprach die Reihenfolge der Vorstellungen, wie sie in der Seele des Redenden und des Hörenden hervortreten. Was aber die sprechende Stimme in ihren Modulationen ausdrückt, sind jene seelischen Stimmungen, die die Einzelvorstellung oder den ganzen Gedanken begleiten. Weil und insoweit diese Stimmungen von der Welt der Objecte abhängig sind, kann ihr Ausdruck mittelbar zugleich objective Bedeutung erlangen: was mich sehr erschreckt, wird sehr schrecklich sein; was mich zögern macht, wird noch im Ungewissen schweben. Wo aber der Ton meiner Stimme fragend, bittend, befehlend ist, da muss er als unmittelbare Äusserung syntaktischer Kategorien anerkannt werden.

Wir haben es hier mit einem Gegenstande zu thun, zu dessen Bearbeitung die Wissenschaft noch sehr schlecht vorbereitet ist. Jede Sprache gestattet in ihrem Laut- und Tonwesen einen gewissen Spielraum, dessen Schranken man ||377|| einhalten, dessen Freiheiten man verständig gebrauchen muss, um richtig zu sprechen. Denn auch die Freiheit ist nicht regellos; sie ist ein Zugeständniss, das die Sprache der Eigenart und der jeweiligen Stimmung des Redners macht, gleichviel ob es sich um den Stil oder um die Laut- und Tonerzeugung handelt. In letzterem Sinne nenne ich ihre Benutzung Ausspracheweise oder Stimmungsmimik. Letzteren Ausdruck wähle ich im Gegensatze zu jener Mimik, die äussere Gegenstände oder Vorgänge nachbildet und in der Onomatopöie ihre sprachliche Vertreterin hat. Der erstere Name soll die äussere Erscheinung, der zweite den Inhalt und die Wirkung dieser Erscheinung bezeichnen.

Die Grenzen und die zulässigen Äusserungen jener Freiheit sind nach Sprache und Mundart sehr verschieden, und die einschlägigen Erscheinungen sind oft leichter mündlich nachzuahmen und mit dem Gehöre aufzufassen, als mit unsern graphischen Mitteln zu beschreiben. Alles das, was man unter dem französischen Namen *accent* begreift, gehört hierher: Höhe und Beugung des Tones, Rhythmus, Art der Lauterzeugung. Aber auch das Negative, die Pause, ist bedeutsam.

Man hat wohl den Ton der Rede in musikalischen Noten ausdrücken wollen, spricht man doch von singender Rede. In dem Sinne aber dürfte kaum eine Sprache oder Mundart singend sein, dass die Notenschrift ihren Zweck ganz erfüllte. Gelänge es, den Redeton eines Thüringers, |362| Schweizers oder Italieners mittels eines Saiteninstrumentes genau wiederzugeben, so wäre der unnennbaren unreinen Mitteltöne kein Ende, und man würde schwerlich mehr von der melodischen Sprache der Italiener reden. Auch ist es gewiss kein Zufall,

398 that it is those completely unmusical people, who can only distinguish a choral from a waltz at most by the rhythm, who have an astoundingly fine ear for linguistic tones and can convincingly imitate the intonation of foreign dialects. In Chinese, depending on the dialect, every word has a specific tone, a lexical tone, which can be flat, rising, falling, cut short, and then either high or low. This sets narrow limits on the rhetorical accent, but still the emotion makes itself felt very clearly, shows its strength and its type, partly through the tempo of the speech and partly through the sharper, at times screeching, stress. But in singing the lexical tone is still clearly perceptible. This speaks also for the independence of lexical tone from musical tone, since the song does not become dissonant because of it.

The expression of mood through tone of voice is something that man shares with all animals that vocalize. It is therefore no surprise that tone of voice is among the most widely spread means of expression in language. We say that the language of this or that region sounds aggressive, blustering, snappy, phlegmatic, melancholic; we observe certain, particularly common melodies and rhythms in the language. Almost as quickly, we also understand its individual rhetorical nuances. We might want to say: there is a common language of emotion, which finds its place in each individual language.

And at times it finds a truly spacious place for itself. In the Saxon dialect the 'hard and soft letters' *p* and *b*, *t* and *d*, and in certain cases also *k* and *g* are not distinguished; *ö*, *ü*, and *eu* are robbed of their *u* component and sound like *e*, *i* and *ei* (*ai*); they are only spoken purely in certain exclamations and calls. But the Saxon loves to paint with his voice, and when he speaks of dull and deep things he can make his voice dull and deep, and then it might happen that he tells us about a *schröcklüche, tüfe Fünsternüss* (*schkreckliche, tiefe Finsterniss*) 'terrible, deep darkness'. We might also hear him speak of *feste Kruntsätze* (*feste Grundsätze*) 'solid principles' in his decisiveness, and then perhaps, softened up a bit, of a *gleenes Gnespichen* (*kleines Knöspchen*) 'a little bud [of doubt]'. His dialect lets the emotional stimuli rule right into the sound system of the language. We should not mock him for this too hastily. The process is essentially formal, just as the mood of the speaker is a forming force. We might want to compare the process with sound imitation (onomatopoeia). Only this represents an external object according to the aural impressions it makes on a person. In our case, by contrast, the inner, emotional behaviour of the speaker finds symbolic expression in sound and tone as the object of speech. Theory requires that we keep these things separate, but in practice they join together in many ways. The moods are dependent

398 dass oft jene ganz unmusikalischen Menschen, die einen Choral von einem Walzer höchstens nach dem Rhythmus unterscheiden, für die Sprachtöne ein erstaunlich feines Gehör besitzen und den Tonfall fremder Dialekte täuschend nachahmen. Im Chinesischen haftet, je nach der Mundart, jedem Worte ein bestimmter Ton an, der gleichmässig gezogen, steigend oder fallend, kurz abgebrochen und dann wieder hoch oder tief sein kann. Dem rhetorischen Accente sind dadurch engere Schranken gesetzt, und doch giebt sich dabei die Gemüthserregung, ihre Stärke und ihre Art sehr deutlich zu erkennen, theils an dem Tempo der Rede, theils an der schärferen, zuweilen kreischenden Betonung. Beim Gesange aber bleibt der Wortton deutlich vernehmbar. Auch dies spricht für seine Selbständigkeit dem musikalischen Tone gegenüber; denn das Lied wird dadurch nicht misstönend.

Die Äusserung der Stimmung durch den Ton der Stimme theilt der Mensch mit dem stimmbegabten Thiere; kein Wunder also, dass sie zu den allverbreiteten Ausdrucksmitteln der Sprache gehört. Wir sagen wohl, die Sprache der oder jener Landschaft klinge zankend, polternd, schnippisch, phlegmatisch, melancholisch; wir beobachten an ihr gewisse, ihr besonders geläufige Tonfolgen und ||378|| Rhythmen. Fast ebenso schnell aber verstehen wir auch ihre jeweiligen rhetorischen Abschattungen. Man möchte sagen: es giebt eine allgemeine Sprache des Gemüthes, die in jeder Einzelsprache ihre Stätte findet.

Und zuweilen findet sie eine recht geräumige Stätte. Im obersächsischen Dialekte werden die „harten und weichen Buchstaben" *p* und *b*, *t* und *d*, in gewissen Fällen auch *k* und *g* nicht unterschieden; *ö*, *ü* und *eu* klingen, ihres *u*-Bestandtheils beraubt, wie *e*, *i* und *ei* (*ai*); nur in gewissen Aus- und Zurufen werden sie rein ausgesprochen. Der Sachse malt aber gern mit der Stimme, und wenn er von dumpfen und tiefen Dingen redet, so kann seine Stimme dumpf und tief werden, und dann kann es ihm geschehen, dass er etwa von einer „schröcklüchen, tüfen Fünsternüss" erzählt. Auch das kann man hören, dass er jetzt in Entschiedenheit von „festen Kruntsätzen" und dann wieder, weich gestimmt, von einem „gleenen Gnespichen" (kleinen Knöspchen) spricht. Soweit, bis in's innerste Lautwesen hinein, lässt seine Mundart die ge|363|müthlichen Regungen walten. Man spotte nicht vorschnell. Im Grunde genommen ist der Hergang doch echt formal, so gewiss die Stim- ächt 1891
mung des Redenden eine formende Macht ist. Man mag den Hergang mit der Schallnachahmung (Onomatopöie) vergleichen. Allein diese stellt ein äusseres Object dar nach dem Gehörseindrucke, den es auf den Menschen macht. In unserm Falle dagegen findet das innerliche, gemüthliche Verhalten des Redenden zum Gegenstande der Rede in Laut und Ton symbolischen Ausdruck. Die Theorie nöthigt uns, diese Dinge auseinanderzuhalten; in der Praxis werden sie sich vielfach miteinander vermählen. Denn die Stimmungen sind

399 on the impressions through which they are aroused, and consequently form the thing that arouses them itself. The naive mind will put the cause A in the place of the response B and vice versa – it will treat them as two sides of an equation: the child will call a thing that once hurt him 'hurty-hurt'. In a reversal of representations, we speak of the worm of conscience, since the worm is a creature that gnaws and bores, and a guilty conscience claims to feel something like a gnawing, boring pain. But there are different properties that are characteristic for the cause A; that is, different effects on our sensation: a dog does not only bark, it also bites. And the same property B can be noticed in different subjects: an owl howls, but so does the wind, as well as a wailing person. In the oscillating game of the soul the representation of biting may at one moment join with that of barking, at another the representation of the wind perhaps with that of the owl. A large object – such as a falling tree or lowing cattle – causes a dull, deep tone. And now I will join the representation of such a tone to everything that is large. The larger my opponent, the more fearsome he will be – therefore I join a representation of a large, dull-toned object with the representation of fear and speak in a dull voice when I am talking about something fearsome, and in falsetto when I am speaking of something really tiny. My language will become soft when I have a sense of softness. All of this can be individual and momentary; it can, however, also insert itself into the construction of a language as an essential element.

In *Chilean* diminutives are formed through the weakening of consonants. Havestadt (1883 [1775]), *Chilidugu* I, p. 135, offers the following examples: '*cuchani* pro *cutani*, aegrotat' (*cuchani* for *cutani*, he or she is ill); '*amochuiu* pro *amotuiu*, eamus nos duo' (*amochuiu* for *amotuiu*, let us two go); '*cuse* pro *cuye*, anus' (*cus* for *cuye*, old woman); '*vochum* pro *votm*, filiolus' (*vochum* for *votm*, little son); '*siu* pro *riu*, carduelis' (*siu* for *riu*, goldfink). I would like to add a few more examples to those of symbolic vocalism that were given earlier. German: *quaken* (the sound made by ducks, frogs), *quäken* (children), *quieken* (piglets); *trappeln* (many feet), *trippeln* (many small feet); *knarren* (wooden doors), *knurren* (dogs, stomach), *knirschen* (snow). Batta: *džarar, džirir, džurur* = 'crawl'; *džarar* is used in general, *džirir* of small creatures, *džurur* of large or feared creatures (V.D. Tuuk [1864], *Tobasche Spraakkunst*, p. 88). In *Kunama, a* indicates the first person, *e* the second person, and *i* the third person (Reinisch [1881], *Kunama-Sprache*, pp. 17-18). In *Grebo* or *Kru* the first and second persons of pronouns are distinguished only by stress. (Payne [1864], *Grebo Grammar*, pp. 19ff. Auer [1870], *Elements of the Gĕdebo Language*, pp. 14ff. Christaller [1890] in *Zeitschrift für afrikanische Sprachen* III, pp. 5-6).

399 abhängig von den Eindrücken, durch die sie erregt werden, mithin von dem Erreger selbst. Nun setzt der naive Geist den Erreger A an Stelle der Erregung B und umgekehrt, – behandelt sie wie die zwei Seiten einer Gleichung: das Kind nennt einen Gegenstand, der ihm einmal weh gethan hat, fortan Weh-weh; und wir reden in umgekehrter Vorstellungsweise, vom Wurme des Gewissens, weil der Wurm ein nagendes, bohrendes Thier ist, und ein schuldbeladenes Gewissen etwas wie einen nagenden, bohrenden Schmerz zu empfinden meint. Nun aber sind dem Erreger A verschiedene Eigenschaften charakteristisch, also verschiedene Einwirkungen auf unser Empfinden: der Hund bellt nicht nur, sondern er beisst auch. Und dieselbe Eigenschaft B kommt verschiedenen Subjecten zu: Die Eule heult, aber auch der Wind, auch der wehklagende Mensch. Jetzt mag sich im flatternden Spiele der Seele die Vorstellung des Beissens mit der des Bellens, die des Windes etwa mit der der Eule verknüpfen. Ein grosser Gegenstand, etwa ein fallender Baum, ein brüllendes Rind, verursacht einen dumpfen, tiefen Ton. Nun verknüpfe ich die Vorstellung eines solchen Tones mit allem Grossen. Je grösser mein Gegner, desto fürchterlicher wird er sein: Darum verknüpfe ich mit der Vorstellung der Furcht die eines grossen, dumpftönenden Gegenstandes, ||379|| rede in dumpfer Stimme, wenn ich von Fürchterlichem rede, und im Fisteltone, wenn es sich um etwas recht Winziges handelt; meine Sprache wird weich, wenn mir selbst weich zu Sinne ist. Alles dies kann individuell und momentan sein; es kann sich aber dem Baue einer Sprache als wesentliches Element einfügen.

Im Chilenischen werden Diminutive durch Erweichung von Consonanten gebildet. Havestadt, Chilidugu I, pag. 135 führt als Beispiele an: cuchani pro cutani, aegrotat; amochuiu pro amotuiu, eamus nos duo; cuse pro cuye, anus; vochum pro votm, filiolus; siu pro riu, carduelis. Den früher gegebenen Beispielen von symbolischem Vocalismus will ich noch einige hinzufügen. Deutsch: quaken, quäken, quieken, – Deutsch: 1891
trappeln, trippeln, – knarren, knurren, knirschen. Batta: *džarar*, *džirir*, *džurur* = kriechen, *džarar* allgemein, *džirir* von kleinen Thieren, *džurur* von grossen oder gefürchteten Thieren gebraucht (V. D. TUUK, Tobasche v. d. Tuuk, 1891 and 1901
Spraakkunst. S. 88). Im Kunama bezeichnet *a* die erste, *e* die zweite und *i* die dritte Person (REINISCH, Kunama-Sprache S. 17, 18). Im Grebo oder Kru sind die ersten und zweiten Personen der Fürwörter nur durch die Betonung unterschieden. (PAYNE, Grebo Grammar, p. 19 flg. AUER, Elements of. the Gĕdebo Language, p. 14 flg. CHRISTALLER in Ztschr. f. afrik. Spr. III, S. 5, 6).

Here we encounter for a second time the sound-symbolic feeling, and indeed as a creative and shaping force. It is to be assumed that this force, in the degree to which and the way in which its use was solidified in the language, came in part as a vestige and in part according to certain analogies into the service of the regulated stem-building and formal grammatical apparatus.

Wir begegnen hier zum zweiten Male dem lautsymbolischen Gefühle, und zwar als einer zeugenden und gestaltenden Kraft. Es ist anzunehmen, dass diese Kraft in dem Masse und nach der Art, wie sich die Sprachen gebrauchsmässig festigten, theils verkümmert, theils nach bestimmten Analogien in den Dienst des geregelten Stammbildungs- und grammatischen Formenwesens genommen wurde.

400 In the primitive proto-language it would have been able to unfold freely, in wild, gypsy-like freedom, on the spur of the moment.

But it will have left traces of its action everywhere. Among these I count the phenomena of accent in the Indo-European languages: in Greek the stressed ἔστιν 'is' as contrasted with the enclitic ἐστιν; the adverbs ἄπο 'from it', ἄνα 'onto it', ἔπι 'on it' as contrasted with the oxytonic prepositions;[17] the active θεοτόκος 'bearing a god', contrasted with the passive θεότοκος 'born of god' – in English *a présent* contrasted with *to presént*. In Japanese the stress shows whether two substantives that have been united in a compound are in an attributive or co-ordinate relation to one another. In Chinese words of a nominal meaning can serve as *neutra intransitiva* – that is, intransitives that take a complement – with the sense of to be or become something in relation to the complement. As a rule these are pronounced in the falling tone, which tends to be marked with the acute accent: *haò* (*hǎo*) 'good', *haó* (*hào* 好) 'to be good to someone', 'love someone'; *siēn* (*xiān*) 'earlier', 'before', *sién* (*xiàn* 先) 'proceed'; *wâng* (*wáng*) 'king', *wáng* (*wàng* 王) 'become king of...' This is of course generally assumed not to be original, but rather the remnant of an affix that has since disappeared. However, it represents the return to an original process.

Indeed this process must be one of the most original. It is not only common to all humans – an excited person uses it in rather the same way as a barking or howling dog. We are now at the point where we should turn our attention back to agglutination theory. It is quite improbable that there could still be a language today with completely unchangeable so-called root-words. But that there was once such a language or a few of this type is, as far as I can see, at the moment neither provable nor to be rejected as something completely impossible. But that such a language represented the oldest form of human speech is unthinkable. No, it must have been thinkable, since it has in fact been thought – but thinkable only if we abstract away from language that which gives it its soul, and can imagine that there is nothing more to life than metabolism.

VI. Interplay of word order and expression of mood

If we now think about how these two formal means of the earliest proto-language would have interacted, we gain an image of surprising freshness. Speech still moved in a linear fashion, the representations following one after the other, adding to each another cumulatively,

17 'Oxytonic' renders *oxytonirt*, a past participle derived from *oxytonon*, which in the system of Greek accents describes a word that has an acute accent on the final syllable, which is stressed.

400 In der Ursprache durfte sie sich noch frei entfalten, in wilder, zigeunerischer Freiheit nach der Eingebung des Augenblickes.

Spuren ihres Wirkens aber dürfte sie überall hinterlassen haben. Ich rechne dahin die Accenterscheinungen indogermanischer Sprachen, im Griechischen das betonte ἔστιν gegenüber dem enklitischen ἐστιν, die Adverbien ἄπο, ἄνα, ἔπι gegenüber den oxytonirten Präpositionen, das aktive θεοτόκος, einen Gott gebärend, gegenüber dem passiven θεότοκος, von Gott geboren, – im Englischen *a présent* gegenüber *to* |364| *presént.* Im Japanischen zeigt die Betonung an, ob zwei zu einem Compositum vereinigte Substantive im Attributiv- oder Coordinationsverhältnisse zu einander stehen. zueinander 1891
Im Chinesischen können Wörter von nominaler Bedeutung als Neutra transitiva fungiren mit der Bedeutung: das und das sein oder werden in Beziehung auf das Object. In der Regel werden sie dann im fallenden Tone ausgesprochen, den man durch den Acut zu bezeichnen pflegt: *haò*, gut: *haó*, Einem gut sein, ihn lieben; *siēn*, früher, vorn: *sién*, vorangehen; *wâng*, König: *wáng*, König werden über ... Das ist nun zwar aller Vermuthung nach nicht ursprünglich sondern Nachwirkung eines verschwundenen Affixes. Aber es ist eine Rückkehr zu einem ursprünglichen Verfahren.

||380||

In der That muss dies Verfahren eines der allerursprünglichsten sein: es ist nicht nur gemeinmenschlich, sondern der erregte Mensch übt es auch ziemlich in derselben Weise, wie der bellende oder heulende Hund. Jetzt ist es wohl an der Zeit, die Blicke noch einmal zurückzuwenden, zur Agglutinationstheorie. Dass es jetzt noch eine Sprache mit ganz unwandelbaren so genannten Wurzelwörtern gebe, ist nicht wahrscheinlich. Dass es früher einmal eine oder mehrere solche gegeben habe, ist, soviel ich sehe, zur Zeit weder erweisbar noch als etwas schlechthin Unmögliches zu widerlegen. Dass aber eine solche Sprache die älteste Form menschlicher Rede dargestellt habe, ist schlechthin undenkbar. Nein, denkbar muss es doch sein; denn es ist wirklich gedacht worden, – denkbar aber nur, wenn man von der Sprache das hinwegabstrahirt, was sie beseelt, und sich unter Leben nichts Besseres als Stoffwechsel vorzustellen weiss.

VI. Zusammenwirken des Stellungsgesetzes und der Stimmungsmimik.

Jetzt denken wir uns jene beiden Formalmittel der Ursprache in ihrem Zusammenwirken, so gewinnen wir ein Bild von überraschender Frische. Immer zwar bewegt sich die Rede linear, die Vorstellungen folgen aufeinander, summiren sich wohl,

401 and not thrusting themselves forward at the same time, as would be the case with looking at a picture. But it is no smooth, unbroken line, nor is it continuous. Modulations of the sound and tonal structure might show reinforcement, climaxes, abrupt or weak forms of the most different kinds; the rhythm – fast or slow, fluent or halting, with longer or shorter pauses – indicates the degree to which the elements of speech are to be joined together or separated from one another. All of this still occurs within the unrestricted freedom that I earlier described as being 'gypsy-like'. But this language was expressive, spoken by excited people to excitable people – it was expressive even when the listener could not see the facial expressions and gestures of the speaker. At this time there was only a small number of strictly defined grammatical categories that were available: a few formal words, the psychological subject and predicate, communicative, exclamatory and interrogative speech, the linking or separating of sentence elements, the predominant importance of a few of these elements. Everything else will have been fluid. But the possibility of a more solid shape was given. The urge for such a shape will also not have been lacking anywhere, even if this urge manifested itself differently in terms of its strength and direction. Representing these differences simultaneously in an exhaustive and organic-systematic way would be the task of a comprehensive general grammar.

Another fact would explain itself at this point, if it would prove to be true. Carl Abel, based on observations made on Egyptian, assumes that in the oldest human speech the same sounds often, if not always, combined opposite meanings ('opposite sense') in themselves. If this was indeed the case then such a language would have either been insufficient for its purpose as a medium of communication or it must have been complemented by mimicry; that is, by gestures or modulation in sound, tone and rhythm. If the latter was the case then the Egyptologists may decide that the so-called ideographic elements of hieroglyphic writing were in the end significant also for reading them, similar to our punctuation, but in a different sense. By the way, the fact that the ironic form of speech is age-old is something that we can hardly doubt. The question is only whether irony was the basis for this 'opposite sense'.

The closer a language is to this original state, the rawer we will consider it to be, regardless of whether it has simply got stuck at this stage of development, or whether it has degenerated to this state of rawness. Of course this rawness is highly compatible with a rich mass of sensual vividness, poetic sense and temperamental warmth, with all the advantages of naivety: it is only schooled thought that does not find what it wants in unsettled language. How does language come to be settled, and thought to be schooled?

401 drängen sich aber nicht gleichzeitig auf, wie bei der Betrachtung eines Bildes. Aber es ist keine gleichmässige, gerade Linie, auch keine ununterbrochene. Jetzt zeigen Modulationen der Laut- und Tonbildung Verstärkungen, Höhepunkte, schroffe oder weiche Formen der verschiedensten Art; jetzt deutet der Rhythmus, ob schnell oder langsam, ob fliessend oder abgebrochen, deuten längere oder kürzere Pausen an, in welchem Masse die Redeglieder miteinander zu verknüpfen oder voneinander zu trennen seien. Noch geschieht Alles in jener schrankenlosen Freiheit, die ich vorhin die zigeunermässige nannte. Ausdrucksvoll war aber diese Sprache, von erregten Menschen zu erregbaren geredet, – ausdrucksvoll selbst dann, wenn der Hörende die Mienen und Gesten des Redenden nicht sah. Festbegrenzte grammatische Kategorien waren erst in kleiner Anzahl vorhanden: einige Formenwörter, das psychologische Subject und Prädicat, die mittheilende, ausrufende, fragende Rede, die Verbindung oder Trennung von Satzgliedern, die vorwiegende Wichtigkeit einzelner derselben. Alles Andere mochte noch flüssig sein. Die Möglichkeit aber zu festerer Gestaltung war gegeben. Auch wird der Drang zu solcher nirgends ganz gefehlt haben, wenn er sich gleich nach Stärke und |365| Richtung verschieden äusserte. Diese Verschiedenheiten zugleich in erschöpfender und in organisch-systematischer Weise darzustellen, wäre die Aufgabe einer vollständigen allgemeinen Grammatik.

Noch eine andere Thatsache würde sich nun erklären, wenn sie selbst erwiesen würde. Carl Abel nimmt, zunächst auf Beobachtungen im Ägyptischen ||381|| gestützt, an, dass in der ältesten menschlichen Rede die gleichen Laute oft, wo nicht immer entgegengesetzte Bedeutungen („Gegensinn“) in sich vereinigt hätten. Wäre dem so gewesen, so hätte jene Sprache entweder ihren Zweck als Mittheilungsmittel verfehlt, oder sie musste auf mimischem Wege, durch Gesten oder durch Modulationen in Laut, Ton und Rhythmus ergänzt werden. War Letzteres der Fall, so mögen die Ägyptologen entscheiden, ob nicht am Ende doch die sogenannten ideographischen Elemente der Hieroglyphenschrift auch für die Vortragsweise bedeutsam waren, ähnlich, doch in anderem Sinne, wie unsre Interpunktionen. Dass übrigens die ironische Redeform uralt sei, wird man kaum bezweifeln können. Die Frage ist nur, ob jenem Gegensinne eine Ironie zu Grunde gelegen habe.

Aegyptischen 1891
gestützt 1891
Aegyptologen 1891
so genannten 1891

Je näher eine Sprache jenem Urzustande steht, für desto roher werden wir sie erklären, gleichviel ob sie von Hause aus in ihrer Entwickelung zurückgeblieben oder durch rückläufige Entwickelung verroht ist. Nun verträgt sich diese Rohheit recht wohl mit einem reichen Masse sinnlicher Anschaulichkeit, poetischer Sinnigkeit und gemüthlicher Wärme, mit allen Vorzügen der Naivität: nur das geschulte Denken findet in der unstäten Sprache nicht seine Rechnung. Wie kommt die Sprache zur Stätigung, das Denken zur Schulung?

402
VII. Classification of words by conceptual categories. Grammatical parts of speech.

We need not concern ourselves here with the question of innate ideas. To declare an idea innate is to declare that it cannot be explained. If primitive men already had fully developed representations of justice and injustice, then they would have either stayed for all time in a state of bestial anarchy or they would have felt the squalor of this state and known how to escape it through the creation of a social order. If our prehistoric ancestors did not possess the logical categories from the start then they either would have had to persevere in a state of animal stupidity or they would have learnt to eavesdrop on the world to hear its conceptual categories. We know how things have come to be, even when we do not know where they have come from. Every deed pays witness to a corresponding ability, and man has proved his logical ability through deeds in the realm of law and custom, religion and aesthetics, even if he has done this to different degrees, depending on his race and kind.

Man names things according to some conspicuous property; that is, he replaces the subject with a predicate. The names of animals in children's language are obvious examples of this. If it is a predicate of action, perhaps an imitation of sound, such as *miaow* = 'cat', then it is clear that this predicate turns a temporary feature into an enduring one. This same cat, which just miaowed, now catches a mouse, and will sleep after that. If it is an adjectival predicate, one of colour, shape, size or similar, then the same predicate could be applied to many things of different kinds; and things which otherwise are quite similar can from this perspective seem very different. Black is the colour of crows, moles, the hair on the head of humans; black is also the colour of coal and a starless night. But the grey hooded crow is the same as the black crow in every way except colour, and black head hair can over the years turn grey. When we are grown up we remember how we were once small and observed growth in animals and plants. In short, man needed only to observe the world to become aware of the difference between thing, property and action.

But for language this difference would have at first been only a material one, and for this reason not very significant. At one moment people would have said of a pebble that it is white, of the owl that it screeches,

402
VII. Classification der Wörter nach Begriffskategorien. Grammatische Redetheile.

Mit der Frage nach den angeborenen Ideen brauchen wir uns hier nicht zu beschäftigen. Eine Idee für angeboren erklären, heisst erklären, dass sie unerklärbar sei. Haben die Urmenschen die Vorstellungen von Recht und Unrecht nicht fix und fertig mitgebracht, so blieben sie entweder auf alle Zeiten im Zustande viehischer Anarchie, oder sie empfanden das Elend dieses Zustandes und wussten sich ihm durch Schaffung einer gesellschaftlichen Ordnung zu entringen. Haben unsere Urahnen die |366| logischen Kategorien nicht von Hause aus besessen, so beharrten sie entweder im Zustande thierischer Dummheit, oder sie lernten es, der Welt die Weltbegriffe abzulauschen. Wir wissen, wie die Dinge gekommen sind, wenn wir auch nicht wissen, woher sie gekommen sind. Jede That zeugt von einer entsprechenden Befähigung, und die logische Befähigung hat die Menschheit durch die That bewiesen, wie die rechtlich-sittliche, wie die religiöse, wie die ästhetische, wenn auch nach Rasse und Volksart in verschiedenem Grade.

Der Mensch nennt die Dinge nach irgendeiner hervorragenden Eigenschaft, das heisst: er ersetzt das Subject durch ein Prädicat. Die Namen der Thiere in der Kindersprache sind naheliegende Beispiele hierfür. Es sei ein Prädicat der Bethätigung, etwa eine Schallnachahmung, etwa Miau = die Katze: so leuchtet ein, dass dies Prädicat ein Vorübergehendes für ein Dauerndes setzt. Dieselbe Katze, die eben miaute, fängt jetzt eine Maus, wird hernach schlafen. Es sei ||382|| ein adjectivisches Prädicat, eines der Farbe, Gestalt, Grösse u. dergl., so kommt das gleiche Prädicat sehr verschiedenartigen Dingen zu; und Dinge, die sonst einander sehr ähneln, können gerade in dieser Hinsicht sehr verschieden sein. Schwarz ist die Saatkrähe, der Maulwurf, das Haupthaar eines Menschen; schwarz ist auch die Kohle und die Nacht bei sternenlosem Himmel. Aber der schwarzen Saatkrähe gleicht die graue Nebelkrähe in allen Stücken bis auf die Farbe, und die schwarzen Haupthaare können über's Jahr ergraut sein. Wer jetzt erwachsen ist, entsinnt sich, wie er ehemals klein war, und hat das Wachsthum auch beim Thiere und bei der Pflanze beobachtet. Kurz, der Mensch brauchte nur die Welt zu betrachten, um des Unterschiedes zwischen Ding, Eigenschaft und Thätigkeit inne zu werden.

dgl., 1891

Für die Sprache mochte allerdings der Unterschied zunächst nur ein materieller, daher unerheblicher sein. Jetzt sagte man von dem Kiesel aus, dass er weiss sei, von der Eule, dass sie schreie,

403 and then again of the white thing that it is a pebble, of the howling thing that it is an owl. But soon these habits of thought will have become established: in some places people will have loved to make the object, the thing into the psychological subject, and in others the action or property. In the first case they put the main emphasis on the substance, and in the second on its attributes. And all the other modulations of the sound form and stress may have sprung forth from the soul differently, may have affected the listener differently, depending on whether they related to the constant thing or its properties or the fleeting action. But whatever form words were available will have naturally preferred to combine with one particular word category than with another. A 'hither, thither', for example, demands a corresponding verbal complement. And now human inertia wanted – as it always will – the habitual to become the rule, for the exceptions to be kept aside for especially pressing cases or even for them to disappear. In this way matter conditioned form, matter found its expression in form: the logical categories pushed to shape themselves into grammatical categories, to sort themselves into parts of speech.

The fact that this did not happen everywhere in the same decisive, thoroughgoing way is no surprise. It could rather seem strange to us that the division, where it has gone to completion, has not kept to the same boundaries everywhere, since certainly, it would seem, the requirements of logic are everywhere and at all times the same. The contradiction solves itself rather easily.

First of all there are innumerable cases where the choice between nominal and verbal representation is given by the nature of the thing. This is the case whenever action is an essential part of the thing. An *animal*, human or non-human, can only be thought of as living; living in that common sense which excludes plant life means to be an *animal* and vice versa. A king who does not rule is no longer a king; the name demands the power, the power demands its exercise. One stands and falls with the other. Furthermore, many properties of things are in the same way effects that more or less strongly act on our senses: what is sour bites, what is bright dazzles, what is loud deafens, what is beautiful is appealing, and so on. Furthermore: someone who is strong conquers us, someone who is sly tricks us, and so on. Other properties contain temporary states of the subject itself, which will also manifest themselves: sick, hungry, quiet, happy and many other adjectives can be appropriately replaced by verbs related in sense, depending on the inner linguistic form.

But secondly also in the case of the inner linguistic form it is true that the habitual pushes for an expansion of its territory, and seeks where possible

403 – jetzt wieder von dem Weissen, dass es ein Kiesel, von dem Heulenden, dass es eine Eule sei. Bald aber mussten sich Denkgewohnheiten einbürgern: hier liebte man es, den Gegenstand, das Ding zum psychologischen Subjecte zu machen, – dort die Bethätigung oder Eigenschaft. Hier legte man den Hauptnachdruck auf die Substanz, dort auf ihre Attribute. Und alle jene sonstigen Modulationen der Lautbildung und Betonung mochten anders der Seele entquellen, anders auf den Hörer wirken, jenachdem sie |367| dem stätigbleibenden Dinge oder seiner Eigenschaft oder der flüchtigen Handlung galten. Was aber etwa an Formwörtern vorhanden war, das verband sich naturgemäss mit der einen Wortkategorie lieber, als mit der anderen. Ein „hierher, dorthin" z. B. verlangt geradezu eine entsprechende verbale Ergänzung. Und nun wollte es die menschliche Trägheit, – und sie will es immer, – dass das Gewohnte zur Regel wurde, die Ausnahmen für besonders dringliche Fälle vorbehalten blieben oder gar verschwanden. So bedingte der Stoff die Form, fand in ihr seinen Ausdruck: die logischen Kategorien drängten dazu, sich in grammatische Classen zu gestalten, in Redetheile zu sondern.

Dass dies nicht überall in gleich entschiedener, durchgreifender Weise geschehen ist, wird nun nicht Wunder nehmen. Eher könnte uns das befremden, dass die Scheidung, wo sie vollzogen ist, nicht überall die gleichen Grenzen eingehalten hat. Denn allerdings, so scheint es, sind die Anforderungen der Logik jederzeit und jeden Orts die gleichen. Der Widerspruch löst sich ziemlich leicht.

Erstens giebt es unzählige Fälle, wo die Wahl zwischen nominaler und verbaler Vorstellungsweise durch die Natur der Sache gegeben ist. So überall da, wo die Bethätigung zum Wesen des Dinges gehört. Das *animal*, Mensch oder Thier, kann nur lebend gedacht werden; leben in jenem gewöhnlichen Sinne, der das pflanzliche Leben ausschliesst, heisst ein *animal* sein, und umgekehrt. Ein König, der nicht herrscht, ist kein König mehr; der Name verlangt die ||383|| Macht, die Macht verlangt die Ausübung, Eins steht und fällt mit dem Anderen. So ferner sind viele Eigenschaften der Dinge doch mehr oder minder stark auf unsere Sinne fallende Wirkungen: das Saure Sauere 1891
beisst, das Helle blendet, das Laute betäubt, das Schöne gefällt u. s. w. Ferner: der Starke bezwingt, der Schlaue betrügt uns u. s. f. Andere Eigenschaften enthalten vorübergehende Zustände des Subjectes selbst, die sich doch auch äussern werden: krank, hungrig, ruhig, froh und viele andere Adjectiva können passend durch sinnverwandte Verba ersetzt werden, je nach der inneren Sprachform.

Zweitens aber gilt gerade auch von der inneren Sprachform jener Satz, dass das Gewöhnlichere nach Gebietserweiterung, wo möglich nach

sole rule. In this way there are languages with a predominantly nominal conception and others that are predominantly verbal. In addition, the verbal concept permits two kinds of conceptions.

Alleinherrschaft drängt. So giebt es Sprachen mit vorwiegend nominaler, und solche mit vorwiegend verbaler Auffassung. Zudem lässt der Verbalbegriff seinem Wesen nach zweierlei Auffassungsweisen zu.

404 First of all, the verbal predicate, in contrast to the nominal predicate, is more fleeting and contingent, dependent on time and circumstances, hence it is the bearer of temporal and modal forms when a language possesses these. And then it presents, in this same opposition, an activity of the subject, a subjective action. If this point of view predominates then it is quite plausible that we will attribute a personality, a will even to subjects that can have no will or whose actions are unintentional. It is simply naive to personify the inanimate as soon as it affects us more intensely, to have the same feelings towards a nut that falls on our head from a tree as we would have towards someone who hits us on the head.[18] This is how we might explain the fact that many languages combine the verbal predicate with subjective pronominal elements. And when in some languages the third person is often left unmarked, this is quite understandable, since, first of all, the subject in the third person is often already named separately (a substantive), and secondly the subject is often an inanimate thing, while the 'I' and 'you' apply only to people. This apparent neglect of the third person can be seen, for example, in Basque, Turkish, Buryat, Kott, Aleutian, Salish and Nahuatl (Mexican). In this way American languages prefer the verbal perspective and allow it to rule even in cases where we would think the predicate could only be nominal. In Maya the sentence 'I am your son' has the form:

a– meχen– en
your son I

In Nahuatl:

ni– mo– piltzin
I your son

And widely spread, in almost all languages of the western part of the world, is the changing of a predicative adjective into a finite verb.

But that other aspect of the verbal concept, as a more unstable and contingent phenomenon, is the one that predominates in the languages of the world. Modal and temporal forms, at times minimal, but often developed into wonderful fullness and fineness, are among the most common grammatical furnishings. It so happens that even languages that do without pronominal conjugation

18 Gabelentz's original text contains a pun here: the comparison is between a nut that falls on our head from a tree and a person who gives us a *Kopfnuss* (lit. 'head-nut'), a slap across the head, also used metaphorically to mean a brainteaser.

404 Einmal |368| nämlich ist das verbale Prädicat im Gegensatze zum nominalen mehr flüchtig und bedingt, von Zeit und Umständen abhängig, daher, wo die Sprache Temporal- und Modalformen hat, vorzugsweise Träger solcher Formen. Und dann besagt es, wiederum im gleichen Gegensatze, eine Thatäusserung des Subjectes, eine subjective Bethätigung. Waltet dieser Gesichtspunkt vor, so liegt es nahe, der Energie eine Persönlichkeit, einen Willen unterzulegen auch da, wo das Subject nicht willensbegabt, oder seine Äusserung keine gewollte ist. Es ist ja lediglich naiv, das Unbelebte, sobald es uns heftiger berührt, zu personificiren und gegen eine Nuss, die vom Baume uns auf den Kopf fällt, ähnlich zu empfinden, wie gegen einen Menschen, der uns eine Kopfnuss versetzt. So wird es zu erklären sein, dass viele Sprachen das verbale Prädicat mit subjectiven Pronominalelementen verbinden. Und wenn dabei in manchen Sprachen die dritte Person leer ausgeht, so ist dies wohl erklärlich. Denn erstens ist das Subject in dritter Person oft auch ohnehin genannt (ein Substantivum), und zweitens ist dieses Subject oft ein unbelebtes Ding, während das Ich und das Du nur persönlichen Wesen zukommen. Diese scheinbare Benachtheiligung der dritten Person beobachtet man z. B. im Baskischen, Türkischen, Burjätischen, Kottischen, Aleutischen, im Selisch und im Nahuatl (Mexicanischen). In diesem Sinne huldigen zumal amerikanische Sprachen der verbalen Anschauungsweise und lassen sie auch da walten, wo nach unseren Begriffen das Prädicat nur nominal sein könnte. Der Satz: „Ich bin Dein Sohn" heisst im Maya:

Unbelebte
1891

a– meχen– en
Dein Sohn ich

im Nahuatl:

ni– mo– piltzin
ich Dein Sohn

– und weit, fast allverbreitet in den Sprachen des westlichen Erdtheiles ist die Verwandlung des prädicativen Adjectivums in ein Verbum finitum. ||384||

Jene andere Seite des Verbalbegriffes als einer mehr vorübergehenden und bedingten Erscheinung ist aber in der Sprachenwelt weitaus die vorherrschende. Modal- und Temporalformen, zuweilen nur dürftig, oft in wunderbarer Fülle und Feinheit entwickelt, gehören zum gewöhnlichsten grammatischen Hausrathe. Es geschieht, dass auch Sprachen, die der pronominalen Conjugation entbehren,

405 often nudge from this perspective the nominal predicates closer to the verbal predicates. In this way adjectival conjugations in Japanese and Korean run parallel to the verbal conjugations.

The opposing tendency, the predominantly nominal view of verbal (energetic) predicates, can also manifest itself in two ways, depending on whether the action is seen more as a property or even as a possession of the subject. In the first case the form of expression is adjectival, in the second substantive; in the first case it might be: 'He is coming', in the second 'his coming'. The periphrastic forms of our conjugations offer ample examples for the former case. Sanskrit *dātā-'smi* 'Giver am I' = I will give; *dātā* 'giver' = he will give; Latin *amamini* 'you (pl.) are loved', according to its form = φιλούμενοι 'those who are loved'; Slavic *dalŭ, dala, dalo* 'he, she, it has given', actually 'one having given'; English *I am reading* etc. A convergence with the other perspective shows itself wherever we use the auxiliary word *have*, as in some perfects and in the future of the modern Romance languages. But it appears more forcefully wherever possessive elements represent the pronominal conjugations. Classical examples for this are offered by the more form-rich languages of the Malay family, with their preference for passive turns of phrase: 'You (are) my captive' instead of 'I take you captive'. But we encounter more or less clear traces of such possessive conjugations in many other places as well. To select a few obvious examples, this is what occurs in the objective conjugation of Hungarian and in the perfect conjugation of the Semitic and Hamitic languages. We must, however, at this point warn of an obvious fallacy. We speak of *possessive conjugations* whenever the subjective personal elements are similar to the possessive affixes of the substantives: there the action is not thought of as a manifestation of the force of the subject, but rather as something that is thought of as having become a possession of the subject; the representation of energy and subjectivity is lacking. But even the genitive, which the personal pronouns replace, is not necessarily possessive: 'my pictures' could be either pictures that I own or pictures that I have painted and which have been transferred into the possession of someone else. And this is not to mention a third possibility, that I am the object represented in the picture. The weight therefore falls on the point that the nominal verb is thought of as something rigid, or is not formally distinguished from the rigid noun. Here indications of tense and mood are sufficient to differentiate the 'living word', as the Chinese call the verb, from the 'dead' noun. But then we will have to check from case to case whether the so-called genitive in the language under consideration is consistently an adnominal case.

405 in dieser Hinsicht die nomi|369|nalen Prädicate den verbalen nahe rücken. So laufen im Japanischen und Koreanischen den verbalen Conjugationen adjectivische parallel.

Auch die entgegengesetzte Neigung, die vorwiegend nominale Anschauung verbaler (energischer) Prädicate, kann sich in zweifacher Weise äussern, je nachdem die Bethätigung mehr als eine Eigenschaft oder gar als ein Besitz des Subjectes angesehen wird. Im ersteren Falle ist die Ausdrucksweise adjectivisch, im anderen substantivisch; im ersteren Falle mag es heissen: „Er (ist) kommend", – im zweiten: „Sein Kommen". Für Jenes liefern die periphrastischen Formen unserer Conjugationen Beispiele die Hülle und Fülle: sanskrit: *dātā-'smi*, Geber bin ich = ich werde geben; *dātā*, Geber = er wird geben; lateinisch: *amamini*, der Form nach = φιλούμενοι; slavisch: *dalŭ*, *dala*, *dalo*, er, sie, es hat gegeben, eigentlich: gegeben habender, englisch: *I am reading* u. s. w. Eine Annäherung zu der anderen Anschauung zeigt sich schon da, wo wir das Hülfswort „haben" anwenden, wie in einem Theile der Perfecta und in dem neuromanischen Futurum. Kräftiger aber tritt sie da hervor, wo Possessivelemente die pronominale Conjugation vertreten. Classische Beispiele hierfür liefern die formenreicheren Sprachen des malaischen Stammes mit ihrer Vorliebe für passive Wendungen: „Du (bist) mein Gefangener", statt: „Ich nehme dich gefangen". Aber auch sonst vieler Orten begegnet man mehr oder minder deutlichen Spuren solcher Possessivconjugationen. So, um näherliegende Beispiele zu wählen, in der Objectivconjugation des Magyarischen und in der Perfectconjugation der semitischen und hamitischen Sprachen. Doch mag gleich an dieser Stelle vor einem naheliegenden Fehlschlusse gewarnt werden. Man redet von Possessivconjugationen, so oft die subjectiven Personalelemente den Possessivaffixen der Substantiva gleichen: Da sei eben die That nicht als Kraftäusserung des Subjectes, sondern als ein ihm irgendwie gewordener Besitz gedacht; es fehle die Vorstellung der Energie und Subjectivität. Aber auch der Genitiv, den die Possessivpronomina ersetzen, ist nicht nothwendig possessiv: „meine Bilder" können sowohl Bilder, die ich besitze, als auch solche sein, die ich gemalt habe, und die vielleicht in fremden Besitz übergegangen sind, – von einer dritten Möglichkeit, dass ich selbst das dargestellte Object bin, zu geschweigen. Also nur darauf fällt das Gewicht, dass das Verbum nominal, wie etwas Starres gedacht, oder von dem starren Nomen formell nicht unterschieden wird. Da genügen nun doch schon Tempus- und Moduszeichen um das „Lebewort", wie die Chinesen das Verbum nennen, vor dem „todten" Nomen herauszuheben. Dann aber auch wird von Fall zu Fall zu ||385|| prüfen sein, ob denn der sogenannte Genitiv in der betreffenden Sprache ein für allemale ein adnominaler Casus ist.

406
The original repertoire of material words would seem to be exhausted by the categories of substantive, adjective and verb. The origin of number words is unclear almost everywhere. The general assumption is that number words were not originally a unique word class, and the few facts that can be established seem to support this assumption. Some adverbs, above all the deictic and interrogative ones, may indeed be among the oldest formal words. But other adverbs, those that are more descriptive, are purely material. And something similar will have been true of those relational expressions that we call prepositions and postpositions, as well as of the conjunctions, of course. Examples for all of this are readily available to everyone.

VIII. Possibility – Rule – Law

In the preceding discussion we have come to know a series of forces in the soul and physical means that, it would seem, offered themselves to language right from the beginning and completely of their own accord. The purpose and effects to which they would have to join together were also clear: one expressed itself in the order of words, another in the modulation of sound and tone. Then we saw how the outer world itself guided man to a classification of his representations and concepts, and how these could influence the shape of his speech. There were only few means that offered themselves to him, and the mental needs that made use of them would still have been raw and poor. It was enough that a majority of the factors were already present which could join together in limitless diversity. It was need alone that decided, only the nature of the thing that restrained the possibilities. Whoever thinks of a grammar as simply an ordered collection of regulations and prohibitions might call this a state without grammar. I have a different view. Every language clothes its matter in forms, even if these are only syntactic and sound-symbolic. And every one of these forms has its own range of effects and meanings, even if these are remote and vague. And all of these forms together constitute a system, even if it is just a simple one; and it is this system that I call grammar. If we consider the most extreme case – everything was possible, any kind of formation was allowed – then here every kind of formation bore its special meaning in itself; the two were not completely equivalent, and each was therefore more or less necessary.

406

In den Kategorien des Substantivums, Adjectivums und Verbums dürfte sich der ursprüngliche Vorrath an Stoffwörtern erschöpfen. Die Herkunft der Zahlwörter ist fast überall dunkel. Die Vermuthung aber und das wenige Thatsächliche, was sich feststellen lässt, spricht nicht dafür, dass sie von Hause aus eine besondere Wortgattung gebildet hätten. Von den Adverbien mag ein Theil, zumal die deutenden und fragenden, zum ältesten Schatze der Formwörter gehören; ein anderer, der mehr beschreibende, aber ist rein stofflich. Und Ähnliches wird von jenen Beziehungsausdrücken gelten, die wir Prä- und Postpositionen nennen, sowie natürlich erst recht von den Conjunctionen. Beispiele für alles dies sind Jedem zur Hand.

festtstellen [in den Berichtigungen, S. 502: feststellen] 1891

|370|

VIII. Möglichkeit – Regel – Gesetz.

Wir haben im Bisherigen eine Reihe von seelischen Mächten und von physischen Mitteln kennen gelernt, die sich, so schien es, der Sprache von Anfang an und ganz von selbst boten. Auch in welchem Sinne und in welchen Wirkungen sie sich verbinden mussten, war wohl einleuchtend: ein Anderes drückte sich in der Wortstellung, ein Anderes in jenen Modulationen der Laut- und Tonbildung aus. Nun zeigte es sich ferner, wie die Aussenwelt selbst den Menschen zu einer Classification seiner Vorstellungen und Begriffe anleiten, und wie dies die Gestaltung seiner Rede beeinflussen konnte. Es waren noch wenige Mittel, die sich ihm boten, und die geistigen Bedürfnisse, die sich ihrer bedienten, mochten noch roh und armselig sein. Genug, dass schon eine Mehrheit der Factoren vorlag, die sich von Hause aus in schrankenloser Mannigfaltigkeit verbinden konnten. Nur das Bedürfniss entschied, nur die Natur der Sache setzte den Möglichkeiten Mass und Ziel. Wer sich unter einer Grammatik nur eine geordnete Sammlung von Vorschriften und Verboten denken kann, der mag diesen Zustand einen grammatiklosen nennen. Ich bin anderer Meinung. Jede Sprache kleidet ihren Stoff in Formen, wären es auch nur syntaktische und lautmimische. Und jede dieser Formen hat ihren Wirkungs- und Bedeutungskreis, sei dieser noch so weit und vag. Und alle diese Formen bilden zusammen ein System, wenn auch ein noch so einfaches; und dieses System nenne ich eben die Grammatik. Wir setzen den äussersten Fall: Alles war möglich, jederlei Formung war erlaubt: so trug doch jede dieser Formungsweisen ihre besondere Bedeutung in sich; es waren ihrer nicht zwei ganz gleichwerthig, jede war daher bedingt nothwendig.

mit 1891

Mannichfaltigkeit 1891

erlaubt; 1891

Zwei 1891

407 The expression was determined by what was to be expressed; if it was chosen differently, then it said something different. Even if this grammar could be summarized on a single page, it still remains a grammar.

It is easy to see what forces pushed for further development, for the grammatical elaboration of language. Firstly, an increased mental life will have continually assigned new, higher tasks to language. And secondly that familiar tendency to raise the habitual to sole rule will have gradually dammed the riverbed of language.

Even the poorest language grants certain freedoms to speech. The speaker has the choice as to whether he wants to clothe the thought that appears before him in this or that form, if it is not one of those everyday matters for which constant forms tend to establish themselves quickly. He has the choice, but we can hardly say that he chooses: he reaches out blindly, he uses his native language with instinctive sureness. He reaches out blindly, but he will always reach where his mental capacity, habit and his current mood direct him. In this way there are three forces that are determinative: two that are constant, habit and individual capacity, and one that is momentary, the current mood. Only the last two of these belong exclusively to the individual: what is habitual for him depends largely on those whose speech he usually hears; that is, on his closest linguistic companions. This factor has a narrowing effect, since the habitual will, as a rule, be preferred, indeed it would always be preferred and finally reach sole rule, if those individual and momentary forces did not from time to time widen the narrowing riverbed of the language. These forces break through the limits of the everyday, create in it – not always grammatically – exceptions, surprising structures, which because of this have a doubly emphatic effect. In this way language at one moment maintains the freedom it already has and at another gains a new additional freedom. We can observe all of this in the present, and confidently assume it in the most distant past.

We will leave the question as to the original unity or diversity of human language unanswered. But if we imagine that all language emerged at one point of the earth, under a single horde of primitive people, then there would still be no need for a Tower of Babel to explain the fundamental diversity of languages. There was very little that was common; further individual development had plenty of room to move. The horde then broke up, linguistic intercourse among the individual groups ceased.

407 Der Ausdruck wurde durch das Auszudrückende bestimmt; wurde er anders gewählt, so besagte er Anderes. Lasse sich diese Grammatik auf einer Druckseite zusammenfassen: eine Grammatik bleibt sie darum doch. ||386||

Es ist leicht einzusehen, welche Mächte hinfort zur Weiterentwickelung, zum grammatischen Ausbaue der Sprache drängten. Erstens setzte ein gesteigertes Geistesleben der Sprache immer neue, höhere Aufgaben. Und zweitens dämmte jene bekannte Neigung, das Gewöhnliche zur Alleinherrschaft zu erheben, das Flussbett der Sprache allgemach ein.

Selbst die ärmste Sprache wird der Rede gewisse Freiheiten gestatten. Der Redende hat die Wahl, ob er den ihm vorschwebenden Gedanken in diese oder jene Form kleiden will, – es müsste sich denn um eine jener Alltäglichkeiten handeln, für die sich schnell ständige Formen einzustellen pflegen. Er hat die Wahl; aber man kann kaum sagen, dass er wähle: er greift eben blindlinks zu, er handhabt seine Muttersprache mit instinctiver Sicherheit. Er greift blindlinks zu; aber allemal wird er dahin greifen, wohin ihn seine Geistesanlagen, seine Gewohnheit und seine augenblickliche Stimmung weist. So sind es drei Mächte, die hier bestimmend wirken: zwei ständige: die Gewohnheit und die individuelle Anlage, und eine momentane: die jeweilige Stimmung. Nur die beiden Letzteren gehören ausschliesslich dem Einzelnen an; was ihm gewohnt ist, hängt zum guten Theile von denen ab, deren Rede er zu hören pflegt, das heisst von seinen nächsten Sprachgenossen. Dieser Factor wirkt einengend; denn das Gewöhnliche wird in der Regel bevorzugt, ja es würde allemal bevorzugt werden und schliesslich zur Alleinherrschaft gelangen, wenn nicht jene individuellen und momentanen Mächte das sich verengende Strombett der Sprache hin und wieder erweiterten. Sie durchbrechen die Schranken des Alltäglichen, schaffen in diesem – nicht immer im grammatischen Sinne – Ausnahmen, überraschende Gebilde, die darum doppelt nachdrücklich wirken mögen. So wird der Sprache bald eine schon vorhandene Freiheit erhalten, bald eine neue hinzugewonnen. Das Alles können wir in der Gegenwart beobachten, in der frühesten Vergangenheit zuversichtlich vermuthen.

Wir lassen die Frage nach der ursprünglichen Einheit oder Mehrheit der menschlichen Sprache unentschieden. Aber wir setzen einmal den Fall, es wäre alle Sprache an einem einzigen Punkte der Erde, unter einer einzigen Horde von Urmenschen entstanden: so bedurfte es noch keines babylonischen Thurmes, um die tiefgehende Mannigfaltigkeit der Sprachen zu erklären. Des Allgemeinsamen war doch nur wenig, die individuelle Weiterentwickelung hatte breiten Spielraum. Nun theilte sich die Horde, der Sprachverkehr unter den einzelnen Schwärmen hörte auf.

408 Depending on where they found their new home, new stimuli and tasks awaited the emigrants, which caused, or rather demanded, new habits of thought and life. And the language with which they had to process all of this had more gristle than solid bone, it was malleable because unshaped. How quickly the inherited traits will have disappeared, and indeed how few of them will there have been in the first place! And if we now imagine how dependent primitive man will have been on the natural world that surrounded him, how different the individual hordes were, and within these, how every member of the tribe will have been raised to the same degree in the school of such a life, some to a life of peaceful, comfortable enjoyment, others to privation, hard work and struggle, then we might guess how varied their world views would become and likewise the languages in which these world views were reflected.

It has been a matter of great controversy as to whether it is justified to compare the state of our prehistoric ancestors with that of the present-day so-called primitive peoples. In the former case immaturity pregnant with future promise, in the latter hopeless degeneration; not children but geriatrics. But we also speak of child-like geriatrics and know that the comparison is pertinent. Neither the savage nor the child is free. Both love inventing all kinds of formal constraints to further restrict the modest range of freedom that their own weakness and the overwhelming power of the outer world permits to their will: strictly regulated games, festivals, sayings, songs, rituals, manners, in short prescriptions and proscriptions of all kinds, often of the most fantastic kind. They are like the orthopaedic splints and bindings that support a weak-boned body, they are in fact bindings and splints for weak minds and continue to be worn out of dear habit by those who have overcome their weakness. In this way limits of the usual, pleasing or permitted will have been set to natural arbitrariness early in the use of language. Special customs developed in every language community and, if human language was originally one, then the evidence of this unity will have been cleared away very quickly.

But these stimuli will have also had a strengthening effect; they showed the way and broke paths where the needs of communicating thoughts could only assign purposes. In this way both forces promoted and attracted each other, and language became the governess of the spirit of the people, which was for its part the creator of language. It was an exchange; they will have mirrored each other, and we have to continue to see them in this way.

408 Der Auswandernden harrten, je nach den Stätten, wo sie ihr neues Heim fanden, neue Anregungen und Aufgaben, die neue Denk- und Lebensgewohnheiten bedingten, verlangten. Und die Sprache, mit der sie alles dies bearbeiten sollten, hatte doch mehr Knorpel, als festes Gebein, war um so bildsamer, je ungebildeter sie war. Wie schnell mochten sich da die verwandschaftlichen Züge verwischen, wie wenige waren es im Grunde gewesen. Und nun stelle man sich vor, wie abhängig der Urmensch von der ihn umgebenden Natur war, wie verschieden die einzelnen Horden, und innerhalb dieser, wie gleichmässig alle Stammesgenossen in der Schule eines solchen Lebens erzogen wurden, die Einen zu friedlich bequemem Genuss, die Anderen zu Entbehrung, zu harter Arbeit und Kampf: so wird man ahnen, wie mannigfaltig sich die ||387|| Weltanschauungen gestalten mussten und die Sprachen, in denen sich diese Weltanschauungen spiegelten.

|371|

Man hat viel darüber gestritten, ob es berechtigt sei, die Zustände unserer Urahnen mit denen der heutigen sogenannten Naturvölker zu vergleichen: so genannten 1891
dort zukunftsschwangere Unreife, hier hoffnungslose Verkommenheit, – hat man eingewendet, – es sind nicht Kinder, sondern Greise. Aber wir reden auch von kindischen Greisen und wissen, dass der Vergleich zutrifft. Frei ist nun weder der Wilde noch das Kind. Beide lieben es, den bescheidenen Spielraum, den die eigene Schwäche und die Übermacht der Aussenwelt ihren Willensbethätigungen gestatten, noch weiter einzuengen durch allerhand selbsterfundenen Formelzwang: strammgeregelte Spiele, Feste, Sprüche, Lieder, Ritual, Etiquette, kurz Ge- und Verbote aller Art, oft der wunderlichsten Art. Sie gleichen jenen orthopädischen Schienen und Binden, die einem schwachknochigen Körper Halt geben, – sind eigentlich selbst solche Binden und Schienen für schwache Geister und werden auch von den Erstarkten aus lieber Gewohnheit gern weiter getragen. So mochten denn auch der naturwüchsigen Willkür in der Behandlung der Sprache frühe schon Schranken des Üblichen, Gefälligen, Erlaubten gesetzt werden. In jeder Sprachgemeinde bildeten sich besondere Bräuche; und wenn alle menschliche Sprache ursprünglich eine war, so mochten sich die Spuren dieser Einheit recht schnell verwischen.

Und kräftigend wirkten jene Anregungen doch, sie wiesen und bahnten da Wege, wo die Bedürfnisse des Gedankenverkehres nur Zwecke setzen konnten. So förderten und lockten beide Mächte einander, und die Sprache ward Erzieherin des Volksgeistes, der seinerseits ihr Bildner war. Es war eine Wechselwirkung, die beiden mussten sich gegenbildlich zu einander zueinander 1891
verhalten; und so müssen wir sie fortan betrachten.

Works cited by Gabelentz

Auer, John Gottlieb. 1870. *Elements of the Gĕdebo Language*. Stuttgart: P.E. Mission.

Caldwell, Robert. ²1875. *A Comparative Grammar of the Dravidian or South-Indian Family of Languages*. London: Trübner and Co.

Caspari, Carl Paul. 1880. *Grammaire Arabe*, E. Uricoechea (trans.). Brussels: self-published.

Christaller, Johann Gottlieb. 1890. 'Näheres über die Kru-Sprache'. *Zeitschrift für afrikanische Sprachen* 3, pp. 1-39.

Cicero. 1919. *The Orations of Cicero against Catiline*, Charles Duke Yonge (trans.). London: Bell and Sons.

Gabelentz, Georg von der. 1869. 'Ideen zu einer vergleichenden Syntax. Wort- und Satzstellung'. *Zeitschrift für Völkerpsychologie und Sprachwissenschaft* 6, pp. 376-384.

Gabelentz, Georg von der. 1875. 'Weiteres zur vergleichenden Syntax. Wort- und Satzstellung'. *Zeitschrift für Völkerpsychologie und Sprachwissenschaft* 8, pp. 129-165, 300-338.

Gabelentz, Georg von der. 1887. 'Zur chinesischen Sprache und zur allgemeinen Grammatik'. *Internationale Zeitschrift für allgemeine Sprachwissenschaft* 3, pp. 92-109.

Gabelentz, Georg von der. 1889. 'Stoff und Form in der Sprache'. *Berichte über die Verhandlungen der Königlich-Sächsischen Gesellschaft der Wissenschaften zu Leipzig, philologisch-historische Classe* 41, pp. 185-216.

Gabelentz, Hans Conon von der. 1852. 'Kurze Grammatik der tscherokesischen Sprache'. *Zeitschrift für die Wissenschaft der Sprache* 3, pp. 257-300.

Gabelentz, Hans Conon von der. 1861. 'Über das Passivum', *Abhandlungen der Königlich-Sächsischen Akademie der Wissenschaften zu Leipzig, philologisch-historische Classe* 8, pp. 449-546.

Havestadt, Bernhard. 1883 [1775]. *Chilidúǵu sive Tractatus Linguae Chilenesis*, Julius Platzmann (ed.). Leipzig: Teubner.

Howell, Mortimer Sloper. 1883. *A Grammar of the Classical Arabic Language*. Allahabad: Government Press.

Humboldt, Wilhelm von. 1838. *Über die Kawi-Sprache auf der Insel Java*, Johann Carl Eduard Buschmann (ed.). Berlin: Akademie der Wissenschaften.

Humboldt, Wilhelm von. 1997 [1822]. 'On the origin of grammatical forms and their influence on the development of ideas' [Über das Entstehen der grammatischen Formen und deren Einfluss auf die Ideenentwicklung]. In Humboldt (1997), pp. 23-51. (Originally published in the *Abhandlungen der historisch-philologischen Klasse der Königlichen Akademie der Wissenschaften zu Berlin* 4, pp. 239-260.)

Humboldt, Wilhelm von. 1997 [1827]. 'On the Dual Form' [Über den Dualis]. In Humboldt (1997), pp. 111-136. (Originally published in 1828 as *Ueber den Dualis*. Berlin: Akademie der Wissenschaften.)

Humboldt, Wilhelm von. 1997. *Essays on Language*, Theo Harden & Dan Farrelly (eds.). Frankfurt am Main: Peter Lang.

Humboldt, Wilhelm von. 1999 [1836]. *On Language: on the diversity of human language construction and its influence on the mental development of the human species*, Peter Heath (trans.), Michael Losonsky (ed.). Cambridge: Cambridge University Press. (Originally published as *Über die Verschiedenheit des menschlichen Sprachbaues und ihren Einfluss auf die geistige Entwickelung des Menschengeschlechts*. Berlin: Akademie der Wissenschaften).

Kleinschmidt, Samuel. 1851. *Grammatik der grönländischen Sprache*. Berlin: Reimer.

Müller, Friedrich. 1876–1888. *Grundriss der Sprachwissenschaft*, 4 vols. Vienna: Hölder.

Payne, John. 1864. *Grebo Grammar*. New York: American Tract Society.

Reinisch, Leo. 1881. *Die Kunama-Sprache in Nordost-Afrika*. Vienna: Gerold.

Sacy, Silvestre de. ²1881. *Grammaire Arabe*. Paris: Imprimerie Royale.

Steinthal, H. 1860. *Charakteristik der hauptsächlichsten Typen des Sprachbaues*. Berlin: Dümmler (Second edition 1893, Franz Misteli (ed.)).

Steinthal, H. 1867. *Die Mande-Neger-Sprachen, psychologisch und phonetisch betrachtet*. Berlin: Dümmler.

Steinthal, H. ²1881. *Abriss der Sprachwissenschaft*. Berlin: Dümmler.

Steinthal, H. 1883. *Die sprachphilosophischen Werke Wilhelm's von Humboldt*, 2 vols. Berlin: Dümmler.

Tacitus. 1999. *Germania*, James B. Rives (trans.). Oxford: Clarendon Press.

Tuuk, Herman Neubronner an der. 1864. *Tobasche Spraakkunst*, 2 vols. Amsterdam: Muller.

Wright, William. ³1896. *A Grammar of the Arabic Language*, 2 vols. Cambridge: Cambridge University Press.

Secondary works cited

Ritter, Joachim, Karlfried Gründer, Rudolf Eisler, Günther Bien & Tilman Borsche. 1971–2007. *Historisches Wörterbuch der Philosophie*. Basel: Schwabe.

Index

For Product Safety Concerns and Information please contact our EU
representative GPSR@taylorandfrancis.com
Taylor & Francis Verlag GmbH, Kaufingerstraße 24, 80331 München, Germany

www.ingramcontent.com/pod-product-compliance
Ingram Content Group UK Ltd.
Pitfield, Milton Keynes, MK11 3LW, UK
UKHW020017200926
475702UK00014B/1941
9781041179955